Spectrophotometric Reactions

PRACTICAL SPECTROSCOPY

A SERIES

Edited by Edward G. Brame, Jr.

The CECON Group
Wilmington, Delaware

ADDITIONAL VOLUMES IN PREPARATION

Spectrophotometric Reactions

Irena Němcová
Ludmila Čermáková
Department of Analytical Chemistry
Charles University
Prague, Czech Republic

Jiří Gasparič
Department of Biophysics and Physical Chemistry
Charles University
Hradec Králové, Czech Republic

CRC Press
Taylor & Francis Group
Boca Raton London New York

CRC Press is an imprint of the
Taylor & Francis Group, an **informa** business

First published 1996 by Marcel Dekker, Inc.

Published 2022 by CRC Press
Taylor & Francis Group
6000 Broken Sound Parkway NW, Suite 300
Boca Raton, FL 33487-2742

© 1996 by Taylor & Francis Group, LLC
CRC Press is an imprint of Taylor & Francis Group, an Informa business

No claim to original U.S. Government works

ISBN 13: 978-0-8247-9451-4 (hbk)

**Visit the Taylor & Francis Web site at
http://www.taylorandfrancis.com**

**and the CRC Press Web site at
http://www.crcpress.com**

Library of Congress Cataloging-in-Publication Data

Němcová, Irena.
    Spectrophotometric reactions / Irena Němcová, Ludmila Čermáková,
Jiří Gasparič.
        p.   cm. — (Practical spectroscopy ; v. 22)
    Includes bibliographical references and index.
    ISBN 0-8247-9451-6 (alk. paper)
    1. Spectrum analysis.   2. Chemical reactions.   I. Čermáková,
Ludmila.   II. Gasparič, Jiří.   III. Title.   IV. Series.
QD95.N46   1996
    543′.0852−dc20   96-6075
                                                                   CIP

Preface

Modern analytical methods aim to develop instrumental methods of increased capability. Spectrophotometric reactions still remain the basis of many of them. However, a better knowledge is needed of the conditions and course of such reactions and of the properties of the analytes, reagents, and chromogens involved.

The mechanisms of many spectrophotometric reactions have been investigated. It is our aim to summarize the present knowledge of these reactions and to discuss their common principles from the point of view of both inorganic and organic analysis.

This monograph should cater to a broad audience — from technicians to university students, professors, and researchers in a great variety of institutions — whose aim is not only to determine particular substances, but also to find and evaluate principles and wider relationships of spectrophotometric analytical measurements. We hope the book will provide inspiration in developing new analytical methods.

We wish to express sincere appreciation to all our colleagues for valuable and constructive comments during the preparation of this monograph and to Dr. P. Rychlovský for help in the preparation of the manuscript.

Irena Němcová
Ludmila Čermáková
Jiří Gasparič

Contents

Introduction

Spectrophotometric methods, which are used in a wide variety of disciplines, are the subject of numerous monographs and original papers. These sources are usually directed towards professional users, and often focus on only one particular category of spectrophotometric analysis. The majority of them are prepared from the viewpoint of analytes, and they cover the system of all methods of the compounds to be determined, either inorganic (metals and nonmetals) [1–4] or organic (classified according to their chemical structure and use of the material analyzed, e.g., drugs, foodstuffs, and biological materials) [5–7]. Another group of monographs concerns the characterization of spectrophotometric reagents (primarily organic complexing spectrophotometric agents) [8,9]. Other monographs deal with kinetic analytical spectrophotometric methods [10] or primarily with UV/VIS instrumentation [11–13]. (It has to be mentioned that with the progress in instrumental techniques, e.g., flow injection analysis and high-performance liquid chromatography with spectrophotometric detection, a broader interest in spectrophotometric reactions can be observed.)

The aim of this monograph is to afford a deeper understanding of the principles of spectrophotometric determinations on the basis of the properties of analytes and reagents used. This should help in eliminating random approaches to the selection of methods for spectrophotometric determination of particular analytes. The methods of determination are not classified in terms of the character of the analytes (i.e., into determinations of elements or organic compounds); with such an approach, the description of the methods is often repetitive and leads to mechanical use of the procedures described, without critical assessment of their usefulness. The classification proposed by us, based on the principles of chemical reactions as applied to analysis, tries to broaden the general understanding of the chemistry involved in spectrophotometric methods. This approach, of course, cannot be considered to be a complete survey of all published spectrophotometric methods, although we hope we have succeeded in covering the main typical reactions. Where possible, the references are mainly to reviews and are followed by examples of recent applications. A description of both theoretical and instrumental terms used through the monograph follows in Chapter 1.

REFERENCES

1. M. Malát, *Absorption Inorganic Photometry* (in Czech), Academia, Prague, 1973.
2. F. D. Snell, *Photometric and Fluorimetric Methods of Analysis—Non-Metals*, Wiley, New York, 1981.
3. Z. Marczenko, *Separation and Spectrophotometric Determination of Elements*, Wiley, Chichester, 1986.
4. E. B. Sandell and H. Onishi, *Photometric Determination of Traces of Metals; General Aspects*, Wiley, New York, 1987.
5. E. Sawicki, *Photometric Organic Analysis: Basic Principles and Application*, Parts I, II, Wiley-Interscience, New York, 1970.
6. Z. J. Vejdělek and B. Kakáč. *Farbreaktionen in der spektrophotometrischen Analyse organischer Verbindungen. Band I. Organische Farbreagenzien*, Fischer Verlag, Jena, 1969.
7. Z. J. Vejdělek and B. Kakáč, *Farbreaktionen in der spektrophotometrischen Analyse organischer Verbindungen. Band II. Anorganische Farbreagenzien*, Fischer Verlag, Jena, 1973.
8. K. Burger, *Organic Reagents in Metal Analysis*, Akadémiai Kiadó, Budapest, 1973.
9. Z. Holzbecher, L. Diviš, M. Král, L. Šůcha, and F. Vláčil, *Handbook of Organic Reagents in Inorganic Chemistry*, Ellis Horwood, New York, 1974.
10. D. Perez-Bendito and M. Silva, *Kinetic Methods in Analytical Chemistry*, E. Horwood, Chichester, 1988.
11. E. D. Olsen, *Modern Optical Methods of Analysis*, McGraw-Hill, New York, 1975.
12. L. Sommer, *Analytical Absorption Spectrophotometry in the Visible and Ultraviolet: The Principles*. Akadémiai Kiadó, Budapest, 1989.
13. J. M. Hollas, *Modern Spectroscopy*, 2nd Ed., Wiley, Chichester, 1992.

1

Principles of Analytical Spectrophotometry

A. ABSORPTION SPECTRA IN THE ULTRAVIOLET AND VISIBLE REGIONS

1. Origin of Spectra

a. Interaction of radiation with matter

All substances providing absorption spectra in the ultraviolet (UV) and/or visible (VIS) regions absorb radiation in the wavelength range of 200–800 nm. The absorption phenomenon arises from a substance's ability to change its ground energy state to an excited one. The absorption of UV and VIS radiation causes changes in a substance's electronic states (ΔE_{el})and simultaneously changes in its molecular vibrational (ΔE_{vibr}) and rotational (ΔE_{rot}) states. Energy transitions between the electronic states in a molecule are much more energetically demanding ($\Delta E_{el} \sim 150$–600 kJ mol^{-1}) than the transitions between vibrational ($\Delta E_{vibr} \sim 2$–60 kJ mol^{-1}) and rotational states ($\Delta E_{rot} \sim 3$ kJ mol^{-1}). Therefore, absorption of radiation with sufficient energy for electron transition also implies the excitation of vibrational and rotational levels. The total energy changes (ΔE) are then described by Eq. (1)

$$\Delta E = h\nu = \frac{hc}{\lambda} = \Delta E_{el} + \Delta E_{vibr} + \Delta E_{rot} \tag{1}$$

where

$h = 6.626 \; 10^{-34}$ Js
ν = the radiation frequency
λ = the radiation wavelength
$c = 3 \cdot 10^{10}$ cm s^{-1}.

The excited states of substances are ultra-short-lived, and with transfer of energy to adjacent particles, there is a return to ground state.

The absorption spectrum of a substance observed in a homogeneous medium (e.g., in a solvent) is obtained as a function of the transmitance T or absorbance A (see Section B) on the wavelength [or the frequency or the wave number ($\bar{\nu}$), where $\bar{\nu} = 1/\lambda$] of the monochromatic radiation used. For describing spectra of pure substances, often the plot is used of the value of the molar absorption coefficient ε, which characterizes the signal intensity (see Section B), or log ε vs. λ.

Thus, the measured absorption spectrum of a substance is a set of bands corresponding to the electronic transitions, with a structure of finer bands corresponding to vibrational and rotational transitions. The fine structure of bands is apparent in the absorption spectrum of the substance in gaseous state (Fig. 1).

The electronic transition band is characterized by an absorption peak of Gaussian type that is substance specific according to its maximum wavelength (λ_{max}). In the evaluation of spectrophotometric reactions, the observed substance can be analyte, reagent, or their product.

b.   Electronic transitions in organic compounds

Intramolecular transitions.   A substance's absorption spectrum depends on its atomic grouping (the chromophore), that is, how its electrons are capable of changing their energy state. Absorption in the UV or VIS regions is caused by the transitions of bonding electrons in a molecule from the molecular orbitals σ or π, or the nonbonding electrons n, to the antibonding molecular orbitals σ^* or π^* (according to the MO LCAO

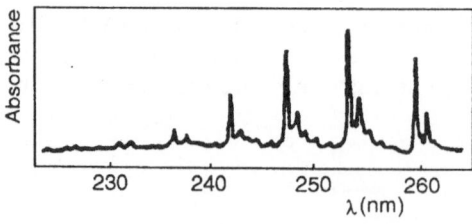

FIGURE 1   Vapor spectrum of benzene.

Hückel theory). In most substances, these transitions occur simultaneously, and the absorption spectrum is composed of the individual absorption bands corresponding to the partial transitions.

The σ–σ* electronic transitions occurring in C−C or C−H bonds require such a great energy that they can be observed upon absorption of radiation of very short wavelength, that is, in the far ultraviolet region (the so-called vacuum ultraviolet region below 200 nm). The spectra of such substances are of no practical importance, but the unsubstituted saturated hydrocarbons, which contain only these single bonds, can be employed as solvents to obtain the spectra of substances in the near ultraviolet or visible region. They include, for instance, *n*-heptane, *n*-hexane, and cyclohexane.

Atoms bearing lone-pair electrons (e.g., S, N, I, Br, Cl) need to be present in the molecule to allow the n–σ* transition. The excitation of these lone-pair electrons from a nonbonding atomic orbital to the antibonding σ* molecular orbitals requires less energy than the excitation of bonding σ electrons, and the spectra formed display an absorption around 200 nm. In substances with strongly electronegative heteroatoms in combination with single bonds, the spectra usually cannot be readily observed, since the bands occur in the region below 200 nm. Some of these substances can serve as suitable spectroscopic solvents (H_2O, alcohols, ethers, amines, halogen compounds $CHCl_3$, CCl_4, ammonia, etc.). The vapor spectrum of methanol shows a λ_{max} at about 183 nm ($\varepsilon \sim 500$), H_2O has a λ_{max} of around 167 nm ($\varepsilon \sim 7000$). The following lower limits apply for the measurement of UV and VIS absorption spectra of compounds in the following solvents: $H_2O \sim 200$ nm, EtOH ~ 209 nm, ether ~ 215 nm, $CHCl_3 \sim 250$, $CCl_4 \sim 260$ nm.

The π–π* transitions are characteristic for compounds with multiple (double or triple) bonds. For instance, the isolated chromophore C=C has an absorption band at $\lambda_{max}=170$ nm. If it is conjugated with another C=C group, the absorption band maximum shifts toward a longer wavelength; there is also an increase in its intensity (bathochromic and hyperchromic shift). Examples are shown in Table 1. In substances with a system of conjugated double bonds, the energy changes (ΔE) are even lower. The more numerous the conjugated double bonds, the more the absorption maximum shifts toward longer wavelengths. These transitions are of importance in aromatics, which for these reasons absorb in the near ultraviolet or visible region of spectrum.

If a molecule with double bonds also contains an atom with a lone electron pair conjugated with them, the less energetically demanding n–π* transition occurs. The substances containing such groups (auxochromes, e.g., C=O, C=S, N=O) have a higher λ_{max} (see Table 1). The n–π* transitions are less intense than the π–π* transitions.

A number of monographs have dealt with the explanation of spectra

TABLE 1 λ_{max} of $\pi-\pi^*$ and $n-\pi^*$ Transitions

Chromophore	λ_{max} of transition (nm)	
	$\pi-\pi^*$	$n-\pi^*$
C=C	170	—
C=C—C=C	220	—
C=C—C=C—C=C	260	—
C≡C	170	—
C=O	166	280
C=N	190	300
N=N	—	340
C=S	—	500
N=O	—	665

of organic substances [1–3]. Tables showing the dependence of ε or log ε on wavelength are contained in [4].

Charge-transfer (CT) transitions. Numerous pairs of organic compounds display an intense absorption in the near ultraviolet (and sometimes visible) region. One of these compounds can be identified as an electron donor (D), and the other as an electron acceptor (A). Their interaction involves electron transfer and the formation of a D–A complex in which these compounds behave as Lewis acids and bases, or as a redox system. The electrons of a strong electron donor solvent can also participate in the electron transfer. If the D–A complex accepts a photon, then a band appears in its electron absorption spectrum, which is formed by the electron transfer, for example, from the ground-bonding π orbital (as well as the nonbonding n orbital) of the donor to the antibonding π^* orbital of the acceptor (intermolecular $\pi-\pi^*$ or $n-\pi^*$ electron transition). (See also Chapter 3, Section B.)

This transition is energetically more advantageous than the electron transitions in the components (donor or acceptor) themselves. The λ_{max} of the resulting band, termed *charge-transfer* (CT) band, is greater than the λ_{max} of the individual D and A components of the complex. Thus, a charge-transfer complex may be manifested by a coloration of originally colorless solutions of these components.

The charge-transfer spectra are well known for a number of organic molecular complexes, examples of which are given in Chapter 3, Section B.

C. Electronic transitions in metal complexes

Metal complexes provide electron spectra [5–8] with characteristic absorption bands that are more or less different from the absorption bands of the complex components, that is, the individual spectra of the metal ion M and ligand L. Primarily the weak organic acids and bases serve as ligands (see Chapter 3). Depending on nature of the bond between M and L, three types of basic electron transitions are considered to explain the absorption spectra of metal complexes, although in some cases it is difficult to distinguish them clearly. They include

> Excitation within the transition metal ions in complex (d–d transitions)
> Excitation related to charge-transfer transitions
> Excitation within the ligands in complex

Sometimes, bands corresponding to other types of transitions can be distinguished in a complex spectrum.

d–d transitions (f–f transitions, respectively). The transition metal complexes with a d^n (f^n, respectively) electron configuration of the central metal ion provide absorption bands corresponding to the d–d (f–f) transitions. The electrons are found at the energy levels formed by a split of the metal ion's basic levels due to the presence of the ligand. The ΔE between these d (f) orbitals is not very great, and thus the transition metal complexes (also their aquo ions) are colored and provide absorption spectra in the visible region. The bands of spectra formed in this way are sometimes denoted as ligand–field absorption bands [7].

The transition metal complexes with a d^1 electron configuration provide a single absorption band. In the complexes with two or more d electrons, more bands can be formed.

From a theoretical point of view, the d–d transition are "forbidden" and thus they have a very low intensity ($\varepsilon \sim 10$–200 for one-electron transitions, and $\varepsilon < 10$ for two-electron transitions). Therefore, their use in spectrophotometric determinations is limited, but they remain of importance in theoretical studies.

Charge-transfer transitions. The explanation of charge transfer bands of metal complexes is associated with a electron transfer between a ligand L and a metal ion M. Radiation absorption is caused by the excitation of electrons from the atom or molecular orbital with a considerably higher electron density to the orbital of another atom or molecule with a lower electron density. Thus, the absorption bands can be formed by the electron transfer from ligand to metal (L–M), or from metal to ligand (M–L). The CT transition intensity is much higher than the intensities of d–d transitions; their $\varepsilon \sim 10^2 - 10^4$. These bands are observed in the complexes of metals that can exist in two oxidation states differing by one, e.g. Fe(III)/Fe(II), Ti(IV)/Ti(III).

The L–M transitions can occur from a σ-bonding orbital or from filled π orbitals of the ligand to the metal ion in a higher oxidation state (which can be reduced by the electron acceptance). For instance, the red color of the ferric thiocyanate complex corresponds to the transfer of an electron from the SCN^- ion to the Fe(III) ion. A simultaneous formation of the CNS radical and Fe(II) ion is assumed. This, in fact, is a photochemical redox reaction. The violet color of MnO_4^- is caused by the charge transfer between the Mn and O ions in tetrahedral arrangement.

M–L transitions occur in complexes where the metal ion is in the lower oxidation state and the ligand contains the π electron system and is capable of back donation. In these complexes, the nonbonding d electron of the metal ion undergoes the transition to the nonbonding orbital of the ligand. Of analytical importance are the tris complexes of 2,2'-bipyridyl, 1,10-phenanthroline, and 8-hydroxyquinoline of Fe(II) (their λ_{max} for the M–L transition are 523, 510, and 581 nm, respectively, and $\varepsilon \sim$ 8,600, 11,200, and 5,000, respectively).

In complexes, the absorption bands of CT transitions are usually manifested in such regions of the spectrum where the radiation is absorbed neither by a free ligand nor by its dissociated anionic form. Substitution on the ligand's π-electron system affects only to a negligible extent the maxima of bands of the corresponding transitions.

Electron transitions within ligands. The absorption bands of complexes associated with electron transitions within ligands are among the analytically important ones, especially in the case of nontransition metal complexes. They are very intense, corresponding to the $\pi-\pi^*$ or $n-\pi^*$ transitions (see Section A.1.b), and depend on the nature of the bond in the complex. If the bond between a metal ion and a ligand is essentially electrostatic, any metal ion affects the spectrum of the ligand similarly as does its dissociation, and the shift of absorption maximum in chelation corresponds to the shifts of λ_{max} in the dissociation scheme of the ligand. The complexation does not much affect the shape of bands or the ε value.

In cases when the bond in complex is strongly covalent, and even the donation of two electron pairs could be anticipated, the energy difference of transitions in the ligand chromophore changes considerably as a result of complex formation.

2.   Effect of Medium on Absorption Spectrum

Chemical changes in the medium are among the most important factors that can result in changes in a substance's spectra (changes in λ_{max} and ε).

Bathochromic, hyperchromic, hypsochromic, or hypochromic shifts may occur.

The effect of solvents on the λ_{max} of absorption bands of an organic compound becomes manifested especially in the case of n-π^* transition. It is possible to distinguish the bands of n-π^* and n-σ^* transitions from those of π-π^* transition, by the measurement of spectra of the compound in different solvents. The maxima of the former transitions move toward a shorter wavelength in a polar solvent, as the nonbonding electron at the ground state are solvated by the polar solvent to a greater extent, and as they require a higher energy for excitation. On the other hand, the bands of π-π^* transitions can be observed at slightly longer wavelengths in such solvents. In nonpolar solvents (hexane, heptane, cyclohexane), spectra with sharper maxima are generally obtained.

In general, it is possible to speak about solvatochromism, a phenomenon whereby the dissolution of a substance is associated with the formation or change of color (see also Chapter 2, Section A), when the stabilization of a nonpolarized or polarized form of a dye occurs due to, for instance, the solvent polarity. For example; 4-nitro-4-dimethylaminoazobenzene is yellow in the nonpolar solvent hexane. With the increasing polarity of solvent, its color changes to yellow-orange and, in ethanol, even to red.

Some colorless or very lightly colored organic compound can form intensely colored solutions by dissolution in mineral acids (for instance, the yellow dibenzalacetone yields a solution of intensely red color in concentrated sulfuric acid). This so-termed halochromism is associated with protonation of the substance molecule (see Chapter 2, Section A).

B. EVALUATION OF SPECTRA

The position of absorption bands in a spectrum in the ultraviolet or visible region is determined by energy absorbed at the electron transition and is related to the structure of the absorbing compound. The determination of absorption wavelength maxima (λ_{max}) is a first step in spectra evaluation. The λ_{max}, together with other information, can serve to identify the compound.

For the purposes of analytical chemistry, it is primarily the quantity of the absorbed radiation that is evaluated, because the quantity is related to the concentration of the analyzed substance (analyte) in the solution.

1. Determination of Analyte Concentration

Radiant flux Φ_0 of monochromatic radiation with the wavelength of λ is weakened by absorption to a radiant flux Φ after passage through the analyte solution. The value of Φ depends on the value of Φ_0, analyte concentration c (mol L^{-1}), and thickness of absorbing layer (cm):

$$\Phi = \Phi_0 \, 10^{-\varepsilon_\lambda lc} \tag{2}$$

Possible losses of the incident radiation flux Φ_0 by reflection or scattering need to be eliminated or minimized.

If $\Phi/\Phi_0 = T$ (where T is the transmittance of the given system), then $0 \leq T \leq 1$, or $0 \leq T \leq 100$ to express $T\% = 100 \, \Phi/\Phi_0$. Introducing the quantity of absorbance $A = -\log T$, we get Eq. (3), the Bouguer-Lambert-Beer law:

$$A = \varepsilon_\lambda lc \tag{3}$$

The value of ε_λ (molar absorption coefficient, molar absorptivity) is used for monochromatic radiation (usually the absorption maximum wavelength is used), in which case it is known as the analyte-characterizing constant. In the SI system its unit is $m^2 \, mol^{-1}$; in analytical chemistry it is usually expressed in $L \, mol^{-1} \, cm^{-1}$. Its value ranges between 10^1 and 10^4, exceptionally up to $10^5 \, L \, mol^{-1} \, cm^{-1}$. Special absorption coefficients can also be defined, if the analyte concentration is expressed in other units. In this monograph, only the values of molar absorption coefficients will be used, and the units will not be shown.

The total radiation absorption under the electron transition considered is proportional to the area under the curve of ε_λ as a function of λ, which is expressed by the integral molar absorption coefficient. This quantity is related to quantities characterizing electron transition probability, for example, oscillator strength. The values of ε_λ are conditional constants, because they depend on experimental conditions, especially on instrumental ones.

According to Eq. (3), there is a linear dependence between the absorbance and the analyte concentration in solution. However, the validity of this rule has certain limitations, and in practice, deviations from this linear relationship have been observed. These deviations can arise from instrumental effects (as shown in Section E), or from a change of chemical equilibria of the analyte in solution. Equilibrium reactions can include acid–base reactions (i.e., the transition between two dissociated forms with different λ_{max} due to the change of H^+ ion concentration), polymerization reactions (the formation of analyte dimers or oligomers, differing again by λ_{max}, at a higher analyte concentration in solution), and the effect of solvent molecules (solvation of ions) and inert salts. When complex formation is used for analyte determination, an important role is played by the ability of weak complexes to dissociate to components that do not absorb at the wavelength used. It has been shown [8] that the error in the absorbance value derived in this way is indirectly proportional to the complex stability constant and ligand concentration.

The most reliable and universal approach for the determination of analyte concentration is utilization of a *calibration plot*, although this approach is relatively time consuming, because it requires the preparation and measurement of a sufficient number of standard solutions of various concentrations under defined optimal conditions (the number of points in a calibration curve depends on whether the measurement is a routine process or whether it involves a development of a new determination method [8], and also on the required precision, the shape of the calibration plot, and the reproducibility of measurement). Results of the measurement are processed either in a graphic way as $A = f(c)$, or by linear regression using the method of least squares, which allows the determination of parameters of the equation

$$A = ac + b \tag{4}$$

where a represents a special absorption coefficient (according to the chosen concentration units), and b is the absorbance of blank. The measurement is performed under a constant thickness of the absorbing layer.

The absorbance of blank can include the absorbance of excess of spectrophotometric reagent and/or the absorbance of solvent, if they absorb at the wavelength selected for the measurement. To get the calibration curve to pass through the origin, the spectrophotometric measurement should be performed against pure solvent or against reagent blank.

It is also possible to evaluate the analyte concentration c_x from its measured absorbance, A_x and the absorbance of a *single standard solution* A_s of a concentration c_s using the Lambert-Beer law:

$$c_x : c_s = A_x : A_s \tag{5}$$

If a sample contains other substances absorbing in the selected wavelength region, it is useful to employ the *standard addition method*, and to determine the analyte concentration graphically or through calculation. Another possibility is to measure against a synthetic solution closely resembling the sample matrix.

Sometimes the *direct calculation* of analyte concentration from the measured value of absorbance is recommended applying the Lambert-Beer law. To obtain the correct value of concentration, it is necessary that the value of ε_λ not be taken from literature, but determined prior to the calculation under the same experimental conditions and using the same instruments that were used to measure absorbance value, because these factors affect the value of ε_λ. Besides, if deviations from the linear dependence occur, it is necessary to check whether the determined value of absorbance lies in the region of validity of the Lambert-Beer law.

A measure of deviation of the measured absorbance values from the linear dependence $A = f(c)$ is the regression coefficient r, the ideal value of which equals 1. If r differs considerably from this value, outlying points may need to be eliminated, necessitating the use of graphical methods or various mathematical procedures [9].

Multicomponent spectral analysis

In cases where the analyzed solution contains more analytes of concentrations c_1 to c_n, absorbing in the same wavelength region, and no interaction occurs among them, then the resulting absorbance at the chosen wavelength is an additive function of the absorbance of individual analytes. Thus, at the unit thickness of absorbing layer,

$$A = \sum_{i=1}^{n} \varepsilon_{\lambda i} c_i$$

(6)

To determine the concentrations of individual components in the solution, it is necessary to perform a series of measurements at appropriately selected wavelengths λ_1 to λ_n [10] (such that one component always absorbs more than the other ones). The absorption coefficients of individual analyzed substances at these wavelengths ($\varepsilon_{\lambda 1}$ to $\varepsilon_{\lambda n}$) need to be determined in advance by the measurement of pure solutions of analyzed substances under the same conditions that apply to the mixture being measured. In this way, n linear equations for n unknown concentrations are obtained for n wavelengths. By solving these equations (for a higher number of analytes, the use of a computer may be necessary), the desired concentration values may be found. An alternative option involves the solution of an overdetermined system, when the measurement is performed at a higher number of wavelengths than that of the analyzed components of the system. To determine c_1 to c_n, graphical methods have been proposed [8].

Derivative spectrophotometry is one of the useful ways to process analytic signals as this technique can be used to distinguish overlapping absorption bands in the spectrum and determine the positions of their λ_{max} and to analyze them quantitatively [8,11,12]. Derivative spectrophotometry is used in multicomponent mixtures of substances to be analyzed and in samples with a significant background absorbance. Differentiating of the absorption spectra may be performed either instrumentally (optically or electronically) or mathematically (using the microcomputer capability of a spectrophotometer), provided that the initial radiant flux Φ_0 is constant.

An example of the spectrum of a two-component mixture with overlapping bands is shown in Figure 2 (derivative of basic curve = 0). At the absorption maximum wavelength, the curves of odd derivatives pass through zero, while the curves of even derivatives display the extreme values (minima and maxima), which can be used for the determination of concentration of both components. The curves of odd derivatives are useful pri-

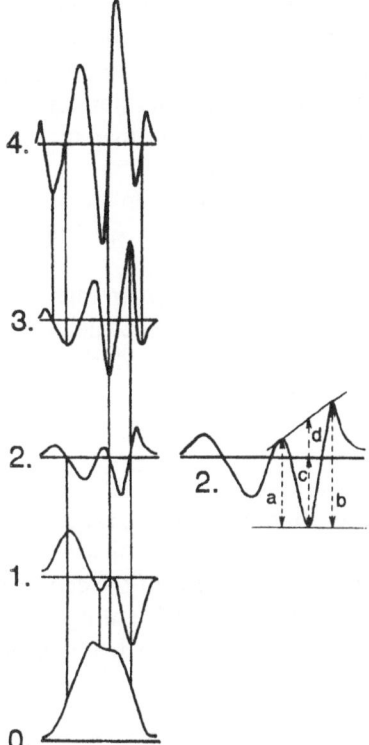

FIGURE 2 Derivative curves of the absorption band formed by two overlapping bands. 1–4, the first to the fourth derivatives.

marily to determine the positions of absorption maxima of the bands that form the superimposed spectrum. The maxima and minima of even derivative curves are used primarily for quantitative analysis. The second derivative curve in Figure 2 shows how concentration may be evaluated.

2. Accuracy, Precision, Sensitivity, and Selectivity

Accuracy is judged by the agreement of measured results with a true value. Because this value is usually unknown, to evaluate the accuracy of a method, it is necessary to compare spectroscopic results with results obtained using other analytical techniques or with the results of analyses of reference materials having about the same composition as the actual sample. To achieve accurate results, it is necessary (1) to use an accurate calibration plot and accurate standards for its preparation, and (2) to work on the same or cross-validated instrument and under the same experimental conditions that were used to prepare the calibration plot. All possible sources of systematic errors need to be considered as well. The best estimate of the true value is the arithmetic average of results, if they are not subject to systematic error. For an accurate method, the arithmetic average approaches the true value.

Precision of measurement can be expressed as the estimate of the standard deviation (s) of a single measurement of absorbance A_i from the average $\overline{A}_i$ [9].

The estimate of standard deviation is a measure of measurement reproducibility. Because the value s depends of the value of the measured variable, it is better to use the relative standard deviation s_r, which is described by the equation

$$s_r = \frac{s}{\overline{A}_i} \tag{7}$$

The measurement precision can be also characterized by the confidence interval [9].

Sensitivity of a measurement is determined by the slope of the calibration graph. It can be expressed as the molar absorption coefficient ε_λ at the wavelength used (usually λ_{max} of the absorption band). Sandell index of sensitivity is also frequently used to express sensitivity [13]:

$$S = \frac{cl}{1000A} \tag{8}$$

where the analyte concentration c is expressed in mg mL^{-1}, and l is in cm.

Selectivity of determination expresses the degree to which other substances interfere with analyte determination by the given method. The interfering substances may be classified as active (ones that interact with analyte, reagent, solvent, etc.) or inert (substances that affect the value of analytic signals without interacting with other substances).

3. Evaluation Techniques

Spectrophotometric measurements can be performed in two basic ways. Most methods used to date are based on measurements in the steady state, that is, after reaching the reaction equilibrium (the quantitative course of a reaction needs to be ensured by an appropriate selection of conditions). It is just as important to perform dynamic measurements, that is, during the course of a chemical reaction. This type of measurement relies on a flow method, termed *flow injection analysis* (FIA) (see Section E), where the absorbance values are read under strictly reproducible conditions before equilibrium is reached. Another category of dynamic measurement includes kinetic methods, in which the change of absorbance value is measured in the course of reaction. The increasing number of dynamic methods has resulted from improvements in measurement instrumentation.

a. Equilibrium (static) measurement

This way of measurement includes all commonly used spectrophotometric methods, in which the evaluation of absorption spectra or of individual band intensity is performed after the constant value of absorbance is reached under the experimentally determined optimal reaction conditions. Using the Lambert-Beer law, the absorbance value measured is then employed to calculate the analyte concentration.

Besides the application for spectrophotometric determination, absorption spectra measured at various H^+ ion concentration, which causes the protonation or deprotonation of some compounds (see Chapter 2) can be also evaluated. At the appropriate wavelength, the dissociation constants of weak (usually organic) acids or bases may be determined from the equation absorbance pH dependence. Similarly, in the reaction between two compounds when a complex is formed the changes of absorption spectra that result from the changing ratio of both components can serve to determine the stability constants of these complexes.

If there are more components in the solution that provide overlapping absorption bands (e.g., several tautomeric or dissociated forms or several complexes of the compound), or if there is only one component that has several absorption bands as the result of several electron transitions from radiation absorption, the resulting spectrum is a sum of spectra of individual forms or transitions. To assign the bands, it is necessary to perform a deconvolution of the spectrum formed, that is, to determine the number of absorption bands and their ε values. Many computer programs have been created for the deconvolution of spectra. Some of these programs are able to evaluate equilibrium constants, as well.

b. Kinetic (dynamic) measurement

Besides measurements using the FIA method (see Section E), dynamic spectrophotometry involves primarily kinetic methods of determination. Kinetic analytical methods [14–18], which are based on the measurement of chemical reaction rate, register the changes in concentration of reacting substances or reaction products in time. Spectrophotometry, i.e. the measurement of absorbance, the value of which is proportional to the concentration of a measured substance, is among the most commonly used instrumental kinetic methods.

The reaction rate v is described as the decrease of reactant (A or B) concentration or as the increase of product (C or D) concentration per unit of time and may be expressed as the derivative of the concentration of reactants or reaction products in time. On the curve of concentration (or absorbance) as a function of time, the rate is represented by the slope of the curve at every point (tg α).

The reaction rate is proportional to the reactant concentrations:

$$v = kA^mB^n \tag{9}$$

where k is the rate constant, and m and n are the reaction orders of individual components. The total reaction order $r = m + n$.

The rate constant (also termed the specific reaction rate) numerically equals the reaction rate at unit concentration of all components involved. It is necessary to keep in mind that to determine which reaction is faster, only rate constants of reactions of the same order can be compared. The values of the rate constant and of the reaction order of individual components can be experimentally determined using differential or integral methods [14–17].

Kinetic analytical methods are usually divided into catalytic and noncatalytic methods. *Catalytic methods* (in homogeneous systems, their use is of major importance) are based on the fact that the rate of catalytic reaction is directly proportional to the catalyst concentration. The analyte (metal, nonmetal, or organic substance) can serve as the catalyst. Catalysts affect the reaction rate even at a very low concentration (the value of which does not change during the reaction). This concentration is usually determined by the differential method. The integral method can be used as well. Catalytic methods allow the determination of the concentration of catalytic substances by several orders of magnitude lower than in equilibrium determinations and are therefore used in trace analysis.

The high sensitivity of catalytic kinetic methods may be further increased by using activators. These include substances that themselves do not catalyze a reaction but that considerably increase its rate in the presence of certain catalysts. Activators act either by affecting the catalyst–substrate interaction or by participating in the catalyst regeneration. Ultimately, they influence the catalytic process indirectly. Since only certain substances that can interact with the catalyst serve as activators, it is possible to increase the catalytic effect of certain ions through the selection of an appropriate activator while suppressing the catalytic effect of interfering ions. In this way, the selectivity of the determination can be also increased. The activation principle can be also applied to determine the activator concentration.

Reaction inhibitors lower the rate of catalyzed reaction. It is assumed that their mode of action is based on reaction with the catalyst, or with one of the reaction components, or even with the activator that affects the catalyzed reaction. In all these cases, the inhibitory effect is proportional to the inhibitor concentration and thus may be used for its determination. Reaction inhibition can be also used in noncatalyzed reactions, where the inhibitor reacts with one of the reaction components.

Catalytic methods can be divided into enzymatic [19,20] and nonenzymatic methods. In clinical laboratories, enzymes (protein catalysts of biochemical reactions) are applied to determine substrates, and reaction activators and inhibitors, and their catalytic effects can be used to determine their concentration. An important property of enzymes is their high specificity. Some enzymes catalyze only one reaction, whereas others are characterized by group specificity (they catalyze only the reactions of substrates with certain functional group) or linkage specificity (they catalyze only the reactions of substances with a certain kind of bond), and some exhibit stereochemical specificity (they catalyze only reaction involving one type of spatial isomers). Enzymatic activity is usually dealt with separately in fields of clinical analyses and so it is not included in this monograph.

Nonenzymatic catalytic methods use classical chemical reactions as do the equilibrium methods. The reactions that are influenced by the presence of a catalyst are named indicator reactions. In kinetic analyses especially, redox indicator reactions are employed. Reactions involving exchange of ligands and complexation reactions are used to a lesser extent. Neutralization reactions are not suitable for kinetic determination due to their high reaction rate.

Noncatalytic methods are used to determine a single component or several components in a analyzed mixture when the reaction rates of these components differ somewhat with a common reagent. These methods do not have as many advantages as catalyzed reactions because it is necessary to use higher concentrations of the substance that is being determined, and thus the relatively lower accuracy of the kinetic method is not compensated by a higher sensitivity of determination. These methods are suitable only for reactions that are too slow for equilibrium determinations or for reactions, in which consecutive reactions occur.

Sensitivity, selectivity, accuracy, and precision of kinetic determination methods, and their comparison with the equilibrium methods are discussed in detail in Ref. [17]. Practical examples of kinetic method applications in individual types of spectrophotometric reactions are presented elsewhere in this volume (Chapter 3, 5).

C. FACTORS INFLUENCING SPECTROPHOTOMETRIC REACTIONS

Spectrophotometric reactions involve transformations of an analyte to products having new absorption bands in the ultraviolet or visible region, thus enabling determination of the original compound. Optimal conditions must be met to enable both the rapid production and the maximum yield of reaction product, and the optimal medium must be selected for the absorbing species to show maximal absorption.

For example, Schiff bases are colorless compounds insoluble in water. However, they dissolve in aqueous mineral acids to form intensely colored protonated forms (see Chapter 6). These bases are formed by the treatment of primary aromatic amines with, for example, p-dimethylaminobenzaldehyde. The formation of colored species is utilized in the determination of amines. Primary aromatic amines can be obtained by reduction of corresponding nitrocompounds, which can be prepared by nitration of aromatic hydrocarbons. All of these compounds can be converted into Schiff bases. Some compounds can be converted into colored products directly; with some compounds it is necessary to perform several steps before the final coloration is reached. The optimal yield of the final product to be evaluated spectrophotometrically depends on several factors, which are physical, physicochemical, or chemical in nature. The most important of these factors include the solvent used, solution acidity, ionic strength of the solution, kind and concentration of electrolytes present, temperature, and presence of interfering substances.

1. Solvent

Water is the solvent most commonly used for spectrophotometric determinations. In some cases, however, to increase the solubility of the organic reagent or the measured reaction product, it is necessary to used mixed solvents (usually formed by water and an organic, water-miscible solvent) or nonaqueous media. Eventually, extraction from the water phase into the organic, water-immiscible solvent can be used.

The choice of solvent can also affect the rate of chemical reaction. For ionic reactions, it has been shown [14] that the increase of solvent-relative permitivity increases the rate constant of reaction between ions of the same charge and decreases the rate constant of reaction between ions of opposite charge. The reaction rate between neutral molecules providing very polar activated complexes increase with the increasing value of relative permittivity. The rate of reaction between ions and neutral molecules that provide weak polar activated complexes does not change much with changes of relative permittivity.

2. Solution Acidity

The solution acidity can have a decisive effect on spectrophotometric reactions. In case of acid–base reactions, a certain concentration of H^+ ions is a prerequisite for the formation of the desired product of the measured substance. In complexation reactions of metal ions with suitable ligands, it is primarily the concentration of H^+ ions that affects the formation of metal hydrocomplexes, which compete with the formation of the desired products. Besides, a number of complexation agents are represented by weak acids or bases, the dissociation or protonation of which depends on the solution pH. In a number of cases, these substances are polybasic acids that can form ligands with different valences based on the solution pH. Even when the dissociation constants of such reagents are known, it is necessary to consider that the concentration of H^+ ions at which the complexation occurs is not identical with the concentration at which the dissociation of the reagent occurs, because the complexation shifts the equilibrium of the reaction $HA = H^+ + A^-$ in favor of the dissociated form such that the complexation occurs at a lower solution acidity. The choice of the optimal solution acidity thus has to be based on the experimental verification of the analyte and reagent properties. The question of suitable acidity is also of major importance in cases when more metal ions that form complexes with the reagent used are present simultaneously in the analyzed solution. If the stability constants of complexes of all metals present differ sufficiently, the optimal acidity for determination of metal ions in the mixture can be found.

In cases when one of the associated components is a complex metal cation or anion or an ion of a weak organic acid or base, spectrophotometric determinations using the formation of ionic associates are affected by the solution acidity in the same way as the complexation reactions described above.

In redox reactions, in which H^+ ions take part, the value of the redox potential (i.e., the oxidation or reduction ability) of reactants depends on the value of $[H^+]$. At the same time, the H^+ ion concentration can also influence the rate constant.

The optimal pH for spectrophotometric reactions can be achieved by the addition of a desired amount of acid or base, or through the use of buffers. In all these cases, it is necessary to verify that these added substances do not enter into competitive reactions with the analyte.

3. Temperature

The effect of temperature on chemical reaction is based primarily on basic thermodynamic relationships. The relationship between the change of the standard Gibbs' energy ΔG^0_T (formerly described as the free enthalpy) and the reaction equilibrium constant K is

$$\Delta G^0_T = -RT \ln K \qquad (10)$$

The change of the equilibrium constant with temperature is described by the equation

$$\frac{d \ln K}{d T} = \frac{\Delta H^0}{RT^2} \tag{11}$$

(where ΔH^0 is the standard enthalpy). If ΔH^0 does not depend on temperature, one can display the dependence of $\ln K$ on $1/T$ and get a straight line with the slope $(-H^0/R)$. For exothermic reactions ($\Delta H < 0$), the equilibrium constant decreases with temperature; for endothermic reactions ($\Delta H > 0$), K increases with temperature.

The kinetics of chemical reactions are significantly affected by temperature as well [14]. The dependence of the rate constant k on the temperature T is described by the Arrhenius equation:

$$k = A \exp(-E/RT) \tag{12}$$

where E is the activation energy (independent of temperature) and A is the frequency factor (which includes the total number of collisions and steric factors).

For many reactions, the rate constant increases 2–3 fold with each 10°C increase in temperature.

4. Spectrophotometric Reagent Concentration

Sufficient reagent concentration shifts the equilibrium reaction toward the formation of measured product. In the case of metal determination by complexation reactions with the reagent in the form of a weak organic acid, the excess of ligand can be achieved either by increasing the reagent concentration (at a constant pH), or by changing the pH (at a constant concentration of excess reagent). At the same time, the reagent concentration affects the reaction kinetics, because the reaction rate depends directly on the concentration of the initial reaction components.

5. Ionic Strength

The ionic strength of a solution can influence all types of spectrophotometric reactions. The effect is greatest when the presence of electrolytes in polar solvents affects the chemical equilibria either of the reagent (dissociation of weak organic acids or bases) or the reaction product (complex formation and stability).

Ionic strength can also affect the rate of chemical reaction [14–16]. The reaction rate increases with increasing ionic strength in the reaction of ions with a common charge, and decreases in the reaction of ions with an opposite charge. The reaction rate of two neutral molecules or of a neutral molecule with an ion is not significantly affected by the ionic strength of the solution. The value of ionic strength has a considerable effect in substance determinations, which utilize the positive effect of surfactants present (see Section C.7).

With respect to the reasons presented, it is recommended that in most substance determinations ionic strength be maintained at a constant optimal value (determined by preliminary study of the spectrophotometric reaction) by the addition of a suitable electrolyte.

6. Interfering Substances

The presence of interfering substances in samples is a serious factor affecting spectrophotometric determinations. In only a few cases is it possible to find a reagent that is completely specific for the substance to be determined or to choose such conditions in which the determination is not influenced negatively by interfering substances. In other cases, before spectrophotometric determination it is necessary either to conduct a suitable type of separation (precipitation or extraction into an organic solvent) or to use a masking reagent.

7. Surfactants

The use of surface-active substances (surfactants, also called tensides) is a very effective way to influence spectrophotometric reactions [20–24]. Surfactants are organic substances, which consist of a nonpolar (hydrophobic) part formed by a long hydrocarbon chain, and a polar (hydrophilic) part. In ionic surfactants, the hydrophilic part of the molecule is either a group carrying a positive charge (quaternary ammonium or pyridinium group, i.e., cationic surfactants) or a negative charge (sulfate or carboxylic group, i.e., anionic surfactants). In nonionic surfactants, there are polyoxyethylene or polyethylene glycol units.

In dilute solutions, surfactants exist in monomeric form (ionic surfactants are dissociated and behave as strong electrolytes). When the critical micelle concentration (cmc) is reached, the aggregates, called micelles, are spontaneously formed. They are usually of spherical or ellipsoidal shape [25]; at high surfactant concentrations they are rod shaped. In water and other polar solvents, so-called normal micelles are formed, in which the surfactant molecules are oriented by the hydrocarbon chain into the center of the micelle and create a nonpolar environment there. The hydrophilic groups are oriented toward the surface of the micelle. The number of surfactant molecules associated in the normal micelles (denoted as the aggregation number) is in the range of 10 to 100. In nonpolar media, inverse micelles are formed, in which the surfactant molecules are oriented in the opposite way.

The cationic surfactants most commonly used in analytical spectrophotometry include cetyltrimethylammonium bromide (CATB): $(CH_3-(CH_2)_{15}N^+(CH_3)_3Br^-$, cmc $= 9.2 \times 10^{-4}$ mol L^{-1}), cetylpyridinium bromide (CPB): $(CH_3(CH_2)_{15}N^+C_5H_5Br^-$, cmc $= 9.0 \times 10^{-4}$ mol L^{-1}), and benzyltetradecyldimethylammonium chloride (Zephiramine: $(CH_3(CH_2)_{13}-N^+(C_6H_5CH_2)(CH_3)_2Cl^-$, cmc $= 3.7 \times 10^{-4}$ mol L^{-1}). The most commonly used anionic surfactant is sodium dodecyl sulphate (SDS): $(CH_3-(CH_2)_{11}SO_4^-Na^+$, cmc $= 8.1 \times 10^{-3}$ mol L^{-1}). Nonionic surfactants include polyoxyethylene(9.5)octylphenol (Triton X-100): $(C_8H_{17}(C_6H_4)O-(CH_2CH_2O)_{9.5}H$, cmc $= 1.0 \times 10^{-4}$ mol L^{-1}) and polyoxyethylene(23)dodecanol (Brij 35): $(CH_3(CH_2)_{11}O(CH_2CH_2O)_{23}H$, cmc $= (6.0–9.1) \times 10^{-5}$ mol L^{-1}).

One of the reasons to use surfactants in spectrophotometry is the ability of micellar surfactant solutions to dissolve some barely soluble compounds and thus allow their determination in an aqueous medium without the need for extraction into an organic solvent. Another reason involves the use of ion associates formed by the dissociated surfactant molecules at c_S < cmc with big organic ions of the analyte having the opposite charge. These associates can either be measured directly in the aqueous medium or, more commonly, they can be extracted into an organic solvent (see Chapter 4).

The ability of surfactants to affect the absorption spectra of metal chelates and acid–base properties of organic metallochromic reagents is most often used. The change of position and height of absorption bands is explained either by the formation of ionic associates and their following solubilization in surfactant micelles or by the change of medium micropolarity in the binding of chelates into the surfactant micelles. Bathochromic or hypsochromic shifts of absorption maxima of chelate bands relative to the band of the reagent used lead, in a number of cases, to the elimination of reagent blank. Hyperchromic shifts increase, sometime enormously, the sensitivity of determination (see Chapter 3).

The effect of surfactants on the acid–base equilibria of organic reagents can markedly affect spectrophotometric determinations; this influence will be discussed in detail in Chapter 2. The effect of surfactants on redox reactions has been studied primarily from a theoretical viewpoint [26].

A special type of surfactant action is so-called micellar catalysis, it is the effect of surfactant micelles on the rate of chemical reactions. Many studies have been devoted to this problem [27–29], and it has been found that the rate constant can be increased by one to three orders of magnitude, or decreased accordingly.

The derivation of relationships for the rate of chemical reaction in micellar solutions is based on the assumption that the observed organic reactant A is partitioned between water and a micellar "pseudophase" in the solution.

In the aqueous phase, a reactant reacts with the rate constant k_v; in the micellar phase, it reacts with the rate constant k_m, giving the same products:

$$M + A \xrightleftharpoons{K} MA$$
$$\downarrow k_v \qquad\quad \downarrow k_m$$
$$\text{products} \qquad \text{products}$$

where

M = micellar phase

MA = micellar phase with the bound reactant

K = binding constant of this interaction.

In simple cases, the electrostatic rule may be applied to estimate the kinetic effect of surfactants. According to the rule, cationic surfactants catalyze the reactions of organic molecules (that are bound and concentrated in micelles) with nucleophilic anions that can be electrostatically attracted to the cationic micelles. Reactions of organic molecules with electrophilic cations are inhibited because of their electrostatic repulsion from cationic micelles. Anionic surfactants catalyze the reactions of micellar-bound molecules with electrophilic cations and inhibit reactions with anions.

The presence of salts in solution usually suppresses the catalytic action of surfactants, because the salt ions cover the surface of micelles and prevent the access of reactants.

D. SOLVENT EXTRACTION IN SPECTROPHOTOMETRY

A number of substances are insoluble in aqueous media. However, by using nonaqueous solvents, their molecular absorption spectra can be measured and used to determine these substances. Organic solvents can affect the solubility of the products of spectrophotometric reactions (e.g., chelates and ionic associates), as well as of the spectrophotometric reagents.

1. Solute–Solvent Interaction

The choice of a suitable solvent as the medium for a spectrophotometric reaction must be based on the knowledge of properties of the reacting substances (analyte and reagent) and the solvent.

The molecular properties of the solute need to be considered (ionic or covalent character, polarity or polarizability, presences of polar, hydrophilic substituents, etc.). Information about solvent behavior depends on its chemical properties, but also on its physical properties, which can play a role in the solute–solvent interaction (protic or aprotic character, solvent polarity defined by the value of relative permitivity ε_r and the dipole moment μ, electron donor/acceptor properties, etc.). Examples of some solvents used in spectrophotometry are listed in Table 2.

Frequently, mixtures of mutually miscible solvents are used. Depending on the ratio of the solvents involved, some properties (e.g., the ε_r values) of such mixtures can differ considerably from the properties of the pure solvents.

The ability of coordinately saturated complexes to bind solvent molecules in outer coordinate sphere—and sometimes also their ability to bind

TABLE 2 Physical Properties of Selected Solvents

Solvent	pK_{SH}	$\varepsilon_r{}^a$	μ^b
Protic (amphiprotic)			
Protogenic			
acetic acid	14.4	6.1	1.7
Protophilic			
formamide	17.0	109.5	3.7
Neutral			
water	14.0	78.5	1.8
methanol	16.7	32.7	1.7
ethanol	19.1	24.5	1.7
n-butanol	21.6	17.5	1.7
Aprotic			
Polar protophobic			
nitromethane	—	35.8	3.5
acetone	—	20.7	2.9
nitrobenzene	—	35.0	4.0
Polar protophilic			
pyridine	—	12.5	2.1
ether	—	4.5	1.2
dimethylformamide	—	36.7	3.9
Inert			
hexane	—	1.9	0
cyclohexane	—	2.0	0
benzene	—	2.3	0
chloroform	—	4.7	1.01

[a]Relative permittivity.
[b]Dipole moment, Debye units.

the solvent in the inner coordinate sphere — is related to solvation properties of the solvent. Examples are shown in Ref. [30].

In most cases, the attractive forces in solute–solvent interactions need to be considered. These are often explained by the existence of weak intermolecular forces (contrary to strong chemical bonds).

Thus, the solution of a substance in a solvent is the result of a number of solute–solvent interactions, making the solvent a very important component of the system, one that affects numerous chemical reactions in the system [30]. For instance, many nonaqueous solvents have a low value of ε_r relative to water and therefore prevent dissociation reactions of molecules. In many cases their application allows the solute being studied to be regarded as ionic associates (see Chapter 4). Another important consideration concerns the effects of solvents upon the absorption spectra of dyes, complexes, and so on.

2. Liquid–Liquid Extraction

In systems of two mutually immiscible liquid phases, the extraction of substances employs a selective transfer of components across an interface. This process is based on differences in solvation, which is reflected by different solubilities of components in these phases. For very low concentrations of component X, the Nernst partition law can be applied and the partition constant K_d can be expressed:

$$K_d = \frac{[X]_2}{[X]_1} \tag{13}$$

where $[X]_2$ and $[X]_1$ are the concentrations of substance X in solvents 2 and 1, respectively. In practical extractions, deviations from this law frequently occur, especially due to mutual miscibility of phases, change that occurs when they are mixed (it is appropriate to use mutually saturated solvents after their previous shaking), entry of solvent molecules into the extracted species, and most importantly side reactions in both phases (dissociation, polymerization, association). Therefore, the partition ratio D_c is used instead of K_d

$$D_c = \frac{c_{X_2}}{c_{X_1}} \tag{14}$$

Its value is determined by the ratio of total analytical concentrations of the partitioned component in given phases. If $D_c \rightarrow K_d \geq 10^2$ [8], the substance X is quantitatively extracted and the value of the conditional molar absorption coefficient approaches the true molar absorption coefficient in the organic phase.

The extraction yield E_A is important in characterizing the extraction process. It expresses the extent of transfer of the substance X from one phase to another (e.g., from aqueous phase 1 to organic phase 2):

$$E_A = \frac{100\, D_c}{D_c + (V_1/V_2)} \tag{15}$$

where V_1 is the volume of the aqueous phase, and V_2 is the volume of the organic phase. The efficiency of a single extraction increases if the ratio V_1/V_2 decreases. However, it is better to extract several times by smaller portions of V_2 than once using large volume. The true extraction yield asymptotically approaches 100% and as a rule, samples are extracted by three smaller portions of solvent (to perform the extraction more than five times does not make sense).The extraction efficiency is often increased by the addition of inorganic salts with polyvalent ion (the so-called salting-out reagent), primarily when there is a decrease of the medium's relative permitivity [31].

The quantitative course of an extraction equilibrium can be described by the extraction equilibrium constant K_{ex}, that is, by the ratio of the concentration of extracted species (e.g., a chelate or ionic associate of AB) in the organic phase 2, and the concentrations of components in phase 1 (e.g., aqueous phase) according to the Eq. (16):

$$K_{ex} = \frac{[AB]_2}{[A^+]_1\,[B^-]_1} \tag{16}$$

provided that B$^-$ is water soluble (e.g., a ligand able to form a complex or the anionic component of an ionic associate). If HB is acid dissolved in the organic phase, other equilibria need to be considered for the aqueous (dissociation) as well as the organic phase (e.g., polymerization).

Extraction spectrophotometry, which is based on the above principles, is used primarily in quantitative analysis. Often selectivity and sensitivity of the determination can be increased by the separation and concentration of product into the nonaqueous phase. Some solvents have the disadvantage of being environmentally hazardous. Hence, in a number of systems it is appropriate to avoid the extraction step by the addition of substances with solubilizing abilities (surfactants), and to perform the determination in the ternary system in an aqueous medium.

For micro- and trace extraction spectrophotometry, reactions often involve the formation of electroneutral complexes (chelates) of metals as well as nonmetals, ionic associates, or solvates. The stoichiometry of absorbing species in an organic solvent usually differs from that in an aqueous medium. Stoichiometry and optimization of individual processes are described in the literature.

The liquid–liquid extraction process is usually performed in shaking flasks or separating funnels. The time and intensity of shaking necessary to establish the equilibrium are characteristic for a given system and are found experimentally. The reproducibility of results depends to a considerable degree on the shaking procedure. Therefore, the use of laboratory shaking machines is recommended.

The extraction process is used either for the isolation and separation of the analyte from the sample to be analyzed or for the isolation of the reaction product (chromogen). In analytical practice, problems can arise by inadequate separation of the aqueous phase (drops of water on the glass walls of the organic layer) and formation of cloudiness (e.g., of chloroform layers). To obtain clear chloroform extracts, it is useful to filter them through a dry filter paper. Sometimes, good results may be achieved by using freshly distilled chloroform or chloroform purified by passing it through a column of alumina. Working at temperatures over 20°C helps bring clear extracts. In manual extraction, the use of separating funnels is a time-consuming operation, and therefore some modifications are often advisable. The whole procedure can be carried out in glass tubes, which may be inserted into the spectrophotometer after the separation of layers [32].

Many studies have dealt with the linking extraction techniques to automated systems [33–36]. Phase separators with a poly(tetrafluoroethylene) porous membrane are employed in the FIA technique (see Section E). This technique is very promising because it offers continuous extraction determinations. Another advantage is offered by the closed-loop system [36], which uses continuous extraction to achieve analyte preconcentration from the aqueous to the organic phase.

E. PROCEDURES

1. Basic Instrumentation

Basic parts of common spectrophotometers for the ultraviolet and visible regions, as well as the most modern instrumentations are described in the literature [3,7,8,17,37–39]. Detailed description of instrumentation is outside the scope of this book. This chapter should be considered as a comment on the standard instrumental techniques.

The analyzed substances are usually measured in the form of solutions in appropriate cells (glass cells for the visible region and quartz cells for the ultraviolet region). If samples are concentrated enough, measurements can be performed on substances in a gaseous state. Techniques have also been proposed to measure the absorbance of samples in a solid state (solid-phase spectrophotometry). For instance, the concentration of a colored product sorbed onto an ion exchanger can be determined by "ion exchanger–phase absorptiometry." In this case it is appropriate to use the term "attenuance"

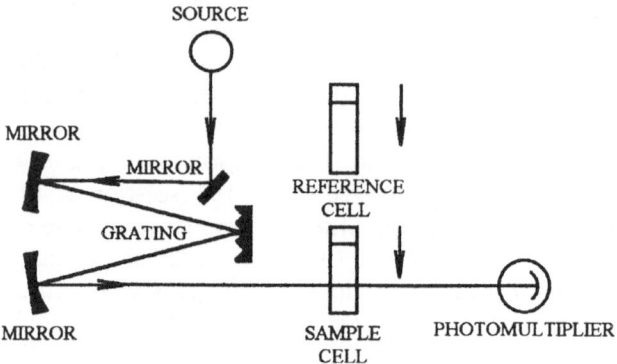

FIGURE 3   Single-beam spectrophotometer.

than "absorbance" because the absorption and dispersion of radiation occur simultaneously. The theoretical background of this method and an overview of metal determinations in the form of sorbed complexes are shown in Refs. [40,41].

Both single-beam (Fig. 3) and double-beam (Fig. 4) spectrophotometers are used. Most instruments are supplemented with a recording device

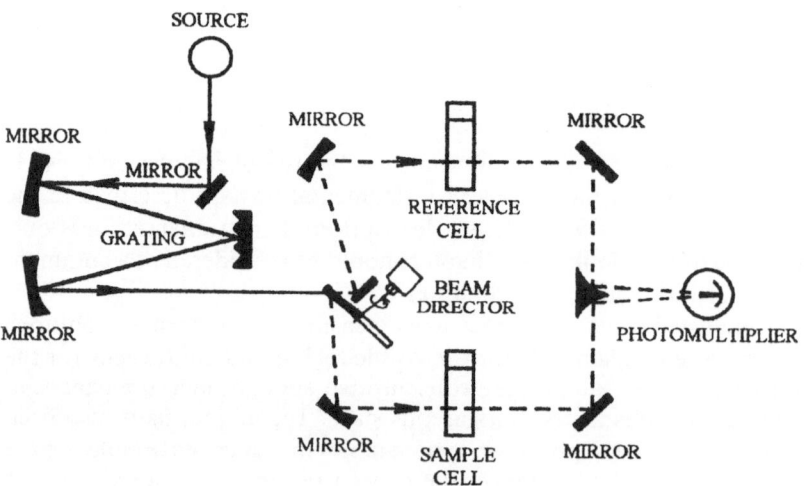

FIGURE 4   Double-beam spectrophotometer.

(chart recorder), digital display, or digital printer. The process of spectrum registration in the entire ultraviolet and visible range generally takes several minutes.

These instruments allow the measurement of absorbance in the range of 0.100–4.000, as well as measurement of log A or transmittance (T). In the course of recording, the scanning speed is often automatically changed according to the spectrum shape. Most instruments allow for calculation of higher-order derivatives of absorption curves. Time dependencies can be also measured at fixed as well as programmed wavelengths. Spectrophotometers are often supplemented with automatic sample handling connected to computers.

A newer way of measurement is represented by diode array spectrophotometers (Fig. 5) [42], in which a single universal detector is replaced by an array of silicon photodiodes (several hundreds of which are used for the range 200–800 nm). The monochromator (a holographic grating) is placed behind the sample, where it separates from the passing radiation individual narrow intervals of wavelengths that hit the photodiodes. The signals of all wavelengths are detected simultaneously with a rapid response rate, so that it takes only a fraction of a second to obtain the entire spectrum. The spectrum is then immediately displayed or stored in the computer's memory. The spectral resolution is determined by the number of diodes in the spectrum range used.

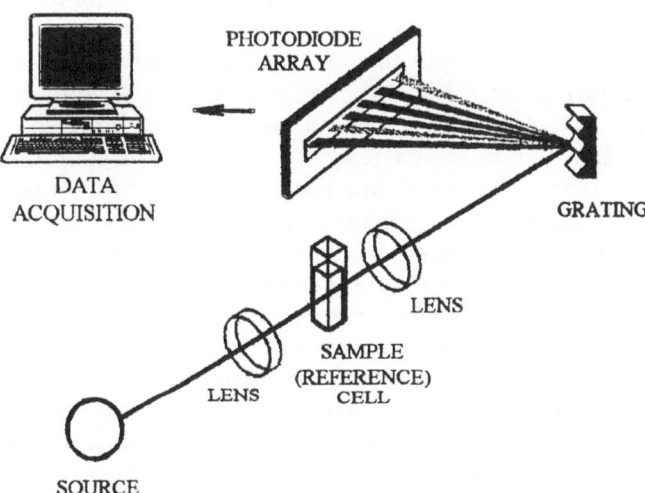

FIGURE 5   Diode array spectrophotometer.

The arrangement shown in Figure 5 enables the measurements of fast changes of spectra caused by chemical reactions. It is therefore especially useful for dynamic measurements. In kinetic measurements it allows simultaneous observation of absorbance-time dependence in reaction mixtures containing several components of a solution that absorb at different wavelengths [43]. The possibility of fast measurement replication, and summation of spectra by computer improves the signal-to-noise ratio.

Another example of modern instrumental techniques employed in spectrophotometry involves the methods of flow analysis, which bring many practical advantages, including greater speed of analysis and primarily the automation of substance determination.

In *continuous flow analysis* (CFA) (Fig. 6), sample, reagent, and other required components of the reaction mixture are aspirated into a mixing coil using a peristaltic multichannel pump (Fig. 6a). Bubbles of introduced air separate individual samples (Fig. 6b). The continuous flow is passed (after deaeration) into the flow cell. The absorbance measurement is usually performed in an equilibrium steady state, and the signal obtained has the shape shown in Figure 6c, with the height h proportional to the analyte concentration. The instrument can be supplemented by equipment for sample filtration, extraction, and dialysis.

This method is suitable for automated measurements of a large series of samples. The multichannel arrangement also allows simultaneous determination of several analytes by various methods. It is used primarily in clinical and environmental analysis.

Flow injection analysis (FIA) (Fig. 7) [44,45] is based on controlled dispersion of a sample within a space and time. An exactly measured volume of sample is injected into a nonsegmented continuous stream of suitable liquid (reagent, buffer, etc.) (Fig. 7a). The sample forms a zone with controlled dispersion in the carrier stream (a concentration gradient is formed) (Fig. 7b). The reaction with a reagent occurs in a reaction coil. The reaction product then moves to a flow cell, where the absorbance is measured. The signal obtained has the shape of a peak (Fig. 7c), in which height h, width W, and area A depend on the analyte concentration. The entire process is highly reproducible provided that all the stages are reproducibly timed. The signal is measured under conditions when chemical equilibrium is not reached. Again, the instrument can be supplemented by equipment for filtration, solvent extraction, and preconcentration.

Another example of this type of measurement is the post column derivatization in HPLC with spectrophotometric detection. Sample components separated by liquid chromatography are treated on-line by a suitable reagent and are determined photometrically.

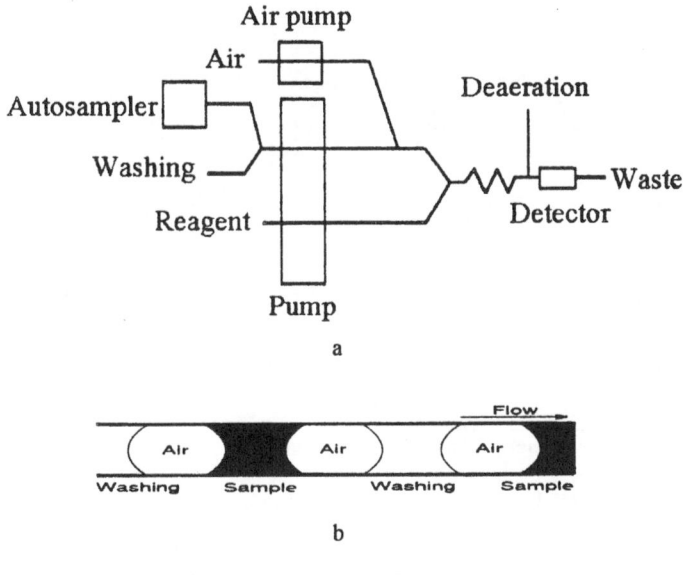

a

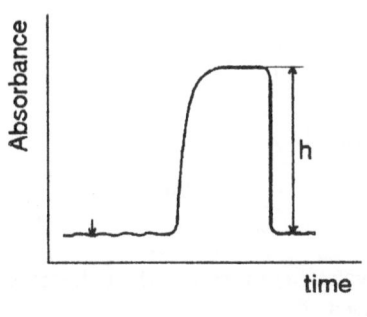

b

c

FIGURE 6 Continuous flow analysis: (a) manifold, (b) schema of the flow after aspiration of two samples, (c) signal.

2. Instrumental Errors of Spectrophotometric Measurements

The accuracy and precision of spectrophotometric measurements are to a considerable degree affected by the quality of the instruments used and by the functioning and correct calibration of their components.

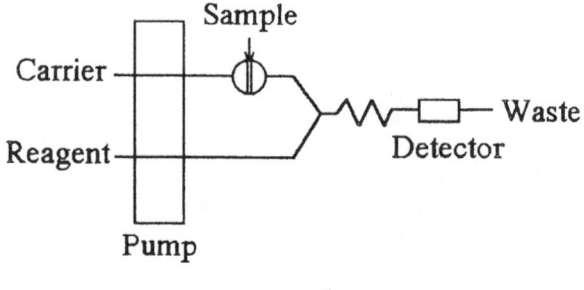

a

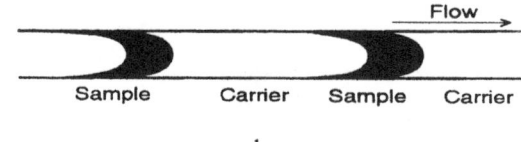

b

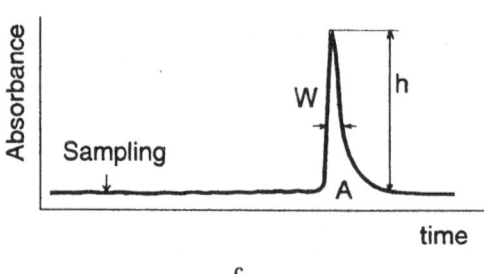

c

FIGURE 7 Flow injection analysis: (a) manifold, (b) schema of the flow after injection of two samples, (c) signal.

Instability of a radiation source causes errors, especially in single-beam instruments, where such instability can induce changes in the observed signal that are comparable to the changes induced by chemical reaction. Instruments rely on voltage and source intensity stabilizers or a feedback system to maintain a constant intensity of the source.

Incorrect adjustment of the monochromator is a frequent source of errors. The calibration of a wavelength scale can be performed by using the spectral lines of discharge tubes [e.g., deuterium lamp (characteristic lines at 374.915 nm, 376.962 nm, 379.687 nm, 383.435 nm, 388.799 nm, 396.899

nm, 410.062 nm, 433.928 nm, 485.999 nm, and 656.100 nm)], the spectra of holmium or didymium filters, or lanthanoid salt solutions. In a similar way, the calibration of an absorbance scale is performed using absorption filters or the solutions of $K_2Cr_2O_7$ (values of ε at 235, 257, 313, and 350 nm are used), Cu(II), or Co(II) salts. For monochromatic radiation there is an optimal monochromator slit width. At excessive width, the spectral resolution is reduced, the fine structure of the spectrum disappears, and the values of absorbance become lower. However, the use of an excessively small width leads to a reduction of determination sensitivity and to a low signal-to-noise ratio, which may induce errors in absorbance readings.

Other errors of spectrophotometric measurement are induced by the formation of stray light, by divergence or convergence of the optical beam, and by scattering of light. These errors usually reduce the value of the signal obtained. In registration spectrophotometers, the speed of spectrum scanning plays a role. At too high a speed, the fine structure of the spectrum is not registered, the shape of absorption bands is distorted, and the value of absorbance is reduced.

Any shortcomings in the function of instrumental components can cause systematic errors of measurement. In most cases, however, these can be removed by the correct adjustment, calibration, and operation of spectrophotometers.

REFERENCES

1. H. H. Jaffé and M. Orchin, *Theory and Application of Ultraviolet Spectroscopy*, Wiley, New York, London, Sydney, 1965.
2. J. Fabian and H. Hartmann, *Light Absorption of Organic Colorants*, Springer-Verlag, Berlin, 1980.
3. J. M. Hollas, *Modern Spectroscopy* 2nd Ed., Wiley, Chichester, 1992.
4. L. Láng, *Absorption Spectra in the Ultraviolet and Visible Region*, Akadémiai Kiadó, Budapest, 1961.
5. D. D. Perrin, *Organic Complexing Reagents: Structure, Behavior and Application to Inorganic Analysis*, Interscience Publishers (Wiley), New York, 1964.
6. K. Burger, *Organic Reagents in Metal Analysis*, Akadémiai Kiadó, Budapest, 1973.
7. E. D. Olsen, *Modern Optical Methods of Analysis*, McGraw-Hill, New York, 1975.
8. L. Sommer, *Analytical Absorption Spectrophotometry in the Visible and Ultraviolet: The Principles*, Akadémiai Kiadó, Budapest, 1989.
9. K. Eckschlager and V. Štěpánek, *Information Theory as Applied to Chemical Analysis*, Wiley, New York, 1979.
10. S. D. Frans and J. M. Harris, Selection of analytical wavelengths for multicomponent spectrophotometric determinations, *Anal. Chem. 57*: 6802 (1985).

11. G. Talsky, L. Mayring, and H. Kreuzer, Feinauflösende UV/VIS-Derivativ-spektrophotometrie höherer Ordnung, *Angew. Chem. 90*: 840 (1978).
12. I. Dol, M. Knochen, and C. Altesor, Enhancement of precision and accuracy in derivative spectrophotometry of highly absorbing samples, *Analyst 116*: 69 (1991).
13. E. B. Sandell, *Colorimetric Determination of Traces of Metals*, Interscience, New York, 1959.
14. K. J. Laidler, *Chemical Kinetics*, 2nd Ed. McGraw-Hill, New York, 1965.
15. H. B. Mark, Jr. and G. A. Rechnitz, *Kinetics in Analytical Chemistry*, Inter-science, New York, 1968.
16. M. Kopanica and V. Stará, Kinetic Methods in Analytical Chemistry, in *Comprehensive Analytical Chemistry*, Vol. XVII, Gy. Svchla, Ed., Elsevier, New York, 1983.
17. D. Perez-Bendito and M. Silva, *Kinetic Methods in Analytical Chemistry*, E. Horwood, Chichester, 1988.
18. H. A. Mottola and D. Perez-Bendito, Kinetic determinations and some kinetic aspects in analytical chemistry, *Anal. Chem. 64*: 407R (1992); *Anal. Chem. 66*: 131R (1994).
19. G. Guilbault, *Enzymatic Methods of Analysis*, Pergamon Press, Oxford, England, 1970.
20. W. L. Hinze, Use of surfactant and micellar systems in analytical chemistry, in *Solution Chemistry of Surfactants*, Vol. 1, K. L. Mittal, Ed., Plenum Press, New York, 1973.
21. L. S. Romsted, Micellar effects on reaction rates and equilibria, in *Surfactants in Solution*, Vol. 2, K. L. Mittal and B. Lindman, Eds., Plenum Press, New York, 1984.
22. E. Pelizetti and E. Pramauro, Analytical applications of organized micellar assemblies, *Anal. Chim. Acta 169*:1 (1985).
23. M. A. Diaz Garcia and A. Sanz-Medel, Dye-surfactants interaction: A review, *Talanta 33*: 255 (1986).
24. L. Čermáková, M. Malát, and I. Němcová, The use of surfactants in analytical chemistry, in *Instrumentation in Analytical Chemistry*, Vol. 2., J. Zýka, Ed., E. Horwood, Chichester, 1994.
25. M. Menger, On the structure of micelles, *Assoc. Chem. Res. 12*: 111 (1979).
26. C. Minero, E. Pramauro, E. Pelizzetti, and D. Meisel, One-electron transfer equilibria and kinetics of *N*-methylphenothiazine in micellar systems, *J. Phys. Chem. 87*: 399 (1983).
27. J. H. Fendler and E. J. Fendler, *Catalysis in Micellar and Macromolecular Systems*, Academic Press, New York, 1975.
28. C. A. Bunton, Micellar catalysis and inhibition, *Purre Appl. Chem. 49*: 969 (1977).
29. A. D. James and B. H. Robinson, Micellar catalysis of metal complex formation; a model system for the study of metal ion reactivity at charged interfaces, *J. Chem. Soc., Farad. Trans. 74*: 10 (1978).
30. K. Burger, *Solvation, Ionic and Complex Formations Reaction in Non-Aqueous Solvents: Experimental Methods for Their Investigation*, Akadémiai Kiadó, Budapest, 1983.

31. Y. Nagaosa, Salting-out of polar solvents from aqueous solution and its application to ion-pair extractions, *Anal. Chim. Acta 120*: 279 (1980).

32. J. Gasparič, A. Barcuchová, and I. Horká, A simplified prodecure for the extraction-photometric determination of organic compounds (in Czech), *Chem. Listy 80*: 1214 (1986).

33. J. Kawase and M. Yamanaka, Continuous solvent-extraction method for the spectrophotometric determination of cationic surfactants, *Analyst 104*: 750 (1979).

34. T. Imasaka, T. Harada, and N. Ishibashi, Fluorimetric determination of gallium with lumogallion by flow injection analysis based on solvent extraction, *Anal. Chim. Acta 129*: 195 (1981).

35. J. A. Sweileh and F. F. Cantwell, Use of peak height for quantification in solvent extraction/flow injection analysis, *Can. J. Chem. 63*: 2559 (1985).

36. R. H. Atallah, J. Ruzicka, and G. D. Christian, Continuous solvent extraction in a closed-loop system, *Anal. Chem. 59*: 2909 (1987).

37. D. L. Andrews, Ed., *Perspectives in Modern Chemical Spectroscopy*, Springer-Verlag, Berlin, 1992.

38. L. G. Hargis and J. A. Howell, Ultraviolet and light absorption spectrometry, *Anal. Chem. 64*: 66R (1992).

39. A. P. Thorne, Fournier transform spectrometry in the ultraviolet, *Anal. Chem. 63*: 57A (1991).

40. K. Yoshimura and H. Waki, Ion-exchanger phase absorptiometry for trace analysis, *Talanta 32*: 345 (1985).

41. M. L. Fernandez-de Cordova, A. Molina Diaz, and M. I. Pascual-Reguera, Determination of trace amounts of cobalt at sub mg L^{-1} level by solid phase spectrophotometry, *Anal. Lett. 5*: 1961 (1992).

42. D. G. Jones, Photodiode array detectors in UV-VIS spectroscopy: Part I, *Anal. Chem. 57*: 1057A (1985), Part II, *Anal. Chem. 57*: 1207A (1985).

43. B. Blanco, J. Gene, H. Iturriada, and S. Maspoch, Application of a photodiode array detector to multi-component determination by flow-injection analysis, *Analyst 112*: 619 (1987).

44. M. Valcárcel and M. D. Lugue de Castro, *Flow Injection Analysis: Principles and Applications*, E. Horwood, Chichester, 1987.

45. J. Růžička and E. H. Hansen, *Flow Injection Analysis*, 2nd Ed., Wiley, New York, 1988.

2

Acid–Base Reactions

A. INTRODUCTION

A vast number of spectrophotometric reactions are carried out in acid or basic media. Establishing optimal conditions ensures that the course of the chemical reaction (i.e., its mechanism and rate) results in the formation of the highest yield of chromogen. The acidity and basicity of the medium can further affect the properties of the chromogens in cases when they undergo different acid–base equilibria (dissociation or protonation). The theory of acid–base reactions is based on the Brönsted-Löwry theory of proton transfer between the solute and the solvent (SH), which has a defined autoprotolytic constant K_{SH} (see Table 2 in Chapter 1).

$$SH + HA \rightleftharpoons SH_2^+ + A^- \tag{1}$$
$$B + SH \rightleftharpoons BH^+ + S^- \tag{2}$$

Acids and bases are classified as strong or weak in a given medium according to their ability to donate or accept protons [Eqs. (1) and (2)]. Their strength is expressed by the value of the dissociation constants of acids K_a or bases K_b for aqueous media as by defined Eqs. (3) and (4).

$$K_a = \frac{[H_3O^+][A^-]}{[HA]} \tag{3}$$

$$K_b = \frac{[BH^+][OH^-]}{[B]} \tag{4}$$

Conditions for the spectrophotometric reactions cover not only the range of the pH scale, but also the region of extreme acidity (H_0 function) and basicity (H_- function). Some analytes under these conditions themselves undergo dissociation or protonation accompanied by color changes in their absorption spectra that can be utilized for their determination.

Besides acidity and basicity of the solvent medium used, ionic strength and the presence of additional components, especially a surfactant, can strongly affect both the reaction course and the spectral properties of the analyte. The latter changes are caused by the ability of the components to interact with the individual dissociated forms of the chromogen.

Examples of acid–base reactions used for spectrophotometric determinations as well as acid–base properties of different types of chromogens and reagents used in the analysis of metals are discussed in this chapter.

B. ACID–BASE REACTIONS OF ANALYTE

Colorless organic analytes having a large π-electron system can form colored salts (cations, anions) on protonation or deprotonation (halochromism). These cations and anions are deeply colored due to the polymethine chromophore and the free electron pairs of heteroatoms participating in the mesomery of the conjugated π-electron system.

Concentrated sulfuric acid is often used as the reagent to produce colored solutions of the analyte. Protonation or carbanion formation is the mechanism responsible for the coloration in some cases, whereas in many others the mechanism is much more complicated and involves sulfonation, dehydratation, oxidation, condensation, or polymerization.

Triphenylmethyl compounds form the corresponding triphenylmethyl cations [1,2], which can be demonstrated for the case of triphenylmethanol [Eq. (5)]:

$$(C_6H_5)_3C-OH + H_2SO_4 \rightarrow (C_6H_5)_3C^+ + HSO_4^- + H_2O \qquad (5)$$

Similarly, the analogous carbonium ion is formed [2] from 1,1-diphenylethylene [Eq. (6)]:

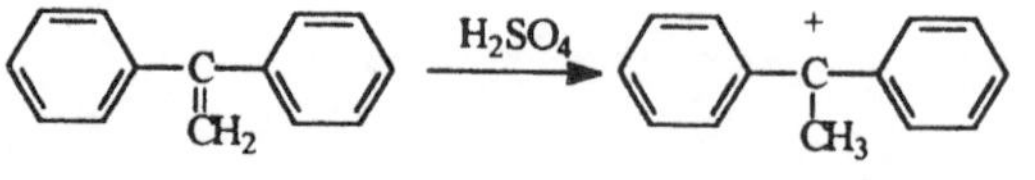

$$(6)$$

The treatment of phenothiazines with concentrated sulfuric acid is accompanied by oxidation and the formation of a radical ion [Eq. (7)] [2]:

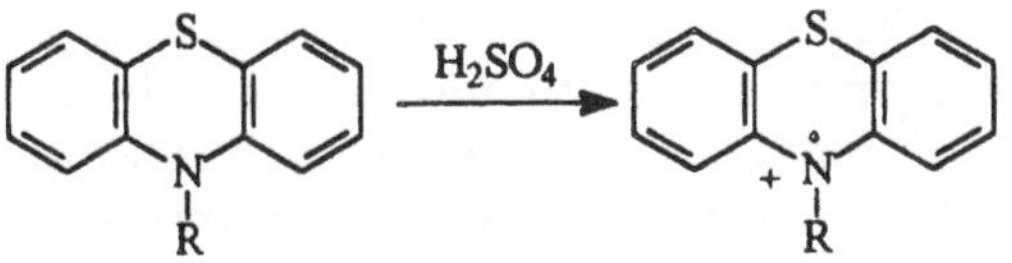

$$ \text{(structure)} \xrightarrow{\text{H}_2\text{SO}_4} \text{(structure)} \qquad (7) $$

Colorations obtained in sulfuric acid solutions are the basis of determinations of carbohydrates, steroids, veratrum alkaloids, pyridoxal, and phenothiazines [1,2]. The determination of steroids is known as the Kober reaction, and much attention has been devoted to the elucidation of the structure of the chromogens thus obtained, [3–11; Eq. (8)]:

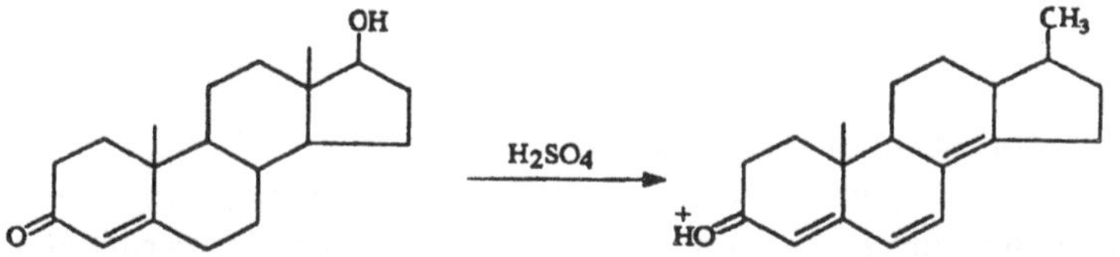

$$ \text{(structure)} \xrightarrow{\text{H}_2\text{SO}_4} \text{(structure)} \qquad (8) $$

Dilute sulfuric acid used to treat five- or six-membered O-heterocyclic compounds causes the formation of colored oxonium salts. An example of the formation of such an oxonium salt is the determination of khellin [2].

Alkali hydroxides are used for determinations of 1,8-dianthrones and 1,8-dianthraquinones. Alkalization causes such solutions to turn red [12]. Analogously, the absorption bands of polynitrophenolates are shifted towards longer wavelengths [13] in comparison with solutions of the corresponding phenols. The determination of picric acid is an example [14,15]. (See also the discussion of nitroso- and nitrophenolate chromogens in Chapter 6.) The mechanism of the reaction of aromatic meta-, di-, and polynitro aromatic compounds with alkalizing agents in polar organic solvents is more complex and is dealt with in Chapter 6. Tetrabutylammonium hydroxide in dimethylformamide can be used instead of alkali hydroxides in aqueous or alcoholic solutions [16].

C. ACID–BASE REACTIONS OF SOME ORGANIC REAGENTS

Acid–base properties are important in characterizing organic reagents and their behavior; they have wide application in spectrophotometric reactions. The determination of dissociation constants (usually termed $-\log K_a = pK_a$) of organic reagents in different media are based on evaluation of spectrophotometric data (including the calculation of thermodynamic constant values [17] and the elimination of experimental effects [18]). The entire reaction scheme has to be solved in cases when the dissociation equilibria may lead to the formation of different tautomers or hydration products. Apart from the structure (basic skeleton type, kind of substituent and its position), of acids and bases, their dissociation constants are also effected by ionic strength, solvent, and the presence of other components, for example, surfactants.

The effect of surfactants on individual dissociated forms of dyes involves a number of partial equilibria [19–25]. It is primarily in the formation of ionic associates with variously dissociated form of the protolyte H_nA at submicellar surfactant concentration, that the electrostatic forces between ions of opposite charge are of major importance. This can be formulated by Eq. (9):

$$H_nA + nS^+ \rightleftharpoons \{nS^+, A^{n-}\} + nH^+ \tag{9}$$

where S^+ is a cation of the surfactant.

Furthermore, it is assumed that the resulting product is solubilized into surfactant micelles with hydrophobic interactions occurring at a supercritical surfactant concentration. This is accompanied by shifts of the absorption band maxima of the dye that are characteristic of the dissociated form of the dye and the surfactant type. As far as interpretating the total scheme of surfactant effects, however, the distribution of the different forms of the dye in the solvent (water) and the micellar pseudophase is usually unknown.

Surfactants also shift the acid–base equilibria, which is manifested by an apparent change of the dissociation or protonation constant of the dye reagent used. In individual cases, this shift is affected to varying degrees by the presence of strong electrolytes that can influence the acid–base equilibrium changes induced by surfactants through the competitive equilibria of ion exchange on the surface of micelles. The change in acid–base equilibrium induced by surfactant action is often almost completely suppressed at high values of ionic strength (see Chapter 1, Section C). Still, surfactant action can result in a improvement of optimal experimental conditions in the spectrophotometric utilization of a dye–surfactant system.

The acid–base behavior of spectrophotometric reagents can be shown by some basic types of dyes, the properties of which have been studied most extensively. They can be divided into nonchelating and chelating dyes.

1. Nonchelating Dyes

These dyes are important acid–base indicators used to determine pH (H_0) values in aqueous as well as nonaqueous media. In many cases, the coloration of their differently dissociated forms is utilized for further spectrophotometric reactions. Nonchelating dyes are used for determinations based on the formation of ionic associates and in redox reactions.

a. Phthaleins and sulfophthaleins

Phthaleine dyes that contain phenol-, naphthol-, cresol-, thymol-, and other groups form a colorless lactone structure in acidic medium and quinoid-colored anions in alkaline medium that can be transferred into a colorless carbinolic base in strongly alkaline medium Eq. (10).

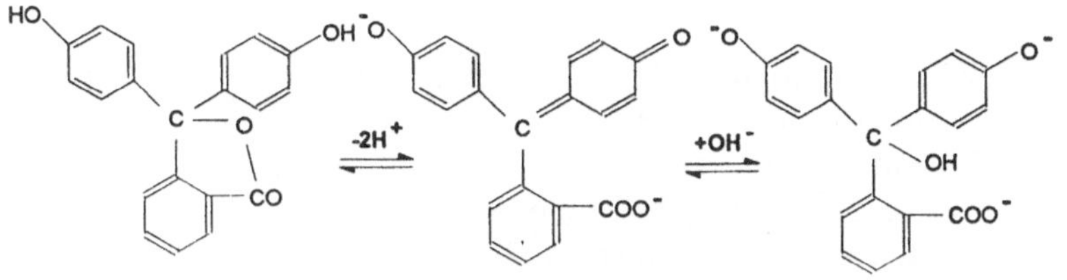

$$(10)$$

In phenolphthalein, the colored anion is purple, while for other compounds shown, the anions range from green to blue. The color transition occurs between pH values of 8 and 9. The most common of these compounds, phenolphthalein, is important as an acid–base indicator and has spectrophotometric applications, for instance, for the spectrophotometric microdetermination of CO_2. In this case, the decrease of dye absorbance in a $Ba(OH)_2$ medium due to CO_3^{2-} or HCO_3^- is measured at $\lambda = 470$–500 nm. The procedure has been modified for carbon determinations in alloys [26].

The sulfophthaleins include a great number of compounds with different subsituents on phenol rings. Among the subsituents are the halogens and the $-CH_3$ and $-CH(CH_3)_2$ groups, as shown in structure (1) and others.

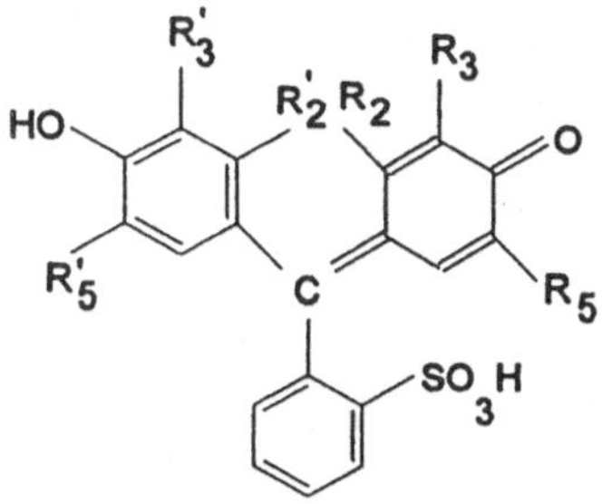

(1)

	R_2, R_2'	R_3, R_3'	R_5, R_5'	pK_a^a
Phenol Red	H	H	H	7.8
Chlorophenol Red	H	Cl	H	6.0
Bromophenol Blue	H	Br	Br	3.8
Bromocresol Green	CH_3	Br	Br	4.6
Thymol Blue	CH_3	H	$CH(CH_3)_2$	8.9
Bromothymol Blue	CH_3	Br	$CH(CH_3)_2$	7.1

^{a}20°C, I = 0.1 mol L^{-1} KCl

The values of pK_a correspond to the $-OH$ group dissociation in the molecule, the dissociation of the $-SO_3H$ group occurs at pH < 2.

Thymol Blue and Bromothymol Blue are used for spectrophotometric determinations of carbon in metals. Phenol Red is used in a similar way for continuous determination of CO_2 in gases [27].

The effect of various types of surfactants on individual dissociated forms of the above dyes and on the change of the pK_a value in media of different ionic strength has been described [28–32]. A comparison of results shows that the absorption spectra (their λ_{max} as well as ε values) of both dissociated forms HA$^-$ and A^{2-}, as well as the pK_a values of the dyes, are affected by the kind and concentration of both the surfactant and the electrolyte that are present.

b.   Basic (cationic) triphenylmethane dyes

These compounds contain different substituted amino groups on the basic skeleton (2).

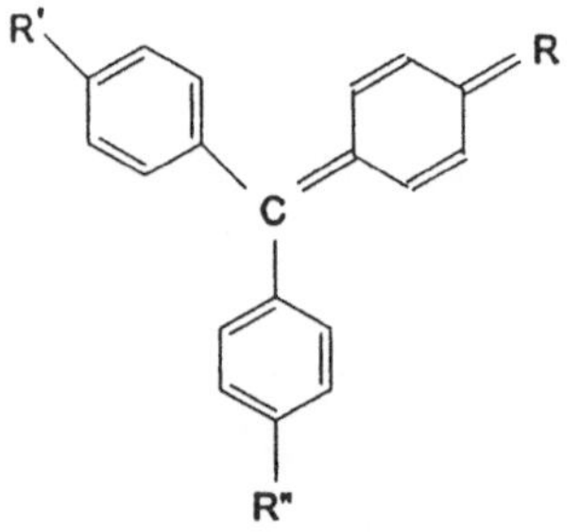

(2)

	R	R′	R″
Malachite Green	$N^+(CH_3)_2$	$N(CH_3)_2$	H
Crystal Violet	$N^+(CH_3)_2$	$N(CH_3)_2$	$N(CH_3)_2$
Methyl Green	$N^+(CH_3)_2$	$N(CH_3)_2$	$N(C_2H_5)_2$
Ethyl Violet	$N^+(CH_3)_2$	$N(C_2H_5)_2$	$N(C_2H_5)_2$

They are acid–base indicators for nonaqueous media and are of importance for redox reactions as well as for formation of ionic associates. Methyl Green and Crystal Violet are used primarily to determine anionic surfactants (λ_{max} = 420 and 630–634 nm for the former dye, and 584–593 nm for the latter dye). Methyl Green possesses importance in spectrophotometric determinations of surfactants containing $-SO_3^-$ or $-O-SO_3^-$ groups, and with a chain length of C_8 or more [33]. The hydrophobicity of Malachite Green as well as of other dyes together with charge changes upon protonation has been evaluated from changes of pK_a values in the presence of inverse micelles [34].

With increasing acidity, the coloration of Crystal Violet, for instance, changes from violet in the basic form, to green, and later to yellow. Equation (11) depicts the course of the protonation.

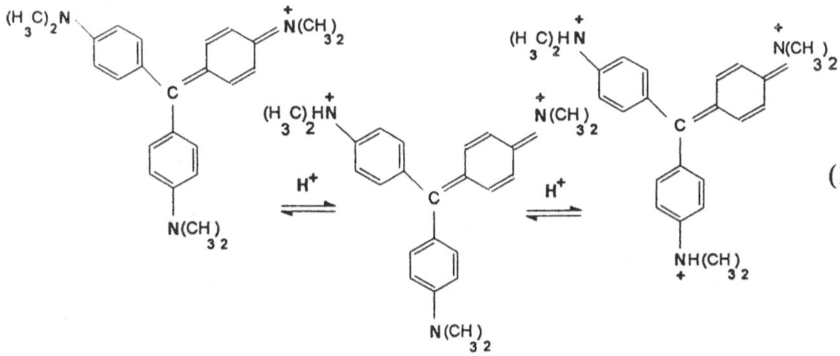

$$(11)$$

C. Azo dyes

Simple azo dyes can occur in two colored acid–base forms, depending on the solution pH. They have a color change usually from red (acidic medium) to yellow (alkaline medium). The compounds belonging to this group include water-insoluble substances, such as *p*-aminoazobenzene or Dimethyl Yellow (i.e., *p*-dimethylaminoazobenzene), as well as water-soluble compounds, such as Methyl Orange (i.e., the sodium salt of 4′-dimethylaminoazobenzene-4-sulfonic acid, $pK_a \sim 3.4$) and Methyl Red (i.e., 4′-dimethylaminoazobenzene-2-carboxylic acid, pK_a ~ 5). The protonation of dimethylaminoazobenzenes, and their derivatives generally occurs simultaneously with their azo-hydrazo tautomerism. An example is shown in the general Eq. (12),

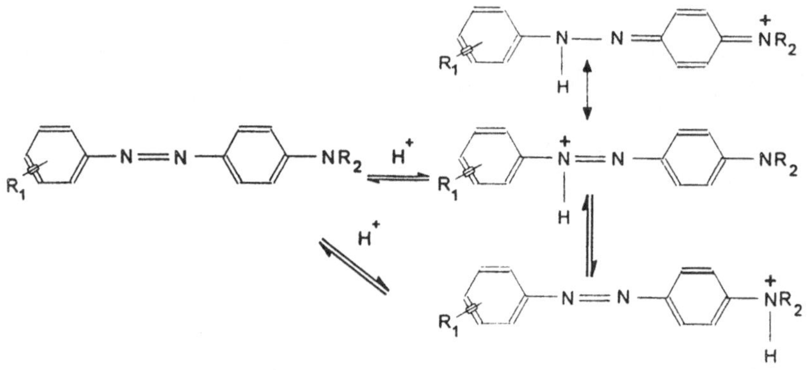

$$(12)$$

where R_1 is a substituent in the phenol ring and R is the CH_3 group.

The change of color that occurs with the acidification of the dye solutions corresponds to the formation of a cation with two resonance hydrazo forms and a tautomeric ammonium ion. Generally, ionic tautomer have a deeper coloration than a neutral dye. The azobenzene behavior in strongly acidic media [35] and the effect of different substituents [36] have been studied. Bishop [37] has showed the effect of subsituents on the pK_a values of some azo indicators, and the effect of medium has been examined [38].

The commonly used acid–base indicators Methyl Orange and Methyl Red are used as model substances to study interactions of azo dyes with synthetic polymers or surfactants. The interactions [39–42] are accompanied by spectral changes in indicator solutions, which allow one to assess, for instance, the effect of solvents [39] or the aggregation of dyes [40]. The entire dissociation scheme covering individual tautomeric forms can be evaluated in the study of the surfactant effect on the pK_a value of these indicators [41,42].

Methyl Orange coloration is employed in boron determination by the FIA method. Boric acid reacts with D-sorbitol to give a complex, the acidity of which is elevated relative to the boric acid itself. Methyl Orange is added to the solution, and the absorbance increase of its acidic form is observed (λ = 520 nm). The increase is directly proportional to the amount of H_3BO_3 [43].

2. Chelating Dyes

These substances are also weak acids or bases. Besides the donor atoms with lone electron pairs (N, O, S), suitable for a chelate ring closure, they contain acidic or basic groups that are dissociable or protonable (usually —OH, —COOH, —NH$_2$, or a substituted amino group). From the viewpoint of assessing their spectrophotometric properties, these compounds contain at least two auxochromic groups in the mutual *o*-position bound on the hydrocarbon skeleton, which is a π-chromophore. When using these organic reagents in chelate-forming reactions, it is necessary, in some cases, to compare the coloration of the individual dissociated forms of the reagent (with the knowledge of its stepwise dissociation constants pK_{a1}, pK_{a2}, etc.) and the coloration of their chelates (with the knowledge of the stability constants of the complexes being formed). In this way, it must be found the optimal solution pH that occurs when the value of the conditional stability constant of the complex (see Chapter 3) is not too low (usually not lower than 10^8) and a sufficiently sharp color contrast is preserved between the metal chelate and the appropriate free form of the reagent.

In the literature, the values of dissociation constants of different metallochromic reagents are usually tabulated together with the values of stability constants of their metal complexes [44,45]. Accompanying the values determined from spectrophotometric data are usually descriptions of absorption spectra (with appropriate data on λ_{max} of characteristic bands and ε values) and the structure of individual dissociated forms of the reagent. In a number of cases, data for the strongly acidic region H$_0$ is also included.

Following are examples of some of the most frequently used reagents from the group of triphenylmethane and gallein dyes, azo dyes, and other compounds.

a. Triphenylmethane dyes and gallein derivatives (xanthene dyes)

Acid–base properties of these groups of dyes are influenced primarily by the kind and position of substituents on the basic skeleton, and this is reflected in the spectrophotometric characteristics of individual dissociated forms of the dyes. They are multiacidic acids, usually containing a sulfo group, which is often already dissociated in the acidic medium.

In a strongly acidic medium, a further protonation can occur with the formation of an H$_{n+1}$A$^+$ cation. With respect to the dye type, the slightly acidic groups (—COOH, —OH) gradually dissociate in a medium with pH 1–13, while individual anionic forms of dyes are gradually formed. Each of these forms is characterized by absorption bands and individual stepwise dissociation constants.

As an example, the structure of the group of Chromazurol S dyes **(3)** is shown,

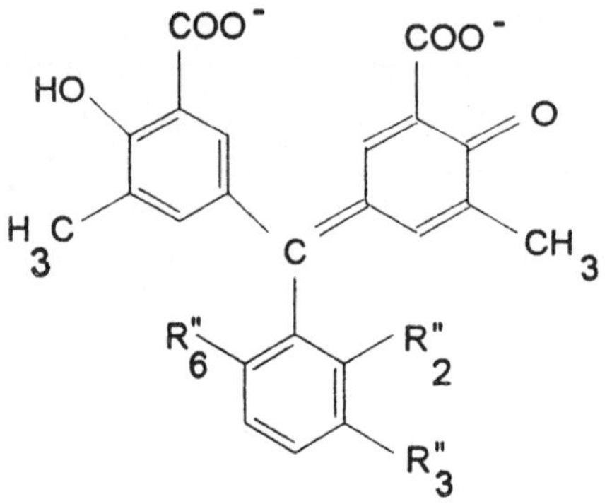

(3)

	R_2''	R_3''	R_6''
Eriochromcyanine R (ECR)	SO_3^-	H	H
Chromazurol S (CAS)	Cl	SO_3^-	Cl
Eriochromazurol B (EAB)	Cl	H	Cl

as well as the structure of the group of pyrogallolsulfogallein derivatives **(4)**.

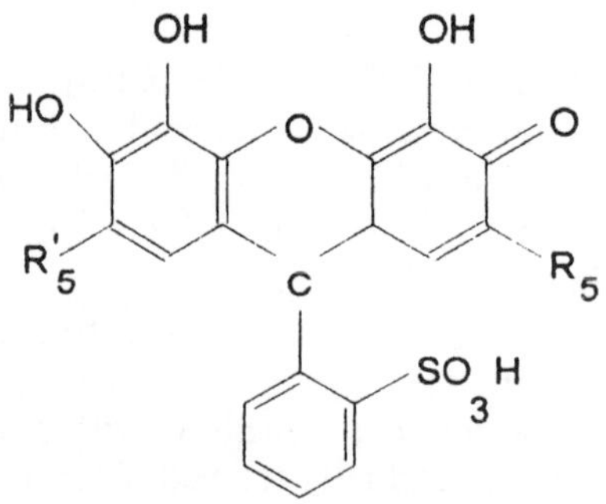

(4)

	R_5	R_5'
Pyrogallol Red (PGR)	H	H
Bromopyrogallol Red (BPR)	Br	Br

A comparison of the effect of structure in the aforementioned dyes is summarized in Table 1, which lists λ_{max} of their individual absorption bands and pK_a values for dissociation [Eq. (13)].

$$H_3A^- \; \underset{\longleftarrow}{\overset{pK_{a2}}{\longrightarrow}} \; H_2A^{2-} \; \underset{\longleftarrow}{\overset{pK_{a3}}{\longrightarrow}} \; HA^{3-} \; \underset{\longleftarrow}{\overset{pK_{a4}}{\longrightarrow}} A^{4-} \qquad (13)$$

The use of dyes in the presence of surfactants often significantly changes spectrophotometric characteristics and the experimental conditions for photometric determinations of metal ions. Therefore, the properties of dye–surfactant systems have been studied with respect to the basic surfactant properties (formation of dye–surfactant ionic associates and monomer–surfactant micelle equilibrium) [50–55]. Apart from changes of absorption bands of individual dissociated forms of a dye, the dye–surfactant interaction is also manifested by a change of dissociation constants for individual dyes, and surfactants. As an example, Figure 1 shows the effect of the cationic surfactant Septonex on the absorption bands of the dissociated form H_3A^-, H_2A^{2-}, and, HA^{3-} of Bromopyrogallol Red. The surfactant has the most pronounced bathochromic and hyperchromic effects on the H_2A^{2-} form.

Table 2 shows the effect of increasing surfactant concentrations on pK_a values of ECR dye. The data confirm that the major effect on protolytic equilibrium is exercised at submicellar surfactant concentrations, when $c_s < cmc$, where c_s is the surfactant concentration and cmc is the critical micellar concentration. At higher surfactant concentrations, there is only a small change in equilibrium constants. In the example shown, the effect on

TABLE 1 Wavelength Maxima (nm) and pK_a Values of Dissociation Forms of Some Dyes

Dye	H_3A^-	pK_{a2}	H_2A^{2-}	pK_{a3}	HA^{3-}	pK_{a4}	A^{4-}	Ref.
ECR	474	2.19	516	5.67	434	11.85	586	46, 47
CAS	470	2.27	492	4.87	427	11.79	599	48
EAB	470	2.68	508	6.07	428	11.95	599	48
PGR	430, 505	6.0	545	9.5	545	12.0	585	49
BPR	433, 530	4.4	556	9.0	556	11.3	598	46, 47
	440, 520	4.39	558	9.06	558	11.31	598	49

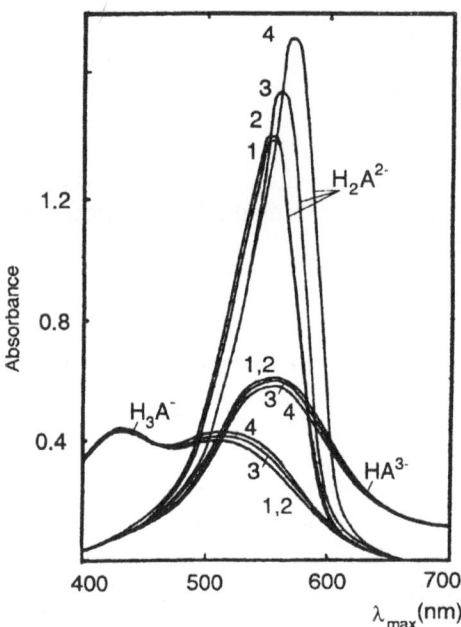

FIGURE 1 Absorption spectra of H_3A^-, H_2A^{2-}, and HA^{3-} forms of Bromopyrogallol Red in the presence of Septonex. $c_{BPR} = 3 \times 10^{-5}$ mol L^{-1}, ionic strength $I = 0.1$ mol L^{-1} (NaNO$_3$), measured against water. Curves $[c_s(\text{mol L}^{-1})]$: 1, 1×10^{-5}; 2, 5×10^{-5}; 3, 5×10^{-4}; 4, 1×10^{-2}.

TABLE 2 $pK_a = f(c_{\text{Septonex}})$ Values for Eriochromcyanine R. Ionic strength $I = 0.2$ mol L^{-1} (NaNO$_3$), $20 \pm 0.5°C$, c_{Septonex} (mol L^{-1})

c_{Septonex}	pK_{a2}	pK_{a3}
0	2.19	5.67
5×10^{-5}	2.15	5.77
2×10^{-4}	1.89	6.31
5×10^{-4}	1.88	6.41
1×10^{-3}	1.67	6.37
3×10^{-3}	1.69	6.35
5×10^{-3}	1.65	6.38

Source: Ref. [47].

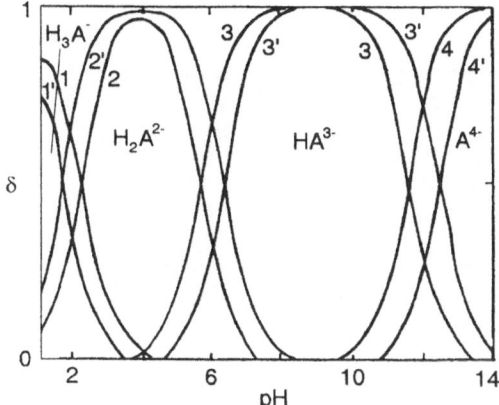

FIGURE 2 Distribution diagram of Eriochromcyanine R in the absence (curves 1 to 4) and in the presence (curves 1′ to 4′) of the surfactant Septonex [47]. c_{ECR} = 4 × 10^{-5} mol L^{-1}, I = 0.2 mol L^{-1} (NaNO$_3$), c_s = 3 × 10^{-3} mol L^{-1}.

individual pK_a constants is also different. With increasing c_s, the pK_{a2} value decreases, the pK_{a3} value increases. This phenomenon, which is observed in a number of dyes, is related to the change of symmetry of boundary resonance structures of stepwise dissociated dye forms caused by the presence of surfactant.

Figure 2 shows the resulting surfactant effect on acid–base equilibria of Eriochromcyanine R. The distribution diagram shows the regions where individual dissociated forms exist in the absence of Septonex and in supercritical concentration (i.e., for solubilized dye forms). These results correspond to the determination of changes of optimal conditions for metal ion determinations using these dyes in the presence of surfactant.

b. Azo dyes

As with simpler azo dyes, *o*- (and also *p*-) hydroxyazo dyes, in which —OH is conjugated with the azo group, can be subject to azo-hydrazo tautomerism. Whereas in unsubstituted azophenols the azo form prevails, in *o*- and *p*-hydroxynaphthylazo compounds, the ketohydrazo form is more stable and always exhibits a bathochromic effect relative to the azo form. In acidic media, the azo group in azobenzenes is subject to protonation when a cation is formed with a bathochromic shift relative to unprotonated compounds. Increasing pH causes dissociation of —OH groups of hydroxyazo dyes, which leads to the formation of the anion [Eq. (14)]:

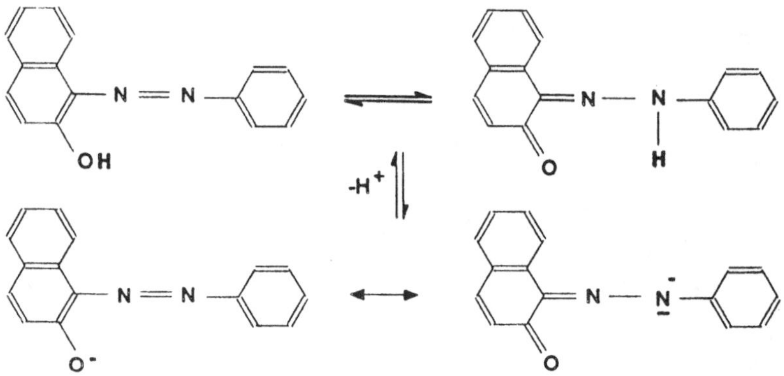

OH –N = N– ⇌ =N — N– O H

–H⁺

O⁻ –N = N– ⟷ =N — N– O

$$(14)$$

The λ_{max} for the anion formed is always longer than for the undissociated species. The study of acid–base behavior of o-hydroxy-substituted heterocyclic azo dyes is related directly to their frequent use in the formation of spectrophotometrically important complexes with metal ions. The acid–base behavior of the often used 4-(2-pyridylazo)resorcinol (PAR) is described by Eq. (15).

$$H_3A^+ \underset{\longleftarrow}{\overset{-H^+}{\longrightarrow}} H_2A \underset{\longleftarrow}{\overset{-H^+}{\longrightarrow}} A^- \underset{\longleftarrow}{\overset{-H^+}{\longrightarrow}} \qquad (15)$$

pK_a: 3.0   5.6   11.9

The H_3A^+ form has $\lambda_{max} = 389$ nm for the azo form, 455 nm for the hydrazo form, 383 nm for H_2A (azo form prevails), 411 nm for HA^- (hydrazo form), and 488 nm for the A^{2-} azo form [56–58].

Depending on the pH of medium (and that of the dye concentration), it is possible to show the formation of different types of complexes in which a variously dissociated form of PAR exists (see Chapter 3) [58]. Absorption spectra of the complexes are compared with the spectra of the deprotonated form of the dye.

The effect of surfactants on numerous spectrophotometric determinations is based on dye–surfactant interaction. The changes of acid–base properties of azo dyes undergoing complexation in the presence of surfactants have been researched in studies dealing with metal ion determinations (see Chapter 3). The changes in PAR acid–base properties in the presence of surfactants (and under the effect of a strong electrolyte) have been documented [56,57].

c. Other substances

Similarly, the acid–base behavior of other organic reagents used in spectrophotometry has been studied, for example, the variously substituted derivatives of 9,10-anthraquinones and 9-anthrones [59], and 1-amino-4-hydroxyanthraquinone [60]. The pK_a values determined by various methods, of some carbohydrazide and thiocarbohydrazide derivatives and dithiocarbamates are given in Refs. [61,62,63], and the pK_a of triazene derivative in Ref. [64].

Various other studies have examined the surfactant effect on acid–base behavior of some complexation dyes, including ferron [65] and other 8-hydroxyquinoline derivatives [66]. The effect of surfactants on acid–base equilibrium of the thiazine dye Azur B [67], and on phenothiazine derivatives [68] have been described.

D. ACID–BASE REACTIONS OF THE REACTION PRODUCT (CHROMOGEN)

Some spectrophotometrically usable chemical reactions result in the formation of colorless or weakly colored products, the protonated or dissociated forms of which, however, are deeply colored. Therefore, the acidification of alkalization, respectively, of the reaction mixture is the last step of such spectrophotometric procedures.

It will be shown later (see Chapter 6) that the condensation of primary aromatic amines with *p*-dimethylaminobenzaldehyde results in the formation of colorless azomethines. Their protonated forms are intensely colored and are the chromogens in many spectrophotometric procedures.

Nitrosophenols or nitrophenols, are formed during the action of nitrous or nitric acid, respectively, on phenols. These compounds are usually colorless or only slightly yellow. However, they can be easily converted into their salts of intense yellow or orange-to-red salts by the addition of alkali hydroxide or ammonia.

Azo dyes obtained by the coupling of phenols with diazotized anilines containing nitro groups are yellow to orange. The color of their solutions turns to a red-to-violet color after the addition of alkali hydroxide. Orange-to-red coupling products obtained with 1-naphthol derivatives also posses properties of acid–base indicators; their color in alkaline media is violet to blue.

REFERENCES

1. Z. J. Vejdělek and B. Kakáč, *Color Reactions in Spectrophotometric Analysis of Organic Compounds: Vol. II, Inorganic Reagents* (in German), VEB Fischer Verlag, Jena, 1973.
2. Z. J. Vejdělek and B Kakáč, *Color Reactions in Spectrophotometric Analysis of Organic Compounds* Supplementary Vol. II, *Inorganic Reagents* (in German), VEB Fischer Verlag, Jena, 1982.

3. M. Kimura, K. Akiyama, and T. Miura, Chromogenic reactions of steroids with strong acids. VI. Behavior of phenolic steroids in concentrated sulfuric acid, *Chem. Pharm. Bull. 22*: 643 (1974).
4. M. Kimura and T. Miura, Chromogenic reactions of steroids with strong acids. VII. Reactions of A-aromatic steroids with concentrated sulfuric acid, *Chem. Pharm. Bull. 24*: 181 (1976).
5. T. Miura and M. Kimura, Chromogenic reactions of steroids with strong acids. VIII. Mechanism of the first stage in the Kober reaction, *Chem. Pharm. Bull. 25*: 1042 (1977).
6. T. Miura and M. Kimura, Chromogenic reaction of steroids with strong acids. IX. Behavior of estrone methyl ether in concentrated sulfuric acid, *Chem. Pharm. Bull. 26*: 171 (1978).
7. T. Miura, H. Takagi, K. Harita, and M. Kimura, Chromogenic reactions of steroids with strong acids. X. Behavior of testosterone in concentrated sulfuric acid, *Chem. Pharm. Bull. 27*: 452 (1979).
8. T. Miura, H. Takagi, and M. Kimura, Chromogenic reactions of steroids with strong acids. XI. Mechanism of chromogenic reaction of testosterone with sulfuric acid, *Chem. Pharm. Bull. 27*: 783 (1979).
9. T. Miura, H. Takagi, K. Adachi, and M. Kimura, Chromogenic reactions of steroids with strong acids. XIV. Proton and carbon-13 nuclear magnetic resonance spectra of testosterone and 17β-hydroxyandrosta-4,6-dien-3-one in sulfuric acid, *Chem. Pharm. Bull. 30*: 3154 (1982).
10. H. Takagi, H. Odai, T. Miura, and M. Kimura, Chromogenic reactions of steroids with strong acids. XVI. Color and fluorescence reaction mechanism of progesterone with sulfuric acid, *Chem. Pharm. Bull. 32*: 2486 (1984).
11. H. Takagi, T. Miura, and M. Kimura, Color reactions of steroids with strong acids. XVII. Carbon-13 NMR spectra of unsaturated oxosteroids in sulfuric acid, *Bunseki Kagaku 33*: 547 (1984).
12. H. Auterhoff and K. Boehme, Borntraeger reaction, *Arch. Pharm. 301*: 793 (1968).
13. T. Abe, Electronic spectra of polynitrophenols and their anions, *Bull. Chem. Soc. Japan 35*: 318 (1962).
14. G. Miller and W. Grabiec-Koska, Colorimetric determination of picric acid in terephthalic acid (in Polish), *Chem. Anal. (Warsaw) 16*: 1267 (1971).
15. S. S. Gitis, Yu. D. Grudtin, A. Ya. Kaminskii, A. V. Ivanov, and S. A. Agapova, Spectrophotometric determination of styphnic and picric acids in 3,3'-dinitrodiphenyl sulfone (in Russian), *Trudy Vses. Nauch-Issled. Proekt, Inst. Monomerov 2*: 139 (1970): *Chem. Abstr. 76*: 30428g (1972).
16. M. I. Walash, A. M. El-Brashy, and M. A. Sultan, Colorimetric determination of some aromatic nitrocompounds of pharmaceutical interest,*Anal. Lett. 26*: 499 (1993).
17. M. Meloun and S. Kotrlý, Determination of thermodynamic dissociation constants and parameters of the extended Debye-Hückel expression: Application for some sulphonephtaleine indicators, *Collect. Czech. Chem. Commun. 42*: 2115 (1977).
18. A. J. Waring, The use of absorbance ratios in pK measurements by spectro-photometric methods, *Anal. Chim. Acta 153*: 213 (1983).

19. M. J. Minch, M. Giaccio, and R. Wolff, Effect of cationic micelles on the acidity of carbon acids and phenols: Electronic and ^{1}H NMR spectral studies of NO_2-carbanions in micelles, *J. Am. Chem. Soc. 97*: 3766 (1975).

20. F. Dorion, G. Charbit, and R. Gaboriaud, Protonation equilibria shifts in aqueous solutions of SDS, *J. Colloid. Interface Sci. 101*: 27 (1984).

21. A. Berthod and C. Saliba, Measures de pH en milieux micellaires, *Analusis 13*: 437 (1985).

22. A. Berthod and J. Georges, Influences de l'effect micellaire sur l'ionisation des indicateurs acide–base, *Nouveau Journal de Chimie 9*: 101 (1985).

23. A. T. Terpko, R. J. Serafin, and M. L. Bucholtz, Conjugate acid–conjugate base equilibrium in reverse micelles, *J. Colloid. Interface Sci. 84*: 202 (1981).

24. F. H. Quina, and H. Chaimovich, Ion exchange in micellar solutions: 1. Conceptual framework for ion exchange in micellar solutions, *J. Phys. Chem. 83*: 1844 (1979).

25. H. Kohara, N. Ishibashi, and T. Masuzaki, The interaction of acidic dye molecule and organic ligand with positive charges on the micelle surface, *Japan Analyst 19*: 467 (1970).

26. J. Juránek, and A. Ambrožová, Kolorimetrisch-chromatographische Ultramikroanalyse von Gasen. III. Bestimmung von Ultramikromengen des Kohlenstoffs im Technischen Eisen und Eisenlegierungen (in German), *Collect. Czech. Chem. Commun. 25*: 2814 (1960).

27. W. D. Maxon and M. J. Johnson, Continuous photometric determination of carbon dioxide in gas streams, *Anal. Chem. 24*: 1541 (1952).

28. H. Kohara, The change in acid strength of acidic dye molecule and organic ligands on the micelle surface, *Japan Analyst 17*: 1147 (1968).

29. N. Funasaki, The dissociation constants of acid–base indicators on the micellar surface of dodecyldimethylamine oxide, *J. Colloid Interface Sci. 60*: 54 (1977).

30. J. Rosendorfová and L. Čermáková, Spectrophotometric study of the interaction of some triphenylmethane dyes and 1-carbethoxypentadecyltrimethylammonium bromide, *Talanta 27*: 705 (1980).

31. V. Kubáň, J. Hedbávný, I. Jančářová, and M. Vrchlabský, Spectrophotometric investigation of interactions of sulfophthalein dyes with surfactants, *Collect. Czech. Chem. Commun. 54*: 622 (1989).

32. V. Kubáň, I. Jančářová, J. Hedbávný, and M. Vrchlabský, Effect of tensides on the optical properties of sulfophthalein dyes, *Collect. Czech. Chem. Commun. 54*: 70 (1989).

33. V. Kubáň and Z. Vavrouch, Determination of anionic tensides in waters by using of extraction spectrophotometric and two-phase titration methods of ion associates, *Folia Facultatis Scientiarum Naturalium Universitatis Purkynianae Brunensis, XXVI Chemia 21*, 11, 1985, p. 5.

34. O. A. El Seoud, A. M. Chinelatto, and M. R. Shimizu, Acid–base indicator equilibria in the presence of Aerosol-OT aggregates in heptane: Ion exchange in reversed micelles, *J. Colloid Interface Sci. 88*: 420 (1982).

35. R. L. Reeves, The protonisation and indicator behavior of some ionic azobenzene in aqueous sulfuric acid, *J. Am. Chem. Soc. 88*: 2240 (1966).

36. H. Mustroph, R. Haessner, and J. Epperlein, Untersuchungen zum UV/VIS-

spectralverhalten von Azofarbstoffen. XI. Allgemeine Zusammenhange zwischen Structur und Farbe protonierter Azofarbstoffe (in German), *J. Prakt. Chemie 326*: 259 (1984).

37. E. Bishop, *Indicators*, Pergamon Press, Oxford, New York, Toronto, Sydney, Branschweig, 1972, p. 85.

38. A. G. Gonzalez, M. A. Herrador, and A. G. Asuero, Acid–base behavior of some substituted azo dyes in aqueous *N,N*-dimethylformamide mixture, *Anal. Chim. Acta 246*: 429 (1991).

39. R. L. Reeves, R. S. Kaiser, M. S. Maggio, E. A. Sylvestre, and W. H. Lawton, Analysis of the visual spectrum of methyl orange in solvents and in hydrophobic binding sites, *Canad. J. Chem. 51*: 628 (1973).

40. T. Takagishi, S. Fujii, and N. Kuroki, Aggregation of methyl orange homologs carrying long alkyl groups in aqueous solution and their binding behaviors to poly (vinylpyrrolidone), *J. Colloid Interface Sci. 94*: 114 (1983).

41. L. Krpejšová, L. Čermáková, and J. Podlahová, A study of the interaction of Triton X-100 with methyl-orange, *Tenside Surfactants Detergents 28*: 366 (1991).

42. J. Jirasová, J. Bílý, and L. Čermáková, The effect of surfactants on the acid-base behaviour of methyl red. Part II. A spectrophotometric comparison of the effect of Triton X-100 and β-cyclodextrin, *Collect. Czech. Chem. Commun. 59*: 322 (1994).

43. K. Nose and M. Zenki, Flow injection spectrophotometric determination of boron with *D*-sorbitol using methyl orange as an indicator, *Analyst 116*: 711 (1991).

44. E. P. Serjeant and B. Dempsey, Ionisation Constants of Organic Acids in Aqueous Solution, *IUPAC Chemical Data Series No. 23*, Pergamon Press, Oxford, 1979.

45. K. Burger, *Organic Reagents in Metal Analysis*, Akadémiai Kiadó, Budapest, 1973.

46. V. Škarydová and L. Čermáková, Spectrophotometric study of the effect of tensides on triphenylmethane dyes, *Collect. Czech. Chem. Commun. 47*: 776 (1982).

47. V. Škarydová and L. Čermáková, The dissociation constants of eriochromcyanine R and bromopyrogallol red in the presence of carbethoxypentadecyltrimethylammonium bromide (Septonex), *Collect. Czech. Chem. Commun. 47*: 1310 (1982).

48. I. Burešová, V. Kubáň, and L. Sommer, Spectrophotometric study of the interaction of chromazurol S and eriochromazurol B with surface active substances — tensides, *Collect. Czech. Chem. Commun. 46*: 1090 (1981).

49. C. Wyganowski, Effect of quaternary ammonium salts on spectrophotometric characteristic of pyrogallol red and bromopyrogallol red, *Microchem. J. 29*: 318 (1984).

50. S. B. Savvin, I. N. Marov, R. K. Chernova, L. M. Kudrjavceva, S. N. Shtykov, and A. B. Sokolov, On the interaction between non-ionic surface-active substances and triphenylmethane group phenolcarboxylic acids (in Russian), *Zh. Anal. Khim. 36*: 1461 (1981).

51. R. K. Chernova, L. N. Kharlamova, K. I. Gur'ev, and I. S. Sergeeva, Quantum-chemical and spectrophotometric study of protolytic equilibria in solutions of catechol violet (in Russian), *Zh. Anal. Khim. 30*: 1065 (1975).

52. S. B. Savvin, R. K. Chernova, I. V. Lobacheva, and G. M. Beloliptseva, Effect of deprotonation of triphenylmethane group reagents in presence of cationic surface-active substances and its influence on complex formation of disulphophenylfluorone and bromopyrogallol red with metal ions (in Russian), *Zh. Anal. Khim. 36*: 1471 (1981).

53. I. Němcová and L. Čermáková, Spectrophotometric study of the interaction of bromopyrogallol red with tensides and the effect of NaNO$_3$, *Collect. Czech. Chem. Commun. 57*: 1658 (1992).

54. I. Němcová and L. Čermáková, Interactions of surfactants and organic spectrophotometric reagents, *Tenside Surfactants Detergents 31*: 331 (1994).

55. J. Fabian and H. Hartmann, *Light Absorption of Organic Colorants: Theoretical Treatment and Empirical Rules*, Springer-Verlag, Berlin, Heidelberg, New York, 1980, p. 77.

56. J. Bílý and L. Čermáková, A spectrophotometric study of the effect of cationic tensides (CPB, CPTB) on the dissociation constants of 4-(2-pyridylazo)-resorcinol: The effect of a strong electrolyte (NaCl), *Anal. Letters 19*: 747 (1986).

57. J. Bílý, L. Čermáková, and J. Knapp, Spectrophotometric study of the effect of sodium dodecyl sulphate on the acid–base behaviour of 4-(2-pyridylazo)-resorcinol, *Collect. Czech. Chem. Commun. 56*: 785 (1991).

58. L. Sommer, *Analytical Absorption Spectrophotometry in the Visible and Ultraviolet: The Principles*, Akadémiai Kiadó, Budapest, 1989, pp. 235 and 238.

59. M. Shamsipur, J. Ghasemi, F. Tammadon, and H. Sharghi, Spectrophotometric determination of acidity constants of some anthraquinones and anthrones in methanol–water mixture, *Talanta 40*: 697 (1993).

60. K. A. Idriss, M. M. Seleim, M. S. Saleh, M. S. Abu-Bakr, and H. Sedaira, Spectrophotometric study of the complexation equilibria of zirconium (IV) with 1-amino-4-hydroxyanthraquinone and the determination of zirconium, *Analyst 113*: 1643 (1988).

61. D. Rosales, G. Gonzalez, and J. L. Gomez Ariza, Asymmetric derivatives of carbohydrazide and thiocarbohydrazide as analytical reagents, *Talanta 32*: 467 (1985).

62. T. Morales, T. Montana, G. Galan, and J. L. Gomez Ariza, Spectrophotometric determination of zinc with 1-[di(2-pyridyl)methylene]-5-salicylidenethiocarbonohydrazide, *Analyst 112*: 467 (1987).

63. A. Hulanicki, Complexation reactions of dithiocarbamates, *Talanta 14*: 1371 (1967).

64. C. Shiti and L. Xu, 1-(4-nitrophenyl)-3-(2-quinolyl) triazene as an extremely sensitive reagent for spectrophotometric determination of mercury, *Talanta 39*: 1395 (1992).

65. K. Goto, S. Taguchi, K. Miyabe, and K. Haruyama, Effect of cationic surfactant on the formation of ferron complexes, *Talanta 29*: 569 (1982).

66. J. L. Beltrán, R. Codony, M. Granados, A. Izquierdo, and M. D. Prat, Acid–

base and distribution equilibria of 5,7-dichloro-2-methyl-8-hydroxyquinoline in Brij-35 micellar media solutions, *Talanta 40*: 157 (1993).

67. M. Dušková and I. Němcová, The effect of tensides on the dissociation of azure B, *Collect. Czech. Chem. Commun. 58*: 1315 (1993).

68. P. Rychlovský, and I. Němcová, The effect of surfactants on the dissociation constants of phenothiazine derivatives: Alkalimetric determination of diethazine and chlorpromazine, *Talanta 35*: 211 (1988).

3

Complexation Reactions

A. METAL COMPLEXES

1. Introduction

Reactions leading to the formation of analytically important complexes (complex cations and anions as well as the formally electroneutral so-called internal complex salts) have been studied since the last century. To date, there has been much literature on the versatile use of these reactions in spectrophotometric determinations of trace amounts of ions (in aqueous as well as nonaqueous media, often following extraction into a nonaqueous solvent) [1–5].

Generally, the equilibrium concentration of particles present in solution following the establishment of equilibrium

$$mM + rH + nL = M_m H_r L_n \tag{1}$$

can be described by a set of stepwise or overall stability constants. For instance, the overall stability constant for the equilibrium β_{mrn} is defined by the relationship

$$\beta_{mrn} = \frac{[M_m H_r L_n]}{[M]^m [H]^r [L]^n} \tag{2}$$

where $[M_m H_r L_n]$ is the molar concentration of the complex, $[M]$ of the metal ion, $[H]$ of the hydrogen ions, and $[L]$ of the ligand. When protonation (usually of a ligand) does not occur, the r coefficient is equal to zero, or it can be negative in cases in which hydroxide anions are being coordinated. The ionic charges are not usually shown and neither are the coordinated molecules of the solvent. It is assumed that all particles in the solution are solvated (which is also implied).

The formation of a complex particle usually occurs through ligand substitution for the coordinated molecules of the solvent; for example, the reaction between ferric and thiocyanate ions in the aqueous solution

$$Fe(III) + SCN^- = FeSCN^{2+} \tag{3}$$

actually describes the reaction

$$Fe(H_2O)_6^{3+}{}_{(aq)} + SCN^-{}_{(aq)} = Fe(H_2O)_5SCN^{2+} + H_2O^* \tag{4}$$

As is evident from the relationship for β_{mrn}, the concentration of polynuclear complexes increases with the concentration of metal ions. Therefore, only mononuclear complexes exist in sufficiently dilute solutions, and for simplicity, the existence of such complexes is most often assumed. In this case, the system can be described by a few constants — either overall $\beta_1, \beta_2, \ldots \beta_n$ or stepwise constants $K_1, K_2, \ldots K_n$, which are defined as follows:

$$M + L \rightleftharpoons ML \qquad K_1 = \frac{[ML]}{[M][L]} \qquad \beta_1 = K_1 \tag{5}$$

$$ML + L \rightleftharpoons ML_2 \qquad K_2 = \frac{[ML_2]}{[ML][L]}$$

$$\beta_2 = \frac{[ML_2]}{[M][L]^2} = K_1 \cdot K_2 \tag{6}$$

$$ML_{n-1} + L \rightleftharpoons ML_n \qquad K_n = \frac{[ML_n]}{[ML_{n-1}][L]}$$

$$\beta_n = \frac{[ML_n]}{[M][L]^n} = K_1 \cdot K_2 \cdot \ldots \cdot K_n \tag{7}$$

where n is the number of totally bound ligands [6–8]. Values of stability constants are listed in a number of monographs and papers [9–13].

Stability of complexes is an important factor in the assessment of analytical selectivity in complex formation. Properties of the complex that is being formed are then related to electronic structure, symmetry, and so

---

*The complex particles are often shown in square brackets, for example, as the complex cation $[Fe(H_2O)_5SCN]^{2+}$. To avoid possible confusion with molar concentrations, which are presented in the square brackets as well, the complex particles in this monograph are not shown in this fashion.

on. In agreement with statistical probability, $K_1 > K_2 > K_3 \ldots > K_n$ for most complexes. Exceptions include complexation systems where the spatial arrangement changes due to the binding of another ligand or when a forced change of central atom electronic configuration from low spin to high spin occurs. The following complexation systems can serve as examples: Hg(II)-Cl$^-$, $K_4 > K_3$; Fe(III)-Cl$^-$, $K_3 > K_2$; Zn(II)-NH$_3$, $K_3 > K_2 \sim K_1$; Ag(I)-NH$_3$, $K_2 > K_1$. Other factors can also change the sequence of K stability constants: the existence of π-bonds, steric hindrance, or intramolecular hydrogen bonds (e.g., in chelates of some analytically important oximes).

Due to the fact that inorganic as well as organic ions or molecules with acid–base properties can serve as ligands in the complexes, complexation reactions are affected considerably by the medium acidity (the solution pH value). The complexation stability constant K becomes the conditional stability constant K', where

$$K' = \frac{K}{\alpha_M \alpha_L}$$

The Schwarzenbach's coefficients α_M and α_L (coefficients of side reactions of ligand L with protons in media of different acidities, expressed using the appropriate dissociation constant pK_a, or of metal ion M with the competitive complexation components of buffers) can be enumerated when equilibrium constants of all important side reactions are known [1,3,7]. In some cases, the change of medium acidity by a few pH units can change the complex stability constant by several orders of magnitude (Fig. 1).

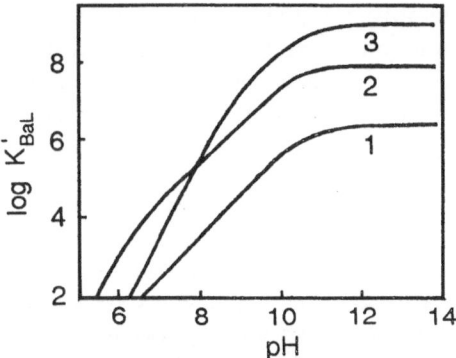

FIGURE 1 Dependence of the conditional stability constants of barium complexes (log K'_{BaL}) on the pH value. L: 1-nitrilotriacetate, 2-ethylenediaminetetraacetate, 3-diethylenetriaminepentaacetate.

   The electronic structures of central ions and ligands, and the symmetry of complexes formed are among the most important factors affecting the stability of complexes.

a. Complexation ability of metals

The metal ions capable of forming complexes with ligands are divided into different groups based on their oxidation state, ionic radius, electronic structure, and polarizability. This articulation is usually based on the stability of metal ion complexes with ligands containing oxygen, nitrogen, or sulfur as donor atoms. This approach to the selectivity prediction of complexation reactions is not too reliable when not combined with a detailed evaluation of ligand donor groups or with further parameters affecting these reactions. The basic theory can be found in the literature [1,3,5,6,14].
   The complexation ability of metal ions can be conveniently evaluated according to *hard* and *soft* Lewis acids and bases. According to this definition, ligands can be considered bases (donors of an electronic pair) and cations can be considered acids (acceptors of an electronic pair). However, it is apparent from this general approach that ligands cannot be put in sequence reflecting their increasing basicity, because the basicity depends on the cation (acid), which is chosen as a standard. It was R. Pearson who proposed to divide acids and bases into two groups denoted as hard and soft. Hard acids preferentially combine with hard bases and soft acids with soft bases (Pearson's principle). This classification has many aspects. With hard particles, the electrostatic interactions occur primarily between the positive charge of the cation and the negative charge (or dipole moment) of the ligand. In aqueous solutions, hydration of particles plays a dominant role, and complex stability is considerably affected by an entropic term (also in unidentate ligands). The interaction between soft particles is of a more covalent character and is affected by the electronic structure of reactants. The decline of enthalpy is the key factor that contributes to the stability of complexes.

b. Complexation ability of ligands

The ability of a ligand to form complexes depends on the kind and number of its donor atoms and their mutual arrangement, the ligand's electronic structure, and steric properties. Based on this viewpoint, ligands are sometimes divided into two groups [15]: those with one or more lone pair(s) of electrons and those without a lone pair of electrons but with π-bonding electrons. To the first group belong ligands that have no vacant orbitals to receive electrons from the metal (e.g., H_2O, NH_3), ligands with vacant orbitals or orbitals that can be vacated to receive p electrons from the metal (e.g., CN^-), and ligands with additional π electrons that can be furnished to vacant metal orbitals (e.g., OH^-, F^-, I^-).

Examples of unidentate oxygen ligands include the H_2O molecule, alcohols, ketones, carboxylic acids, and ethers. Unidentate ligands containing nitrogen donor atom are represented by NH_3 and amines.

If the multidentate ligand (with the same donor atoms, e.g., O, O, or with different donor atoms, e.g., O, N) is coordinated on the solvated cation, a higher number of solvent molecules are released. This increases the number of free molecules in the system, which leads to increased entropy and thus to stabilization of the system. Diamines are bidentate ligands with strong complexation ability (e.g., ethylene diamine), as are some heterocycles (1,10-phenanthroline, 2,2-bipyridine). Of importance are the multidentate chelating agents, which can form stable complexes with ions of almost all metals. These include NTA (nitrilotriacetic acid), EDTA (ethylenediaminetetraacetic acid), and diethylenetriaminepentaacetic acid. The stability of complexes formed increases with the increasing number of ligand donor atoms. Complexes with EDTA anions (6 donor atoms) are usually more stable than complexes with NTA anions (4 donor atoms). Some higher amines such as diethylenetriamine (den), triaminotriethylamine (tren), and triethylenetetramine (trien) are more selective than EDTA (i.e, they do not form complexes with alkali metal ions or alkaline-earth metal ions).

The use of the so-called chelate effect, when one or more five- or six-membered rings are formed as the result of the reaction of metal ions with multidentate polydentate ligands, is important in analytical chemistry and spectrophotometry.

Greater or lesser strain and deformations can occur in complexes because of geometric arrangement of chelate rings, which sometimes leads to a decrease in the number of coordination bonds. The EDTA anion, for instance, contains only 5 coordination sites arranged around a big nickelous cation, and the remaining coordination site is occupied by a water molecule.

Steric effects in ligand molecule can affect the stability of complexes. If the steric factors hinder the electronic orbital overlaps of the central atom and those of the ligand, the complex stability declines. These factors are of importance in chelate-forming ligands, and they also determine the analytical selectivity. For example, the stability constants of 1,10-phenanthroline complexes (e.g., Cd(II), Ni(II), Fe(II)) decrease due to the exchange for the ligand 2,9-dimethyl-1,10-phenanthroline. With the former ligand, Fe(II) forms a tris low spin red complex, while with the latter ligand it forms only a bis complex, which is high spin, colorless, and of a lower stability. Specificity of a number of organic reagents is also explained by steric effects (e.g., cuproine grouping for copper). The metal cations forming octahedral complexes do not react with these reagents.

c. Spatial arrangement of complexes

Another determining factor of complexation equilibrium is the spatial arrangement of complex species; only a certain number of ligands of a given size have a chance to get to such a proximity with the central atom that permits the formation of a bond.

      The complexes of elements with the coordination number 2 (hybrid orbital sp or p^2) have a linear arrangement of complex species (Cu(I), Ag, Au(I), Hg(I), Hg(II)). Complexes with the coordination number 4 have a square planar arrangement (hybrid orbital dsp^2: Ni(II), Cu(II), Pd(II), Pt(II), Au(III)) or form tetrahedrons (hybridization sp^3 or d^3s: Be, Mg, B, Zn, Cd, Pb, Sn(IV), Fe(III), Co(II)). Complexes with the coordination number 6 (d^2sp^3 hybridization) have an octahedral arrangement (Co, Mn, Fe, Cr, Ti, V, Ru(III), Rh(III), etc.). Based on the valence bond theory, the complexes are described as "ionic" (outer sphere, high spin) or "covalent" (inner sphere, low spin).

      During the stepwise formation of the complexes of the metal ion concerned, the change of complex species stereochemistry becomes manifested by an anomaly of the already mentioned dependence of the stepwise stability constants K (when log K_n/K_{n+1} is positive and mostly constant). For instance, $HgCl_2$ is linear, $HgCl_4{}^{2-}$ is tetrahedral. The change of hybridization during the stepwise formation of complexes results in anomalously high value of K_2/K_3 (log $K_1 = 6.7$; log $K_2 = 6.5$; log $K_3 = 0.9$; log $K_4 = 1.0$; $I = 0.5$).

d. Solubility of complexes

The solubility of complex ions in aqueous medium is determined by their ability to orient the polar water molecules around themselves. With the exception of complexes with highly hydrophilic, strongly polar ligands, the electrically neutral complexes are insoluble in water. The associated (often polynuclear) molecules of complexes behave similarly. The water solubility of these products can be influenced through ligand molecule modification by either polar substituents (e.g, groups $-SO_3H$, $-COOH$) or nonpolar groups (e.g., methyl), which lower by steric hindrance the attainment of total coordination saturation of the central atom. Substituents that can be protonated in the acidic medium (amino groups) also increase the solubility of complexes in acidic medium. In coordinatively unsaturated complexes, the formation of complexes with coordinated water molecules or of mixed ligands complexes can have a similar effect (e.g., diacetyldioxime as a specific ligand for Pd(II), contrary to Ni(II), binds hydroxide ion in alkaline medium, while a water-soluble anionic chelate is formed).

Another option to increase the solubility of products is to employ mixed aqueous solvents or to add solubilizing substances (e.g., surfactants).

2. Complexes of Metals with Inorganic Ligands

A number of metals form analytically important complex compounds with inorganic ligands, including halides, thiocyanates, cyanides, and ammonia. Some of them are used for classical spectrophotometric determinations of metals. Determinations following the conversion on peroxocompounds and those using heteropolyacids can be included in this group. Reviews of related studies are presented in a number of monographs [16–18].

a. Halide complexes

Unlike fluoride anions, the other halide anions are more polarizable and form stable complexes with ions that have an unoccupied d shell. Usually, they are colorless or only moderately colored, and if their excess in solutions is not ensured, mixtures of successive complex species are formed, because the values of their stability constants are usually similar. Complexes of chlorides (e.g., Cu(II), Co(II), Fe(III), Ga(III), Au(III), Hg(II)), bromides (e.g, Au(III), Tl(III), In(III), Te(IV)), and iodides (Bi(III), Te(IV), Cd(II), Pd(II), Tl(III), Hg(II)) are of importance in spectrophotometric analysis, because they have sufficiently high values of ε, and thus the maxima of their absorption bands in aqueous media can be used for metal ion determinations. For BiI_4^-, for instance, the $\lambda_{max} = 340$ nm and $\varepsilon = 3.4 \times 10^4$ (the second maximum, $\lambda = 460$ nm, is less sensitive). Te(IV) ions with KI provide the complex anion TeI_6^{2-}, with $\lambda_{max} = 285$ and 335 nm. Sometimes, mixed solvents are used in spectrophotometric analysis to increase sensitivity. The determination of iron in the form of $FeCl_6^{3-}$ can be performed in a dimethylformamide–water mixture at 340 nm or 360 nm (0.1–20.0 ppm Fe).

Complex K_2HgI_4 reacts with ammonia in alkaline medium, and the likely product $(OHg_2NH_2)I$ is formed. To utilize this reaction for ammonia determination, many methods with different alternatives of reagent preparation (Nessler reagent) have been proposed. The yellow or reddish brown coloration of the solution is measured, usually at $\lambda = 400$–435 nm. To determine traces of NH_3 (down to 0.07 ppm), low temperature and an absorption band of $\lambda = 380$ nm are employed. A number of substances can interfere [19].

Much more often, the formation of halide metal complexes is utilized for sensitive extraction determinations of metals when these anionic complexes form ionic associates with different basic, usually triphenylmethane, dyes in nonaqueous medium. The maxima of absorption bands of these dyes are then employed. These reactions have been applied, for instance, in the determination of Tl(III) (TlCl$_4^-$ associate with rhodamine B has a λ_{max} = 560 nm in benzene, and 1-2 ppm Tl can be determined; TlBr$_4^-$ with Crystal Violet can be used in benzene to determine 0.2-0.4 ppm Tl).

Fluorides do not form complexes with metal ions that are directly utilizable in spectrophotometric analysis. An exception is boron determination after the conversion to HBF$_4$ and the subsequent extraction of ionic associate with Methylene Blue (λ_{max} = 660 nm, ε = 6.5 × 10^4, after extraction into dichloroethane, 0.008-0.033 ppm B can be determined). Tantalum determination is based on the same principle.

The major importance of fluorides is their use as masking agents, because they form colorless, usually very stable, complexes with some metal ions (Fe, Zr). This property is used in some cases for indirect determination of fluoride ions based on the decrease in λ_{max} of the colored metal complex after the addition of fluoride.

b. Thiocyanate complexes

Thiocyanate complexes are formed by a number of ions (Ti, V, Nb, Mo, W, Fe, Co, Rh, U, etc.). The formation of colored rhodanide complexes can be used for photometric analysis in aqueous media and mixed solvent media, as well as after extraction of ionic associates of thiocyanate complexes (e.g, with basic dyes, pyridine, tributylamine, diantipyrylmethane), into CHCl$_3$, etc. The sensitivity of individual determinations varies greatly, with the most sensitive ones being the determinations of Ti, Fe, and Nb. The stability of the formed products differs as well. For these reasons, a number of method modifications have occurred, for instance, in the often used Fe(III) determination. Fe(SCN)$^{2+}$ is formed in the acidic aqueous medium. In the presence of solvents miscible with water (ethanol, acetone), the formation of Fe(SCN)$_2^+$ is assumed; in ether it is the formation of Fe(SCN)$_3$. From the extraction by tributylphosphate (TBP), Fe(SCN)$_3$ · 3TBP is formed; from the medium of tri-n-butylammonia, $\{(NH(C_4H_9)_3)_3^+$, Fe(SCN)$_6^{3-}\}$ is extracted. The coloration of these various products is measured at λ = 470-520 nm.

In the determination of W(VI) by KSCN, separation on polyuretane foam has been used [20].

c. Complexes with ammonia

Ammonia forms amine complexes with a number of ions. The complexes of transition metals are colored. Although their stability is sometimes high (for Cu(NH$_3$)$_4^{2+}$, β_4 ~ 5.10^{14}, blue coloration can be measured at λ_{max} = 620 nm but also in the ultraviolet region at 220 nm), practical determinations are not feasible.

d. Complexes with hydrogen peroxide

Hydrogen peroxide forms colored complexes with some transition metals (Cr, Mo, Ce, Ti, V, Nb, U, Ta, W), of which only some have sufficient determination sensitivity to be of importance in spectrophotometric analysis. Examples are the formation of red-brown pervanadate (λ_{max} = 460 nm or 290 nm), yellow perniobic acid (λ_{max} = 435 nm or 342 nm), yellow-to-yellow-red peruranates (λ_{max} = 355 nm in the presence of carbonates, but preferably measured at $\lambda \sim$ 400 nm), or the yellow-to-orange color of pertitanate compounds (measured at around 400 nm). Besides uranium, the peroxocompounds of other elements are formed in media of different acidities. Although the determinations are of relatively little sensitivity (1–100 ppm of metal), a number of studies and practical applications exist.

e. Heteropolycomplexes

These complexes are important primarily for spectrophotometric determinations of Si, P, As, Ge, and other metals that are able to coordinate polymerized ions of molybdate, tungstate, and others. For instance, yellow-colored products derived from appropriate heteropolyacids of the common composition $H_nX(Mo_3O_{10})_4$ (n = 3 for X = P(V) or As(V); n = 4 for X = Si(IV) or Ge(IV)) are formed in acidic medium. Phosphorus determination is performed in the range λ = 310–430 nm (1–15 ppm P; the sensitivity can be increased by the additions of acetate). After extraction into isobutanol, 0.06–1.8 ppm P may be determined.

Using the FIA technique, silicon has been determined in silicates following the isolation of silicic acid (employing an ion exchanger) and the reaction with the molybdate reagent when the yellow silicomolybdic acid is formed [21].

A second use of heteropolyacids involves their ability to be reduced by different reagents ($SnCl_2$, hydrazinium sulfate, and others) to molybdenum blue. These reactions are addressed in Chapter 5.

The third possibility for determination is to use the formation of ionic associates of heteropolyacids (see Chapter 4). Very sensitive determinations ($\varepsilon \sim 10^5$) of silicon, germanium, and phosphorus are possible after extractions of these associates with basic dyes (Rhodamine B, Crystal Violet, and others) [22]. For instance, silicomolybdic acid forms an associate [$4B^+ \cdot Si(Mo_3O_{10})_4^{4-}$], where B^+ is the cation of basic dye.

3. Complexes of Metals with Organic Ligands

A general classification of the most important organic chelating agents can be based on a number of perspectives: chemical structure of the ligand organic molecule, number of rings formed in the chelation (and the number of their members), number and kind of ligand donor atoms. In this section, the organic ligands are classified with respect to the kind of donor atoms. The only exception is the group of azo dyes, where many combinations of possible metal–ligand interactions occur. However, as oxygen and nitrogen are in most cases the main donor atoms of these compounds, the azo dyes as a whole have been placed into this group of ligands.

Examples are given of different types of metal complexes with organic ligands and their use in the spectrophotometric determination of both metals and organic analytes. Besides these direct determinations, indirect determinations of organic analytes are possible. For example, an analyte can be oxidized by a metal ion and the reduced form of the metal then determined using a metalloorganic reagent, or the complex formed with the analyte can be separated and its metal content (determined by using a metalloorganic reagent) used to measure the analyte content.

a.   Ligands with oxygen donor atoms

This group includes ligands of two different types:

 1.   ligands in which the donor atoms are not included in the conjugated π-electronic system of the molecule.
 2.   ligands in which the incorporation of nonbonding oxygen electrons into the conjugated π-electronic system of the ligand occurs in chelation.

On chelate formation, electrons in the first type of ligands remain localized on the oxygen donor atom, and during excitation they can be transfered to the electronic system of metal ions. These ligands can react with metal ions, which are able to accept the excited valence electrons into their orbitals, primarily with the metal ions at a higher oxidation state. Especially the bands corresponding to these π–d transitions are employed analytically.

During radiation absorption, the major absorption band in the second type of ligand is induced by a π–π^* transition, and due to delocalization of electrons in chelation, it is bathochromicaly shifted relative to the π–π^* band of the free ligand. The π–d transition occurs also in the transition metal complexes at the higher oxidation state.

The dissociation of protons from the neutral molecules of organic ligands occurs in ligands containing the hydroxy or carboxy groups in chelation. The ligand anions formed compensate the positive charge of the metal ions (with respect to the metal oxidation state, several chelates can be gradually formed). At the same time, saturation of the metal ion coordination sphere occurs. If the result is the formation of electroneutral, coordinatively saturated chelate, then this chelate is (with minor exceptions) insoluble in water and soluble in nonpolar organic solvents. In these cases, spectrophotometric determination is usually performed following extraction.

The basic types of ligands used in spectrophotometry that form chelates through two oxygen donor atoms are presented along with examples of their use, in the following subsections.

Aliphatic and aromatic polyhydroxy-, hydroxycarboxy-, (poly)hydroxy(poly)carboxy-, and carbonyl compounds. In spectrophotometric determination of metals, primarily aromatic compounds of this type are used as reagents, while aliphatic compounds are used mostly as masking agents. An overview of aromatic ligands of this type, and of their reactions, appears in Refs. [23,24].

Polyphenols and polynaphthols with hydroxyl groups in mutual position ortho-, or peri- (pyrocatechol, pyrogallol), and their sulfoacids, of which the best-known analytical reagents are Tiron (disodium salt of 1,2-dihydroxybenzene-3,5-disulfonic acid) and chromotropic acid (disodium salt of 1,8-dihydroxynaphthalene-3,6-disulfonic acid), provide five- to six-membered chelates **(1)**, **(2)** in reactions with metals. *o*-Hydroxy- and α-hydroxycarboxylic acids (salicylic acid, glycolic acid, mandelic acid) form the chelates **(3)**.

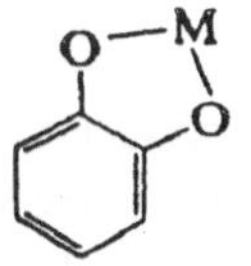

(1) 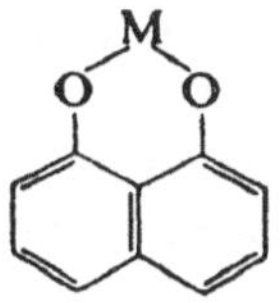   **(2)** 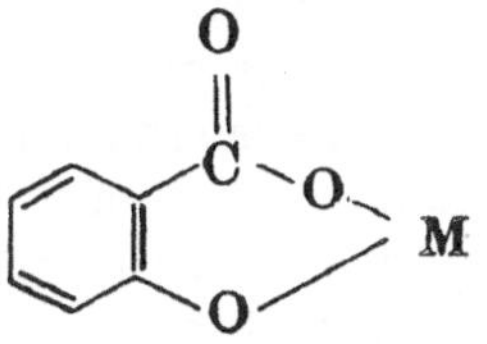   **(3)**

In the case of polyhydroxycarboxylic acids and carbonyl compounds (gallic acid, ascorbic acid, sugars), it is assumed that the carboxyl and carbonyl groups are not involved in the chelate formation. Two adjacent hydroxyl groups participate in the reaction, and thus a five-membered cycle is formed.

The following structures have been suggested for the complexes Mo(VI) with pyrocatechol in slightly acidic medium **(4)** [25], and Ti(IV) with ascorbic acid **(5)** [26,27].

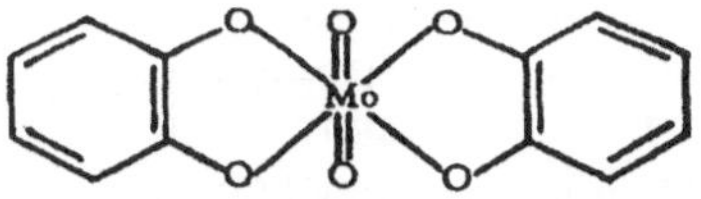

(4) 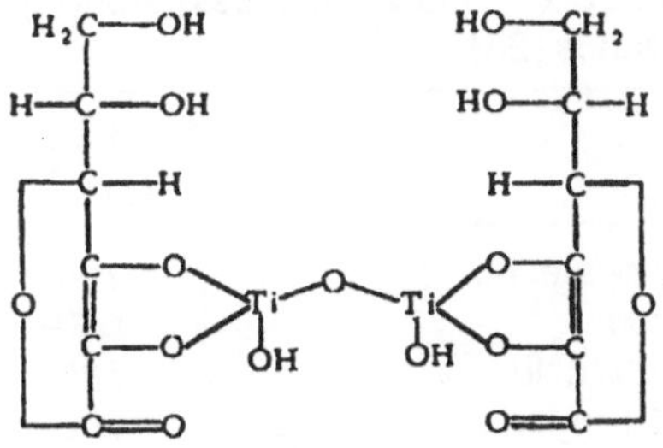   **(5)**

All of the above substances are ligands with poor selectivity, because they form colored chelates with many metals, primarily in a higher oxidation state (e.g., Fe(III), Ti(IV), Ce(IV), Mo(VI), V(VI)). Depending on the ligand concentration and the solution pH, these ligands can provide more complexes. For instance Fe(III) ions form FeL, FeL_2, and FeL_3 complexes. The value of λ_{max} decreases with increasing number of coordinated ligands. Example of chelates of these types with several aromatic ligands are shown in Table 1.

The position of the chelate absorption maxima depends also on the polarity of the M–L bond. The value of λ_{max} decreases with increasing polarity.

Of recent interest have been the reaction of 2,3-dihydroxybenzoic acid with Ti(IV) [28]; the reaction of vanadium in different oxidation states with pyrogallol [29]; and the determinations of Ce, Mo, Ni, V, Sc and rare earths with Tiron [30]. Chromotropic acid has been used for Ti(IV) [31] and boron [32] determination using flow injection analysis (FIA). Complexing with ascorbic acid has been used for U(VI) determination [33], and several azo derivatives of pyrocatechol have also been prepared and proposed for zirconium determination [34].

These complexes can also be used for the determination of analytes with structure analogous to the reagents. Thus, 1,2-diphenols and polyphenols (pyrocatechol, pyrogallol, adrenaline, etc.) can be determined using their reaction with ferric chloride in the presence of triethanolamine or pyridine, with ferrous sulfate in weak alkaline media or in the presence of alkali tartrate, with titanium(IV) sulfate, with cobalt(II) nitrate, with copper(II) and mercury(II) salts, or with molybdate in neutral or slightly acidic media.

Salicylic acid and its derivatives are determined mostly using a reaction with Fe(III) salts; Cu(II) and Fe(II) salts have also been used. Some aliphatic acids (oxalic, citric, tartaric, malonic, malic, L-ascorbic, tartronic)

TABLE 1 Characteristics of Iron(III) Chelates with Some Phenolic Ligands [23]

Ligand	FeL λ_{max} (pH = 2–4)	FeL_2 λ_{max} (pH = 5–7)	FeL_3 λ_{max} (pH = 8–9)
Pyrocatechol	700	580	490–495
Tiron	660–680	560	480
Gallic acid	580	550–560	440–500
2,3-Dihydroxynaphthalene	580–590	560–530	450–460
Chomotropic acid	725	630–640	

decolorize the red Fe(III) complex of 5-nitrosalicylic acid, which can be used for their indirect determination [26,27].

Methods for drug determination based on these reactions have been reviewed [35].

β-Diketones. Chelates of aliphatic β-diketones are derived from the enol form of these substances (6). In this system, π-electrons are delocalized through all atoms, and thus a quasi-aromatic system is formed.

The most common ligand of this type is acetylacetone (2,4-pentanedione, where $R_1 = R_2 = CH_3$). In the spectra of its chelates, the absorption maximum corresponding to the π–π* transition can be found, bathochromicaly shifted by about 25 nm relative to the ligand itself with $\lambda_{max} = 272$ nm (e.g, for the Be(acac)$_2$ chelate, $\lambda_{max} = 295$ nm). For chelates with transition metals, the maximum corresponding to the π–d transition (for the Fe(acac)$_3$ chelate, $\lambda_{max} = 438$ nm) [36] is formed in the visible part of the spectrum.

The six-membered chelates of metals with acetylacetone are insoluble in water, but their extraction into organic solvents has been studied [37]. The extraction spectrophotometric metal determinations (e.g, of Al, Fe, Cr, Mo, Ru) are reviewed in Ref. [4].

Thenoyltrifluoroacetone (7) can be applied in spectrophotometry. Owing to its electron-accepting group $-CF_3$, the value of its pK_a decreases, and thus the extraction is performed in a more acidic medium. It is used for determinations of Ce, Zr, Rh, and Tl [4].

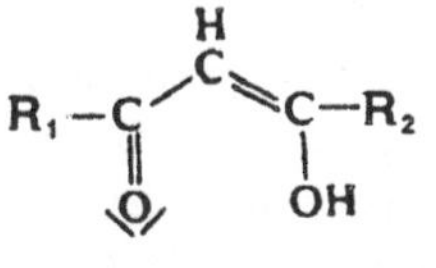

(6) 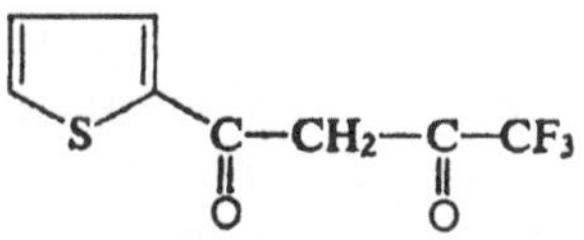   **(7)**

The Fe(III) chelate, which is formed with acetylacetone and similar ketones, can be used for their determination. Aqueous and ethanolic media can be used, or the complex can be extracted into an organic solvent [26,27,38].

Substituted hydroxylamines. Only a few of this large group of substances (in Ref. [39] more than 20 derivatives are described) have proven suitable for practical application in metal determinations. Cupferron (an ammonium salt of nitrosophenylhydroxylamine) (8) and several hydroxamic acids (e.g, benzohydroxamic acid) (9) are the primary reagents used.

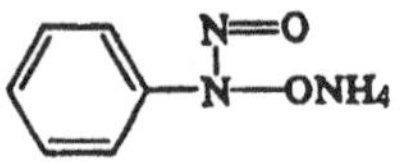

(8) 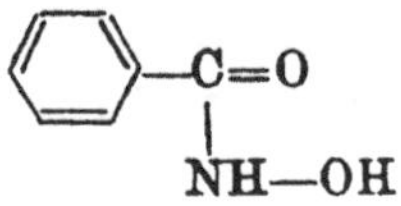   (9) 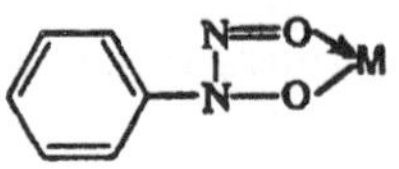   (10)

With the ions of Fe(II), Fe(III), Cu(II), Al(III), Bi(III), Ti(IV), Zr(IV), cupferron forms chelates of the type **(10)**, which are very insoluble in water and are highly soluble in organic solvents. Thus, they are employed in extraction spectrophotometric determinations.

In hydroxamic acid, the effect of aliphatic, substituted aliphatic, aromatic, substituted aromatic, and heterocyclic substituents has been studied [40] with respect to the activity of metal ions, and it has been found that for chelate formation the pH, solvent, and reagent concentration — but not the kind of substituent — are of decisive importance. The extension of conjugation through the introduction of a double bond into the side chain between the carbonyl carbon atom and phenyl group, however, increases molar absorptivity ε.

N-Phenylbenzohydroxamic acid (*N*-benzoyl-*N*-phenylhydroxylamine), which has been used for spectrophotometric determinations of Fe(III), Mo(VI), V(V), U(VI), and others [40], is the most commonly used substance from this group. These metals react with the keto form of the reagent, whereas boron [41] reacts with the enol form.

The complexation of ferric ions with hydroxamic acid is also a very important spectrophotometric reaction for the determination of a number of organic compounds that are easily converted to hydroxamic acid. Thus, hydroxamic acids are readily formed by the reaction of esters, anhydrides, lactones, lactames, and chlorides of carboxylic acid with hydroxylamine in alkaline medium. Amides, imides, and hydrazides are less reactive, but their reaction with hydroxylamine can be catalyzed by nickel(II) chloride or carbodiimide. Carboxylic acids alone must be converted to esters or chlorides prior to reaction with hydroxylamine, or their reaction with hydroxylamine can also be catalyzed by the presence of nickel(II) chloride or carbodiimide. In that case, the reaction proceeds in neutral media. These catalysts can also accelerate the reaction in other cases.

The course of the reaction with hydroxylamine (hydroxylaminolysis) depends on many factors, especially on the character of the analyte molecule and functional groups present in it; on the concentration, mutual ratio, and sequence of the reagents added; and on pH, temperature, reaction time, and solvent used. The most frequently used concentration of hydroxylamine is 1 mol L^{-1}. The concentration of hydroxide ions accelerates the reaction; however, in some cases, it can cause undesired hydrolysis of esters.

The complexation reaction with Fe(III) ions is affected by the type of the Fe(III) salt used, its concentration, medium acidity, and presence of ethanol. The stability of the complex can be affected by an excess of hydroxylamine. The procedure has also been used for the determination of hydroxy groups, which are acetylated prior to the reaction with hydroxylamine.

Many examples of the application of the reaction on compounds having the functional groups mentioned can be found in the literature [42–45]. The procedure has been used for the determination of O- and N-acetyl groups in organic compounds as well. Primary and secondary alcohols must be converted to esters by acetylation with acetic anhydride prior to the reaction with hydroxylamine. The procedure has also been used for the determination of the content of hydroxy groups, which are acetylated prior to the reaction with hydroxylamine [26,27].

Aldehydes can be converted to hydroxamic acids by using the reaction with benzenesulfohydroxamic acid and ketones after their conversion to oximes and their hydrolysis to hydroxylamine and its determination as formylhydroxamic acid [26,27].

Flavones. Flavones are compounds of plant origin and have many analytical applications [46,47]. The best known are quercetin (3,5,7,3′,4′-pentahydroxyflavone) (11), and morin (3,5,7,2′,4′-pentahydroxyflavone).

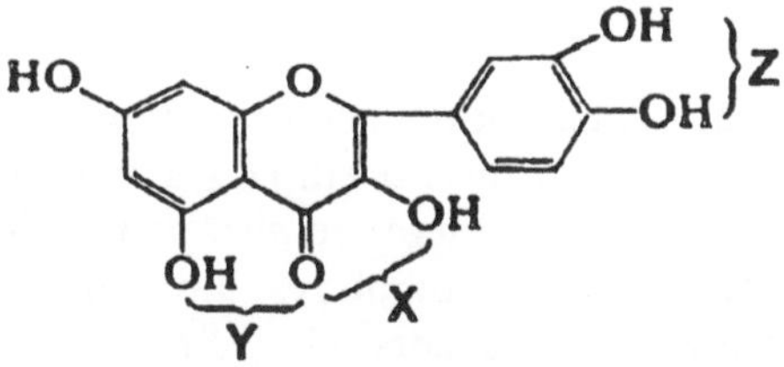

(11)

In the quercetin molecule and in a number of other derivatives, there are several possible coordination centers, which are shown on structure (11). It has been found that the type of chelate that is being formed is metal specific. For instance, aluminum forms only five-membered chelates with atoms of the X center and six-membered chelates with atoms of the Y center; niobium forms chelates of X type in acidic medium, and of Z type in neutral medium [46].

Quercetin provides colored reactions with a number of metals (e.g., Cr(III), AL(III), Sn(IV), Fe(III), lanthanoids, Th(III), Ti, Zr, and U(VI)), and these reactions are used for their determination. Many examples of the application of the determination of flavonols as metal complexes have been described in the literature [26,27,48,49].

Fluorones. Fluorones are dyes with the basic structure **(12)** corresponding to 2,3,7-trihydroxy fluorone (2,3,7-trihydroxyxanthene-6-one). R can be alkyl (methyl, propyl, pentadecyl) or aryl (phenyl; hydroxy-, bromo-, methoxy-, nitro-, dimethylamino-, carboxy-, sulfophenyl; pyridyl, quinolyl, etc.). Structure **(13a,b)** depicts the 1:2 and 1:1 complexes of molybdenum with propyl fluorone [50].

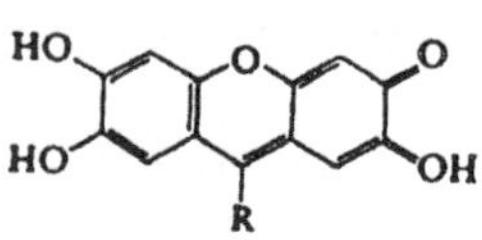

(a)

(b)

(12) 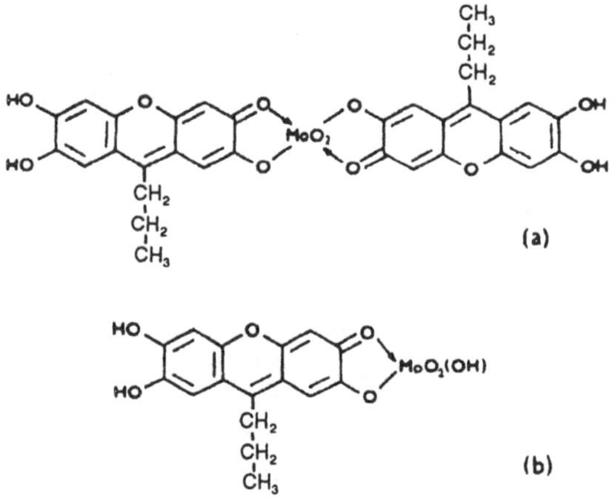   **(13)**

In the spectra of nontransition metal complexes, the absorption bands corresponding to $\pi-\pi^*$ transitions are employed analytically. In the spectra of transition metal complexes, there are also bands corresponding to $\pi-d$ transitions. In the formation of complexes, a bathochromic shift of the $\pi-\pi^*$ transition absorption band occurs relative to the free ligand as a result of the entry of nonbonding electrons of the oxygen donor atom into the conjugated π-electronic system. The greater the effective charge of the metal ion (i.e., the charge related to the ion unit radius), the more localized are the nonbonding electrons on donor atoms (the bond polarity increases), and the smaller the bathochromic shift. For instance, in the sequence Zr(IV), Ti(IV), Sn(IV), Ge(IV), the ion effective charge increases from 3.33 to 4.33, and the λ_{max} of these complexes decreases from 540 nm to 510 nm [36].

From this group, the most commonly used reagent is 9-phenyl fluorone and its derivatives. An overview of their structures and applications is found in Ref. [50]. Phenyl fluorone is most often used in germanium determination (at least 60 studies dealing with the reaction of Ge(IV) with phenyl fluorone have been published). The absorption maximum of the chelate is in the range 500–530 nm; the ε value (in 0.5 M HCl and 20% ethanol) is about 4×10^4.

To prevent the precipitation of the germanium complex from solution, a protective colloid (gelatin, arabic gum, polyvinyl alcohol) is usually applied. The same effect is achieved by using surfactants, when the determination sensitivity is enhanced as well (in the presence of anionic surfactant sodium dodecyl sulfate, $\lambda_{max} = 504$ nm, $\varepsilon = 1.18 \times 10^5$; in cationic surfactant dodecyltrimethylammonium bromide, $\lambda_{max} = 503$ nm, $\varepsilon = 1.72 \times 10^5$) [50]. Using o-chlorophenyl fluorone for germanium determination in practical samples, the value of $\varepsilon = 1.82 \times 10^5$ ($\lambda_{max} = 516$ nm) has been achieved in the presence of cetyltrimethylammonium bromide [51]. An analogous increase in the value of ε occurs in the complex of phenyl fluorone with titanium [52] in the presence of the nonionic surfactant Triton X-305 and OP emulsifier (($CH_3)_3$-C-CH_2-C($CH_2)_2$-C_6H_4(OCH_2-$CH_2)_{10}OH$).

Photometric determinations using extractions into CCl_4, cyclohexanol, and methylisobutylketone have also been performed [50].

Triphenylmethane dyes. These are among the most widespread dyes used in spectrophotometric determinations of metals [50]. When substituted only in the 4 and 4' positions, and sometimes in the 4" position (usually by $-OH$, $-NH_2$, and $-N(alkyl)_2$ groups), and bearing a positive or negative charge depending on the substituent kind and solution pH, these dyes provide ionic associates with counterions of opposite charge (see Chapter 4). Dyes with suitable substituents in positions 3,4, and 3',4' can form chelates with metal ions. Besides the substituents mentioned, the remaining positions of the basic skeleton of triphenylmethane dyes can be occupied by various substituents that affect the position of absorption maximum, acid–base properties, chelate formation, solubility, stability in time, and so on.

With respect to the arrangement of substituents with oxygen donor atoms, these dyes can be divided into several groups: the Pyrocatechol Violet group (**14**), the aluminon group (**15**), and the Eriochromcyanine R group (**16**).

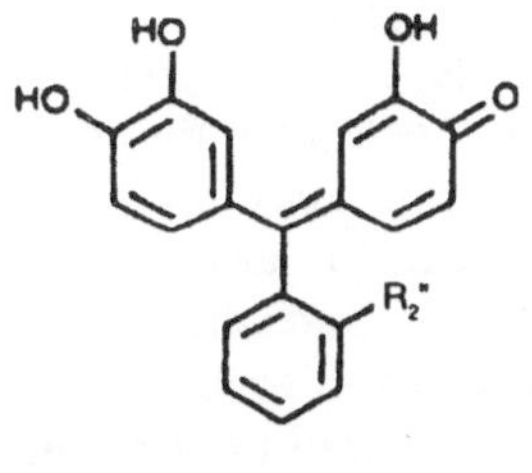

| (14) | (15) | (16) |

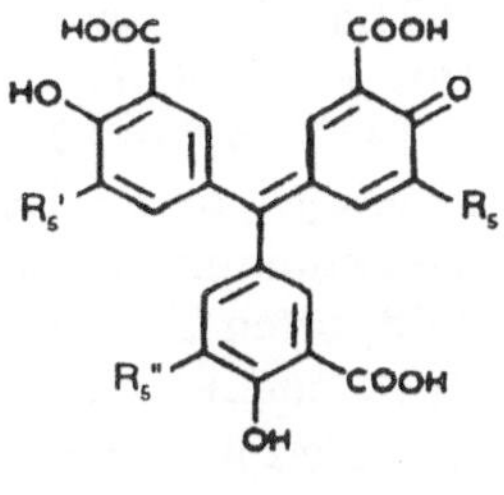

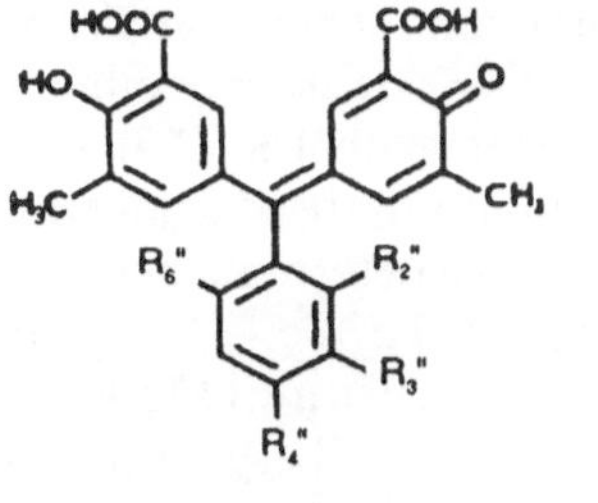

Dyes with the sulfo group on the third ring are well soluble in water; the derivatives with $-SO_3H$ in the position R_2'' are denoted as sulfophthaleins.

Pyrocatechol Violet ((**14**), $R_2'' = SO_3H$) is used primarily to determine trivalent and tetravalent elements (B, Al, Ga, In, Sb, Bi, Sc, Yt, La, lanthanoids, Ge, Sn, Ti, Zr, V, Nb, Ta, Mo, W, Fe), as well as Cu, Zn, Cd, but not all elements of the IIA group. Perhaps most frequently it is used for tin(IV) determination. The chelates that are formed have an absorption maximum of around 550 nm, and the published values of ε are in the range 5 to 8×10^4, depending on the measurement conditions. Measurements are performed at pH = 2–4 in the presence of protective colloids. This was the first system in which it was observed that the presence of gelatin can induce a bathochromic shift of the absorption maximum (in this case up to 640 nm) [53]. Analogous effects have been found in the presence of cationic surfactants (CTAB, CPB): λ_{max} = 660–670 nm (thus the bathochromic shift is 110–120 nm relative to the complex without surfactant); at the same time a hyperchromic shift occurs: ε = 8 to 10×10^4 [50].

The reaction of Pyrocatechol Violet with tin(IV) has also been used in solid-phase spectrophotometry [54], in which the forming complex is bound to an anion-exchange gel and its absorption spectrum measured.

The extraction spectrophotometric determinations of tin with Pyrocatechol Violet in the presence of a third component in the solution have also been suggested (see the discussion on the formation of ionic associates in Chapter 4), for example, extraction into butylalcohol in the presence of 1,3-diphenylquanidine [55].

The intensely colored complexes of Pyrocatechol Violet in the presence of 1,10-phenanthroline have also been studied using the Hückel MO LCAO calculations [56].

The decolorizing of the bismuth and Pyrocatechol Violet complex from EDTA in the presence of the cationic surfactant Septonex has been used to propose a method for EDTA determination [57].

Aluminone ((**15**), $R_5=R_5'=R_5''=H$) and alumocresone ((**15**), $R_5=R_5'=R_5''=CH_3$) are usually used in the presence of protective colloid to determine beryllium and aluminum [50]. In the case of aluminum, the nonionic surfactant Triton X-100 has also been employed [58].

Eriochromcyanine R ((**16**), $R_2''=SO_3H$, $R_3''=R_4''=R_6''=H$) and Chromazurol S ((**16**), $R_2''=R_6''=Cl$, $R_3''=SO_3H$, $R_4''=H$) are the most commonly used triphenylmethane dyes of the last group. They are often used to determine beryllium, and aluminum, as well as Ga, In, lanthanoids, Ti, Zr, Th, Fe, V, Cu, Ru, and Pd, often in the presence of surfactants.

Structure (**17a,b**) shows 1 : 1 and 1 : 2 complexes of Eriochromcyanine R with beryllium [50], and Table 2 lists examples of beryllium determination with Chromazurol S in the presence of cationic surfactants [50].

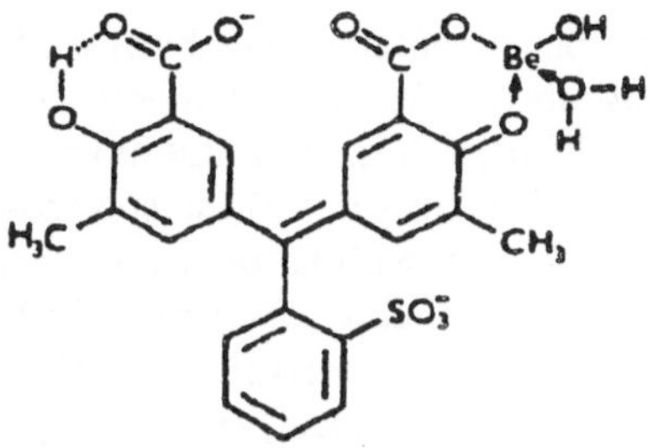

(17a)

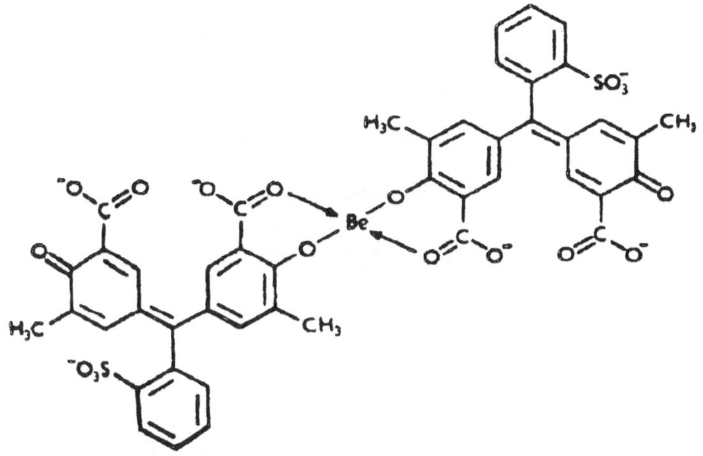

(17b)

TABLE 2 Determination of Be(II) with Chromazurol S and Cationic Surfactants

Tenside	pH$_{opt}$	λ_{max} (nm)	$\varepsilon \times 10^{-4}$	Determination range (mg mL^{-1})
Zephiramine	3.5–4.5	610	9.90	0.02–0.80
Zephiramine	10.0	525	5.35	0–0.08
CTAB	6.6–7.0	596	5.85	0–0.06
CPB	5.0	605	9.6	0–0.06
PolyT[a]	5.8–6.3	615	6.47	0–0.13

Source: From Ref. [50].
[a]Poly(vinylbenzyltriphenylphosphonium chloride).

The reactions of UO_2^{2+} with Chromazurol S and Eriochromcyanine R in the presence of cationic surfactants have been studied in detail [59], as have been the determinations of rare earth elements [60,61].

The determinations of beryllium [62] and aluminum [63] in aqueous medium has been performed using solid-phase spectroscopy (dextran-type anion exchange gel). The reaction with Eriochromcyanine R and derivative spectrophotometry have been applied to determine Cr(III) in steel [64].

Gallein (pyrogallolphthalein) derivatives. These substances **(18)** have properties analogous to fluorones, and to triphenylmethane dyes. They can be also included within the group of xanthene dyes [50].

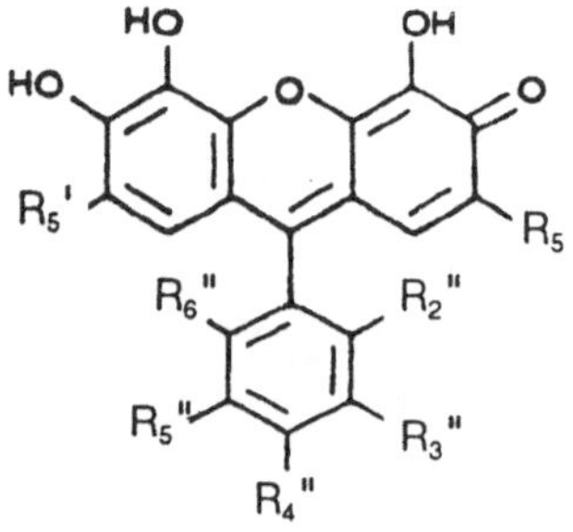

(18)

Among the most frequently used derivatives of this group are Pyrogallol Red ($R_5 = R_5' = R_3'' = R_4'' = R_5'' = R_6'' = H$, $R_2'' = SO_3H$), and Bromopyrogallol Red (BPR) ($R_5 = R_5' = Br$, $R_2'' = SO_3H$, $R_3'' - R_6'' = H$). These reagents are employed to determine Sn, Mo, and W, as well as the elements of group III lanthanoids, and the elements of groups IV and V [50].

In the case of a complex of silver with BPR, it has been found that the introduction of phenanthroline as a third component of the complex induces a significant bathochromic shift of the complex absorption spectrum: for the yellow Ag–BPR complex the $\lambda_{max} = 390$ nm; for the blue Ag(I)–BPR–phenanthroline complex the $\lambda_{max} = 635$ nm [65]. This complex can be also employed for indirect determination of CN^- ions, because these ions cause its discoloration owing to the formation of a more stable complex of Ag(I) with CN^- ions [66].

Derivative spectrophotometry has also been applied to determine praseodymium by using Bromopyrogallol Red [67]. This determination is not affected by high concentrations (up to hundredfold) of other rare earths, with the exception of Yt, La, and Nd.

Hydroxy- and carboxyquinone derivatives. The most widespread spectrophotometric reagents of this group are anthraquinone derivatives: alizarine (1,2-dihydroxyanthraquinone) **(19)**, alizarine S (sodium salt of 1,2-dihydroxyanthraquinone-3-sulfonic acid), and quinalizarine (1,2,5,8-tetrahydroxyanthraquinone). These substances can form chelates of the type **(20)** and **(21)** with metal ions.

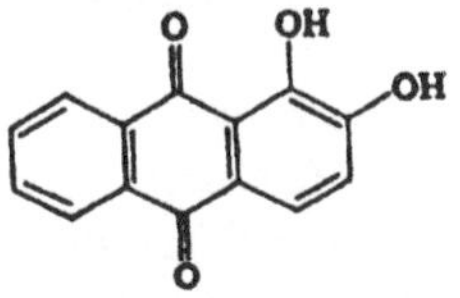

(19) 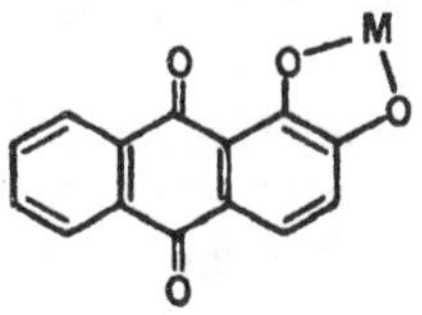   **(20)** 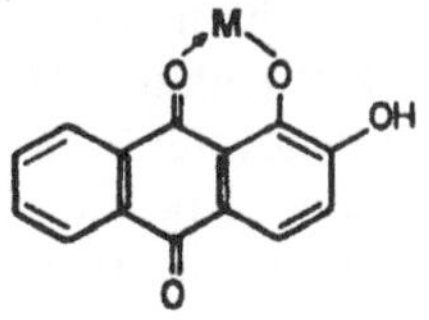   **(21)**

These substances have been used to determine Al, B, Be, Ga, In, Th, Zr, V, and other metals, and their complexes have been studied theoretically [68]. The determination of magnesium with 1,8-dihydroxyanthraquinone [69] and the simultaneous determination of beryllium and magnesium, and of cobalt and nickel in the complexes with 1-hydroxy-2-carboxyanthraquinone, have been performed [70]. These have been done using derivative spectrophotometry, similar to the simultaneous determination of beryllium and aluminum with 5,8-dihydroxy-1,4,-naphthoquinone in the presence of Triton X-100 [71]. The determinations are carried out in ethanol–water alkaline medium.

For lead determination, the mixed-metal complex of calcium and purine (1,2,4-trihydroxyanthraquinone) has been proposed, which has a higher stability, extractability and ε value when compared with the binary complex [72].

The study of the reaction between Zr(IV) and 1-amino-4-hydroxyanthraquinone has confirmed that carbonyl and hydroxyl groups participate in complex formation, whereas the amino group does not [73]. Emodin (1,3,8-trihydroxy-6-methylanthraquinone) has been employed in the determination of magnesium (pH $= 10$, $\lambda_{max} = 553$ nm) [74].

Alizarine and quinalizarine are often used for fluoride determinations, which are based on the decolorization of complexes of these reagents with suitable metals (La, Zr, Th) by fluorides, because the complexes of fluorides with these metals are more stable.

Chloranilic acid (2,5-dichloro-3,6-dihydroxybenzoquinone) is also included among reagents of this type. It is used for the determinations of Al, Ca, Sr, Mo, Zr, and so on, and in the form of barium salt it is used for sulfate determinations. In the latter reaction, $BaSO_4$ and red chloranilate ion with $\lambda_{max} = 530$ nm is formed, the absorbance of which is proportional to the SO_4^{2-} concentration. This well-known reaction has been used for FIA determination [75].

Phenoxazones (22) are less frequently used as reagents with oxygen donor atoms; e.g., they have been used for aluminum determination in the presence of cationic surfactant [76]. Hematoxylin (23) has been suggested for the determination of Ga and In in cetyltrimethylammonium bromide medium, which causes bathochromic shifts of absorption maxima and an increase of ε value [77].

Tetracyclines (24) can undergo complexation with ions of different metals (e.g., Mg, Co, Cu, U, Fe, Zr, Th, Sb, and Zn). The most often used spectrophotometric procedure for the determination of different tetracycline derivatives are based on the formation of complexes with Th, Zr, and Zn salts. The determination with Fe(III) has been carried out using the FIA method [78]. The interaction with Bi(III) salts (Dragendorff reagent) is better described as ion pairing. Oxytetracycline can be determined as a yellow complex with Ce(III) salts [26,27].

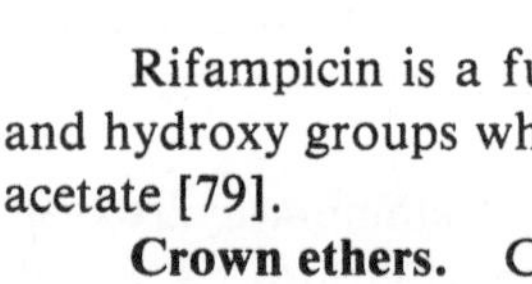

(22) 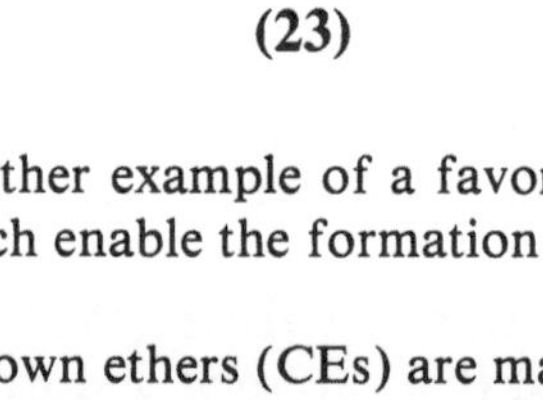   (23) 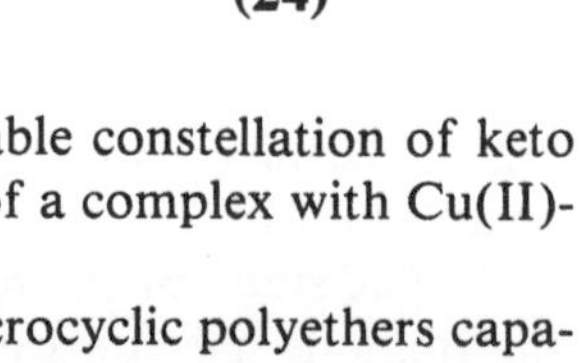   (24)

Rifampicin is a further example of a favorable constellation of keto and hydroxy groups which enable the formation of a complex with Cu(II)-acetate [79].

Crown ethers. Crown ethers (CEs) are macrocyclic polyethers capable of forming stable complexes with alkali metal ions as well as with other ions (Ag(I), Hg(II), La(III), Ce(IV), Tl(I)). They are highly soluble in organic solvents with a low polarity. For some transition metals (Co(II), Ni(II), Cu(II), Zn, Cd, Pb), the use of dimethyl sulfoxide has proved suitable [6]. The most simplest CEs contain repeated sequences of $(-CH_2-CH_2-O)_n$ with a spatial arrangement that allows the inclusion of ions into nucleophilic cavities of various sizes, determined by the number of oxygen atoms. For instance, in the 14-crown-4 compound (25) the cavity diameter is 0.12–0.15 nm; in the 18-crown-6 compound (26), it is 0.26–0.32 nm.

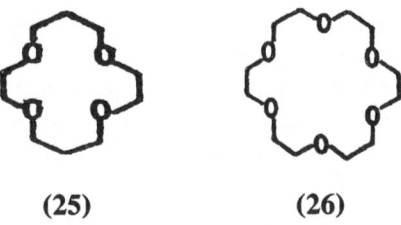

(25)   (26)

Therefore, the reaction selectivity is determined by the size of the reacting ion, and the inclusion complex that is formed is the result of ion–dipole interaction between cation and lone electron pairs of oxygen atoms. For spectrophotometric determinations of metals, colorless cationic complexes with CEs associated with intensely colored anionic dye (e.g., of the sulfophthalein type) are employed. Other alternatives involve the use of specially prepared CE derivatives with an azo or thio group placed into the etheric skeleton, or the use of a double-armed CE or other armed macrocycles [6]. The heteroatoms on the side arm participate in the coordination.

For example, selective determination of potassium with Orange 4-picrylamino-benzo-15-crown after extraction into chloroform ($\lambda_{max} = 560$ nm, $\varepsilon = 4.6 \times 10^3$) has been described [6]. Analytical applications of crown ethers are summarized in Ref. [80].

b. Ligands with nitrogen donor atoms

Ligands of this type often form complexes with metals in lower oxidation state that are able to lose an electron. In these cases, an intense band occurs in absorption spectra which reflects the transition of electrons from metal into free π-molecular orbitals of the ligand (i.e., d–π* transitions). Metals in a higher oxidation state also provide colored complexes, but their bands are not intense enough.

Reactions of metals with the following ligands of this type are the ones mostly commonly used in spectrophotometry:

2,2′-Bipyridyl (27) and 1,10-phenathroline (28).

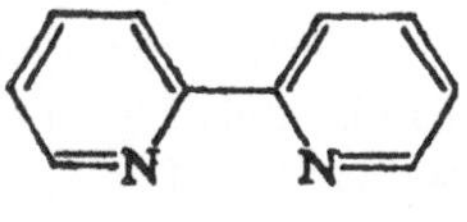

(27) 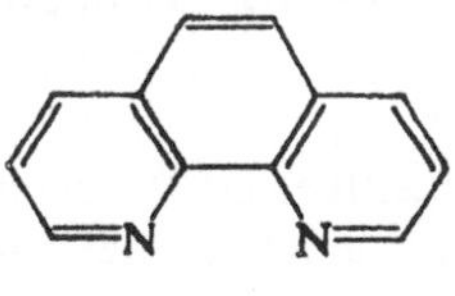   (28)

In the reaction with Fe(II) ions, red-colored complexes of FeL$_3$ type are formed. For Fe(bipy)$_3{}^{2+}$, $\lambda_{max} = 524$ nm; for Fe(phen)$_3{}^{2+}$, $\lambda_{max} = 510$ nm. Cu(I) and Ti(III) provide similar complexes.

The sulfate of tris(phenathroline) complex with Fe(II) is known as the redox indicator ferroin, which undergoes a transition from red to blue (ferriine) during the course of oxidation. The utilization of this reaction in spectrophotometry is presented in Chapter 5.

The formation of the Fe(II) complex with 2,2′-bipyridyl through the substitution of the CN$^-$ anion in Fe(CN)$_6{}^{4-}$ is catalyzed by Ag(I) and Au(III) ions. This reaction has been used to determine trace amounts of the above cations ($\lambda_{max} = 520$ nm) [81].

$$\text{Fe(CN)}_6{}^{4+} + 3(\text{bipy}) \rightleftharpoons \text{Fe(bipy)}_3{}^{2+} + 6\,\text{CN}^- \tag{9}$$

Ligands of this type are also represented by cuproine (2,2′-biquinoline) (29), neocuproine (2,9-dimethyl-1,10-phenanthroline) (30), bathophenanthroline (4,7-diphenyl-1,10-phenanthroline) (31), and bathocuproine (2,9-dimethyl-4,7-diphenyl-1,10-phenanthroline) (32).

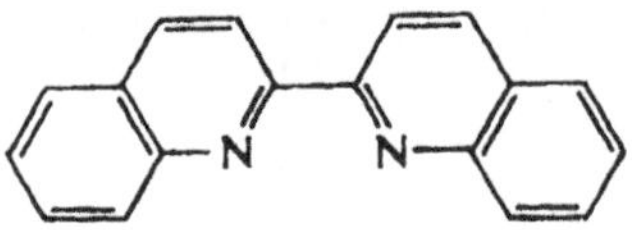

(29) 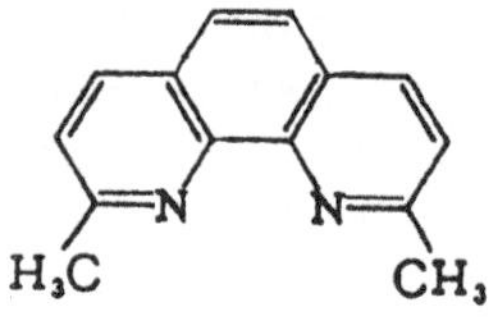   (30)

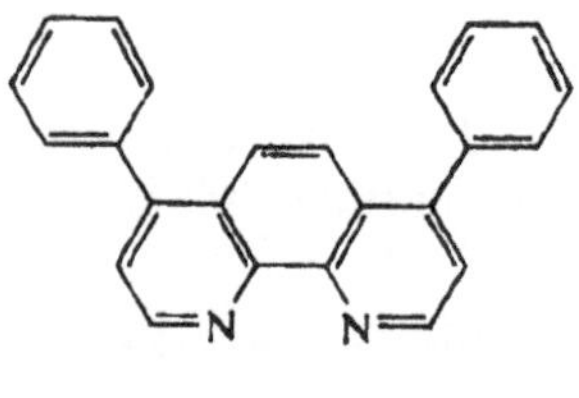

(31) 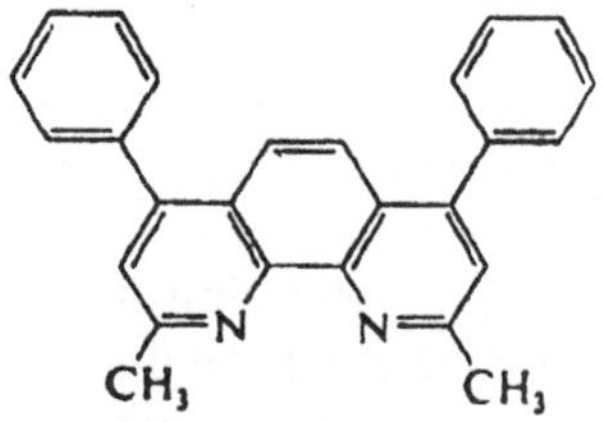   (32)

The Fe(II) complex with bathophenanthroline is insoluble in water (pH = 4–8) but can be extracted into organic solvents (iso-amylalcohol, toluene, nitrobenzene, $\lambda_{max} \sim 530$ nm). Relative to phenanthroline, the determination is more sensitive. Of the related cations, only Cu(I) forms a complex, which is colorless in acidic medium, however; Co(II) also forms a complex, but not one that can be extracted from the acidic medium.

Bathocuproine and neocuproine, in which the −CH$_3$ group acts as a steric hindrance, do not react with Fe(II), but they do with Cu(I) (in the presence of reduction agents, e.g. hydroxylamine hydrochloride). The orange complex of CuL_2 **(33)** is formed in the reaction with bathocuproine.

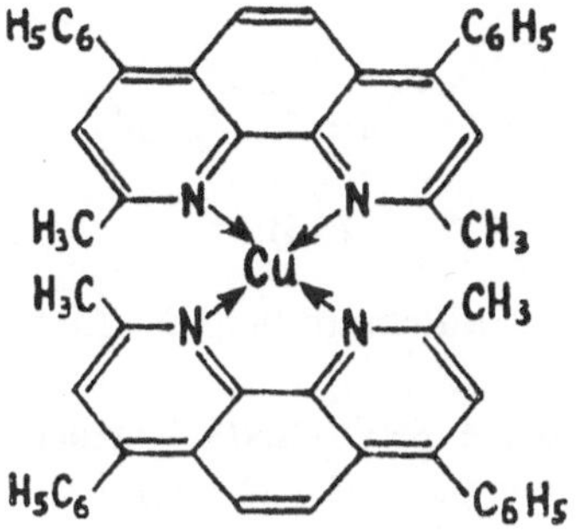

(33)

The above properties have been used to propose a procedure for the simultaneous iron and copper determination using the FIA method [82].

It is also possible to determine Cu(I) by neocuproine in the presence of a large surplus of Cu(II) [83]. This reaction can be also used for the determination of reductants (see Chapter 5).

Triazines. The very sensitive reagent for Fe(II), 3-(4-phenyl-2-pyridyl)-5-phenyl-1,2,4-triazine **(34)**, is included among ligands of this type [84]. The chelate of this ligand is insoluble in water. The sulfonated derivative provides water-soluble chelates of the $Fe(II)L_3$ (λ_{max} = 565 nm) and $Cu(I)L_2$ type (λ_{max} = 454 nm). It has also been used to determine reductants (ascorbic acid, L-cysteine). The Fe(II) and Cu(I) chelates are less stable than with bipyridyl, probably due to electron-acceptor properties of the sulfo group [85].

To determine Fe(II) and Cu(I) in practical samples, tetraammonium salt of 2,4-bis(5,6-diphenyl-1,2,3-triazine-3-yl)pyridinetetrasulfonic acid **(35)** has been used [86], and 3-(2-pyridyl)-5,6-diphenyl-as-triazine and its sulfonated derivative (ferrozine) **(36)** were used for Ru(II) determination [87].

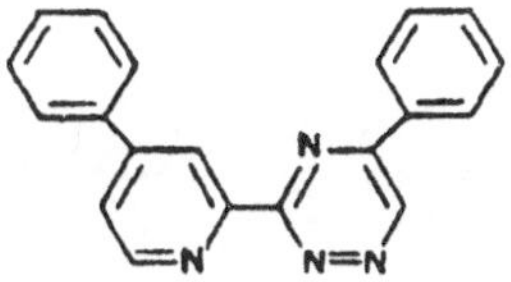

(34)

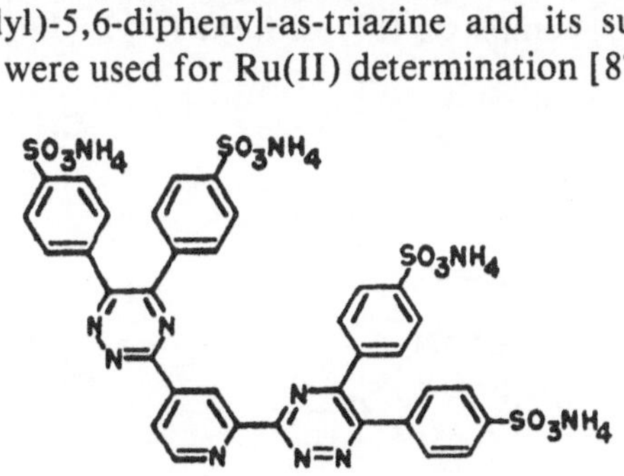

(35)

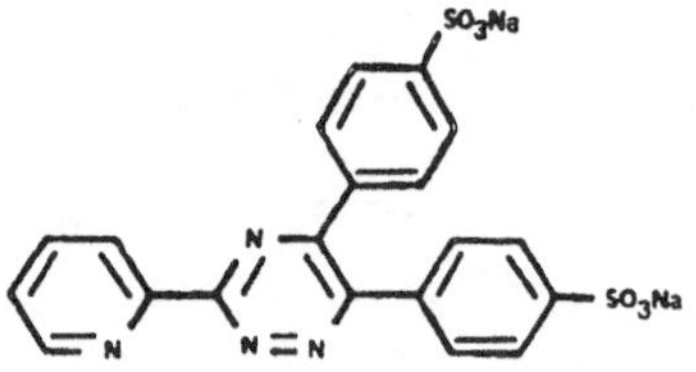

(36)

All these substances act as bidentate ligands, forming tris complexes with the elements studied.

Pentacyanoaminoferrate $Fe(CN)_5NH_3^{3-}$ reacts with ferrozine to form the $Fe(CN)_4(ferrozine)^{2-}$ complex. This reaction is catalyzed by Hg(II), Ag(I), and Au(III), which can be spectrophotometrically determined in this way [88].

2,4,6-tripyridyl-1,3-triazine (37) also belongs to this group. It is a very sensitive reagent for Fe(II) determination [89].

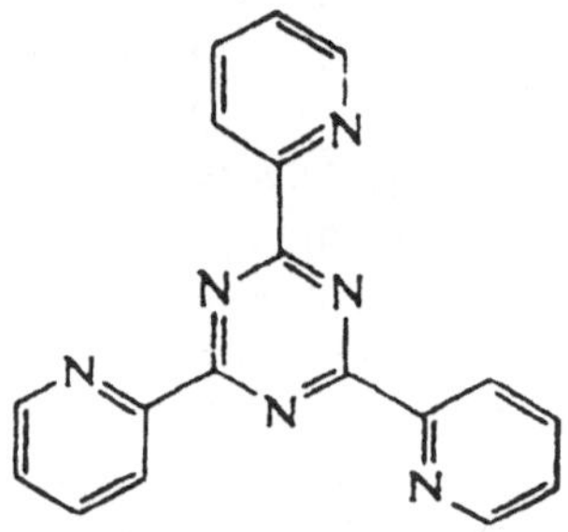

(37)

α-Dioximes (38). These represent another group of substances with two nitrogen donor atoms. The most commonly used reagent of this type is diacetyldioxime (dimethylglyoxime) (38) [90] (R = CH$_3$), which forms colored complexes (39) with Pd(II), Pt(II), Ni(II), Fe(II), and Co(II) [91].

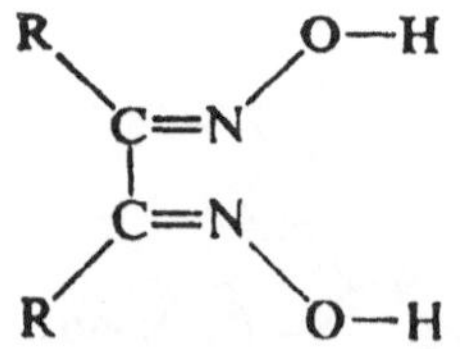

(38) 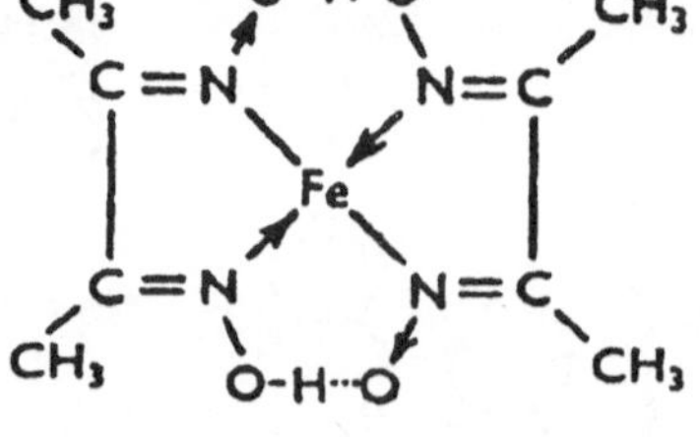   (39)

The ligands are connected by a strong intermolecular bond, which stabilizes the square-planar complex of the ML_2 type. In chelates of metals with coordination number 6, the binding of two other monodentate ligands ($-OH_2$, $-NH_2$, $-$halide) usually occurs in the direction of the z-axis.

The chelate of Ni(II) is insoluble in water, because a bond between nickel atoms is formed, but the other chelates are soluble in water. In these, the central atom binds a molecule of solvent or monodentate ligands. Contrary to other metal ions, the Co(II) complex coordinates iodide anions (λ_{max} = 435 nm); this reaction can therefore be exploited for their selective determination [4].

Diacetyldioxime complexes with Pd(II) ions are very stable, and this reaction is specific for Pd(II) at pH = 1(λ_{max} = 380 nm). A similar situation occurs for α-furyldioxime [92]. The ions of gold and platinum do not interfere, nor do the presence of other metal ions; but CN^- does when the complex is extracted into $CHCl_3$.

With diacetyldioxime, Ni(II) provides a yellow complex extractable into organic solvents. In alkaline oxidation medium, a wine-colored complex is formed that is soluble in water and contains Ni(IV) [93].

Diacetyldioxime is poorly soluble in water (more soluble in ethanol and even more so in diethyl ether and acetone). Cyclohexanedione dioxime (nioxime) **(40)** and cycloheptadionedioxime (heptoxime) are water soluble reagents with similar properties [94].

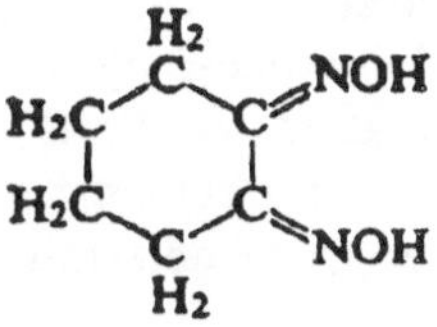

(40) 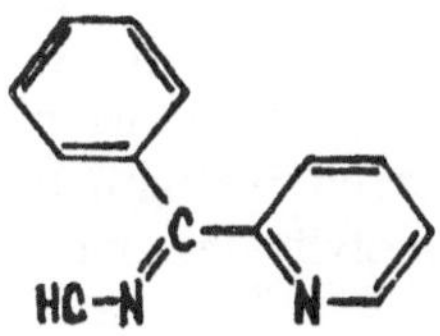   **(41)**

Another ligand of this type is syn-phenyl-2-pyridylketoxime **(41)** [95] which has been proposed for the determination of Fe(II), Co(II), Pd(II), and Cu(I) ions.

The determination of diacetyl can be performed following reaction with hydroxylamine in acidic medium and the conversion of the dioxime formed to this chelate. Acetoin and 2,3-butanediol can be determined in this way after oxidation to diacetyl [26,27].

Heterocyclic hydrazones of heterocyclic aldehydes. Hydrazones are substances with the basic structure **(42)**,

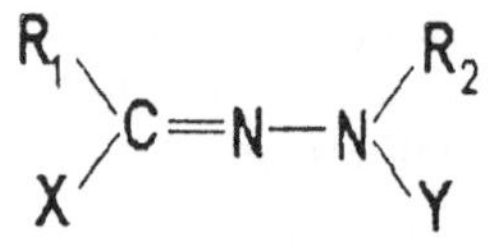

(42)

where R_1, R_2, and Y are H, alkyl, aryl, heterocycles or RCO; and X is H, alkyl, aryl, heterocyclus, OR, SR, CN, SO_2R, or NO_2.

Heterocyclic derivatives have been proposed for analytical practice [96] as derivatives of the basic structure **(43)** (pyridine-2-aldehyde-2'-pyridylhydrazone).

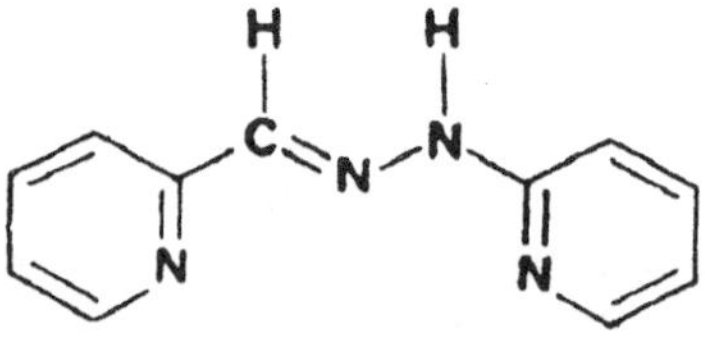

(43)

They are tridentate ligands, and their chelates with Fe(II), Co(II), Ni(II), Zn(II), Cu(II), Pd(II), and so on have been studied theoretically [97–99] as well as for practical applications (di-pyridylketone-2-pyridylhydrazone for V(II) determination [100], its nitro derivative for Fe(II) determination [101], and the derivative with the extended π-electronic system pyridine 2-aldehyde-2-quinolylhydrazone for nickel trace determination in foods [102]).

The traditional analytical applications of hydrazones are listed in Ref. [103]. The structure of the complex of Cu(II) with pyridine-2-aldehyde-2-quinolylhydrazone is shown as an example **(44)**.

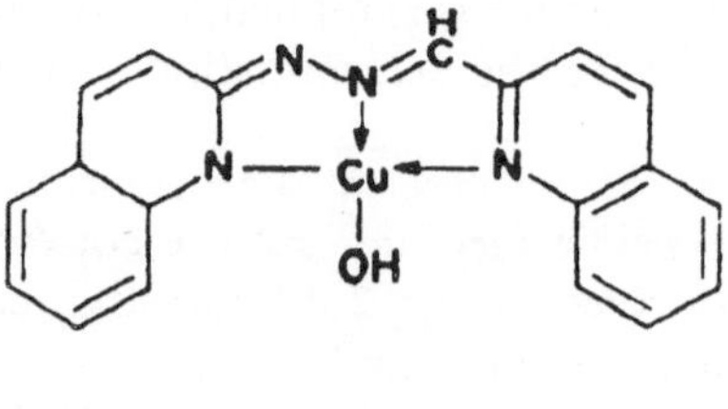

(44)

Triazenes.   These are a new group of spectrophotometric reagents. The simplest substance of this type is 1-(4-nitrophenyl)-3-(2-quinol)triazene **(45)**, which has been suggested for mercury determination in the presence of cetylpyridinium bromide [104].

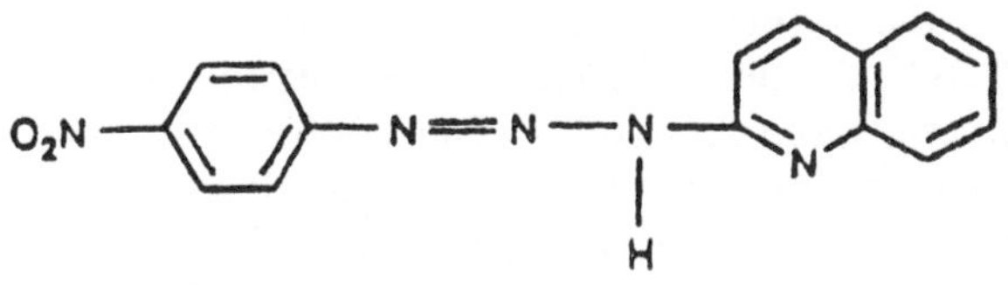

(45)

Cadion, (1-(*p*-nitrophenyl)-3-(*p'*-azobenzene)-triazene) is a similar compound; only the triazene group participates in chelate formation. Ternary cadmium complex with cadion and 1,10-phenanthroline have been proposed for extraction spectrophotometric determination of cadmium (λ_{max} = 490 nm, ε = 9.2 × 10^4), Mg, Hg, and Pd [105].

The extraction of methylmercuric ion complex with cadion has been studied as well [106]. High sensitivity is achieved in silver determination with cadion in the presence of nonionic surfactant Triton X 100 (λ_{max} = 565 nm, ε = 1 × 10^5) [107].

Porphyrin derivatives.   These are tetradentate ligands **(46)**. $\alpha,\beta,\gamma,\delta$-tetrakis(1-methylpyridinium-3-yl)porphine has been employed to determine nanogram quantities of cadmium [108]. The reaction of Mn(II) ions with $\alpha,\beta,\gamma,\delta$-tetrakissulfonatophenylporphine is catalyzed by the ions of Pb(II) and Hg(II), and can be used for their determination [109,101]. The measurement is performed through the observation of absorbance decrease in the reagent absorption maximum (λ = 413 nm) at fixed time.

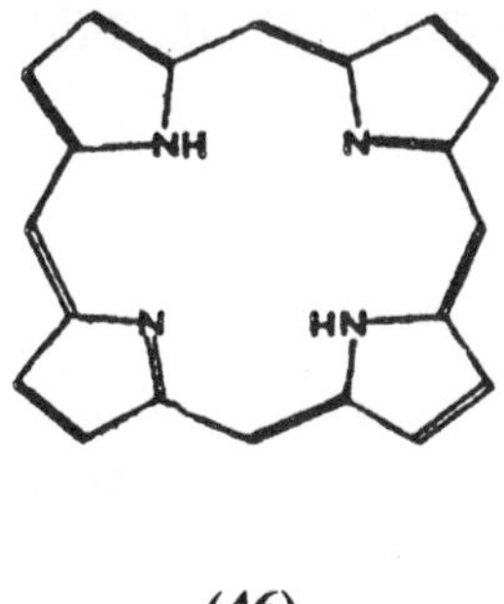

(46)

Biuret reaction.   Peptides and proteins form colored chelates (biuret reaction) with first transition metals (Co, Cu, Ni) in alkaline media. The chelates with Cu(II) ions are red to violet. For a comparative study of the suitability of the metals mentioned see Ref. [111]. The composition of Cu complexes is dependent on reaction conditions, especially on the proportion of the protein, the copper salt, and alkali hydroxide. Mostly *N,N*-chelates are supposed to be formed. The reaction is used to determinate proteins in biological materials and foodstuffs. The sensitivity of the reaction is low and therefore it is suitable for the determination of higher contents of proteins (e.g., in serum). The sensitivity can be increased by measuring in the UV region (310–340 nm). Many modifications of the spectrophotometric procedure have been described in the literature [112].

Conversely, complexation with Cu(II) ions can be used for the determination of biuret (imidodicarbonic diamide) [113].

c.   Ligands with sulfur donor atoms

Ligands of this group form complexes preferentially with "soft" metals, and the properties of the complex formed are strongly affected by the ligand structure. Ligands containing two sulfur donor atoms can be divided into the two following groups:

Aliphatic and aromatic dithiols.   The best known of this group is dithiol (toluene-3,4-dithiol) **(47)**. With metal ions (e.g., Mo(VI) in acidic medium [25]), colored precipitates are formed, which are extracted into organic solvents (for molybdenum it is tri-*n*-butylphosphate [114], isobutylmethylketone [115], and others).

Vicinal dithiols can be determined [26,27] using the formation of complex with manganese(II) acetate. Thiols can all form colored complexes with Co(II) and Cu(II) salts. Their aducts with mercuric chloride are also used for the determination of penicillins and cephalosporines using the reaction with imidazole [116,117].

Another reagent of this group is rubeanic acid (dithiooxamide) **(48)**, used for determinations of Cu(II), Co(II), Pd(II), and Os(VIII).

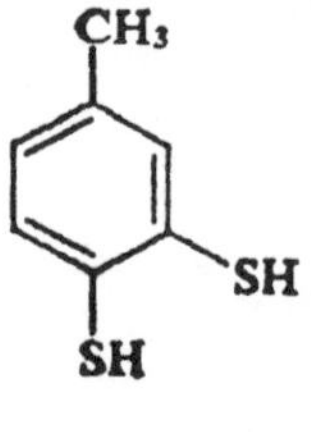

$$H_2N-\underset{\underset{S}{\|}}{C}-\underset{\underset{S}{\|}}{C}-NH_2$$

(47) 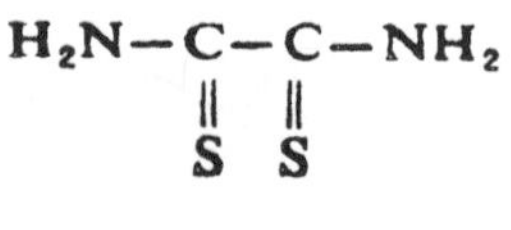   (48)

Dithiocarbamates. These are substances of the basic structure (**49**), in which π-electrons are delocalized across all atoms. (R_1 and R_2 are alkyl or aryl groups.) With metals, these substances form four-membered cycles (**50**) (these are the smallest experimentally confirmed cycles), which are not very stable due to a deformation of ligand bond angles.

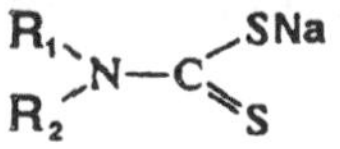

(49) 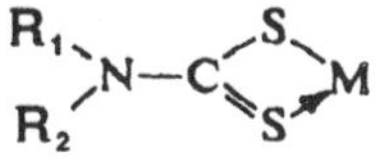　　　　　(50)

The most often used reagents of this group are sodium diethyldithio-carbamate (Kupral), and dimethyldithiocarbamate, respectively. An intense band corresponding to the π–π^* transition (for Cd(II) complex, $\lambda_{max} = 262$ nm) can be found in absorption spectra of its complexes. In nontransition metal complexes, a longer wavelength n-π^* band can also be found (for Ag(I) complex, $\lambda_{max} = 340$ nm). In transition metal complexes π–d transitions also occur (for Cu(II) complex, $\lambda_{max} = 436$ nm) [36].

Hundreds of works and a few reviews [118,119] deal with the metal complexes with dithiocarbamates. According to review [119], dithiocarbamates can be divided into three groups: (1) compounds with hydrophobic substituents, which provide water-insoluble complexes extractable into organic solvents; (2) compounds with hydrophilic substituents ($-OH_1 - $COOH), which provide water-soluble complexes; and (3) compounds with two CS_2 donor groups, which form polymeric chelates insoluble in all solvents. Review [120] deals with disulfides of dithiocarbamates (i.e., with substances R-CS-S-S-CS-R).

Different stabilities of diethylthiocarbamate complexes with metals have been used for determinations of Hg(II) and Ag(I), the insoluble complexes of which are more stable than the Cu(II) complex ($\lambda_{max} = 445$ nm, pH = 8–9). In the presence of the nonionic surfactant Triton X-100, solubilization of insoluble complexes occurs, and thus extraction into organic solvents is not necessary. The absorbance decrease of aqueous solutions is measured directly at 445 nm [121].

Antirheumatics flufenamic and mefenamic acid (anthranilic acid derivatives) have been determined using Cu(II) ions and diethyldithiocarbamate [122]. First, reaction of these acids with $Cu(NH_3)_4SO_4$ was carried out. The complexes formed were extracted into $CHCl_3$ and treated with diethyldithiocarbamate. The complex with Cu(II) is more stable than the complexes of the above acids. Thus, the acids are released from the complex into the solvent and the complex $[(C_2H_5)_2NC(:S)S]_2Cu$ is formed; its absorbance (which is proportional to the acid concentration) is measured at 430 nm.

The complexation reaction can be also used for the determination of dialkyldithiocarbamates. Copper(II) acetate and 0.3 mol L^{-1} HCl are used as reagents [27].

Dialkyldithiocarbamates are easily prepared by the reaction of dialky-lamines or cycloalkylamines with carbon disulfide in neutral or alkaline medium (10):

$$CS_2 + R_2NH + Cu(II) \rightarrow R_2N-C \underset{S}{\overset{S}{\diagdown}} Cu^+ \tag{10}$$

This reaction and subsequent complexation with Cu(II) salts in the presence of diethanolamine or pyridine can be used to determine the amines mentioned [27,123,124], carbon disulfide [125–127], or such compounds that release those compounds on hydrolysis. Thus, technologically important metal dialkyldithiocarbamates or xanthates have been hydrolyzed with hydrochloric acid and determined in this way, Eqs. (11) and (12):

$$RR'N\text{-}CS\text{-}SMX + H^+ \rightarrow RR'NH + CS_2 + MX^+ \tag{11}$$

$$R\text{-}O\text{-}CS\text{-}S\text{-}MX + H^+ \rightarrow R\text{-}OH + CS_2 + MX^+ \tag{12}$$

The procedure is also applicable to organomercury compounds, which decolorize the yellow solution of copper(II) diethyldithiocarbamate in chloroform at pH 4.5 [27].

d.   Ligands with oxygen and nitrogen donor atoms

Aromatic ethanolamines (51).   These (and other molecules containing alcoholic or phenolic hydroxy groups and amino groups) form colored chelates with ions of heavy metals. They react with Cu(II) ions in alkaline medium to form chelates extractable into 1-butanol, 1-pentanol, chloroform, and toluene, and they are suitable for quantitative analysis [26,27]. The determination of ephedrine, which forms a complex **(52)**, can serve as an example [128] (Eq. 13).

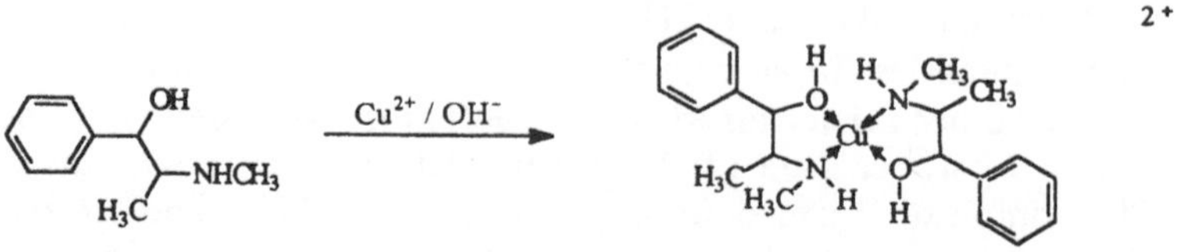

(13)

(51)   **(52)**

8-Hydroxyquinoline (oxine) and its derivatives. Oxine (**53**) is a reagent that forms chelates (**54**) with almost all transition metals, but also with nontransition metals. It forms ML_2 chelates with divalent metals in neutral or slightly acidic medium, ML_3 chelates with trivalent metals at pH ~ 4, and ML_4 chelates with Th(IV) and Zr(IV) in acidic medium.

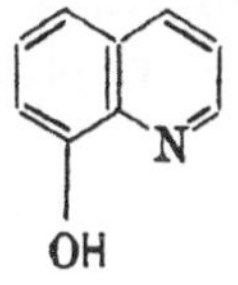

(53) 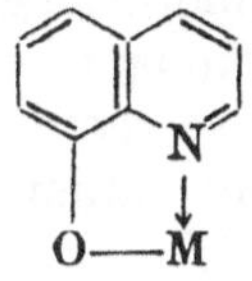   **(54)**

In absorption spectra of the nontransition metal chelates (Al, Bi, Ga), primarily the bands corresponding to π–π* and n–π* transitions in the conjugated π-electronic system of the reagent are used for determinations. Apart from these bands, more intense bands, corresponding to the charge transfer transitions, are formed in the spectra of transition metals [36].

If electroneutral and coordination-saturated chelates, which are insoluble in water, are formed (e.g., with Co(II)), extraction spectrophotometric determination is usually performed [4].

It has been found that the extractability of some valence-saturated, but coordination-unsaturated, chelates can be improved by the addition of quarternary ammonium compounds. In the case of magnesium oxinate, zephiramine has been used; its presence increases ligand coordination and leads to the formation of $\{Mg(Ox)_3, (zeph)\}$ associate [129].

The most selective derivative of this group is 7-iodo-8-hydroxyquinoline-5-sulfonic acid (ferron), which provides colored chelates with Fe(III) and Al(III) ions, contrary to the colorless chelates of other metals. The cationic surfactant cetyltrimethylammonium bromide increases the values of the stability constants of Al and Fe(III) complexes (primarily of the ML_3^{3-} complex), expands the pH range in which complexes can exist, and improves the linearity of calibration curves [130]. Cetylpyridinium bromide also induces bathochromic and hyperchromic shifts of the absorption maximum of the complex of niobium with 8-hydroxyquinoline-5-sulfonic acid, because it favors charge-transfer transitions [131].

A number of arylazo-quinolinols (5-benzeneazo-, 5-ortho-, 5-meta-, 5-para-hydroxyphenylazo) have been prepared, and the effect of substituents on their dissociation constants has been observed in dioxane–water (1 : 1) medium. Because these electron-acceptor substituents increase the acidity of the derivatives, the chelates are formed at lower pH values [132].

The complexation reaction of 8-hydroxyquinoline and its halogenated derivatives can be used for their determination. Fe(III) ions in methylcellosolve medium or Cu(II) ions in aqueous medium with subsequent extraction of the complex into chloroform are used as reagents [26,27,133–137].

The vanadium-8-hydroxyquinoline complex reacts with primary, secondary, and tertiary alcohols as well as with phenols in organic media to form red-to-blue vanadium-oxinate-hydroxycompound complexes, which can be used for the determination of alcohols or phenols [26,27].

Oximes (55).   These represent a very wide group of ligands with oxygen and nitrogen donor atoms. Their spectrophotometric applications are reviewed in Ref. [138].

The simplest compounds of this group are α-hydroxyaldoximes (e.g., salicylaldoxime **(56)**.

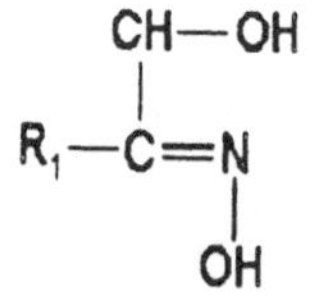

(55) 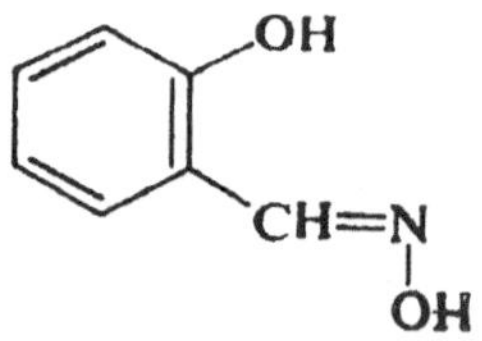   (56)

Chelates with metals are insoluble in water. The chelate with Cu(II), Ni(II), and Pd(II) has the structure **(57)**; the chelate with Mn(II), Fe(II), and Zn(II) has the structure **(58)** [4].

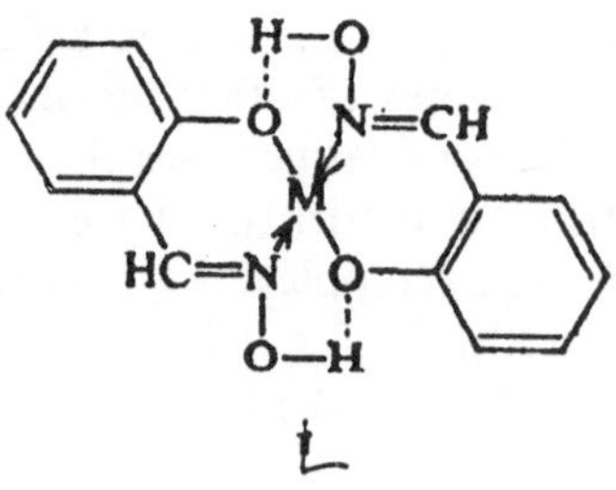

(57) 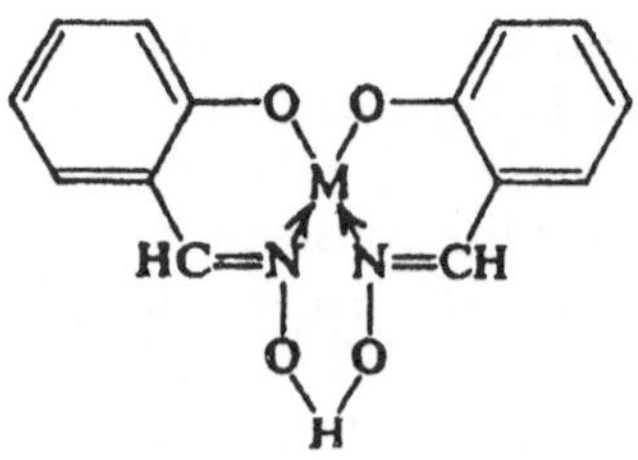   (58)

The absorption band corresponding to the π–π* transition is used for spectrophotometric determination. The chelate formed is a planar formation, which leads to the extension of conjugation and a shift of absorption maximum toward longer wavelengths (for salicylaldoxime, $\lambda_{max} = 300$ nm; for its chelate with Cu(II), $\lambda_{max} = 344$ nm) [36].

β-Benzoinoxime (Cupron) (59) is used for the determinations of Cu(II), Mo(VI), V(V), and W(VI).

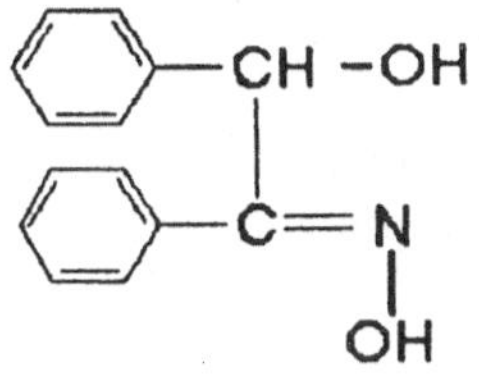

(59)

Different types of oximes (as well as α-dioximes) are widely used for palladium determination (quinoline-2-aldoxime, 2-pyridylaldoxime, salicylaldoxime, acetophenoxime). Their absorption maxima lie in the range 280–430 nm, and the determination is usually performed after extraction into chloroform. In this way, ppm concentrations of palladium can be determined. Palladium has also been determined following extraction with 3-hydroxy-2-methyl-1,4-naphthoquinone-4-oxime into molten naphthalene [139].

Resacetophenone oxime (2,4-dihydroxyacetophenone oxime) forms a water-soluble, yellowish brown complex with Mn(II) (pH = 10.5), with $\lambda_{max} = 380$ nm and stoichiometric composition of 1 : 1 [140].

Azomethines (Schiff bases). These are compounds characterized by an —RC=NR′ group (R and R′ are alkyl, cycloalkyl, aryl, or heterocyclus) [141]. o-(Salicylidene-amino)phenol (60) is a tridentate ligand that provides chelates with divalent cations (Co(II), Mn(II), Ni(II), Zn(II)) in the stoichiometric ratio of 1 : 1. These chelates are difficult to extract into inert, not very polar solvents (benzene, 1,2-dichlorethane). In the presence of tetradecyldimethylammonium chloride (surfactant zephiramine), Ni(II) and Zn(II) react with ligand anion (pH = 11) at a ratio of 1 : 2. The resulting chelate, with a charge of negative two, then reacts with two zephiramine cations and forms a [ML₂, (zeph)₂] complex that is extractable into the above solvents [142].

Another important azomethine ligand is glyoxal bis(2-hydroxyanil) [143], a quadridentate ligand that forms uncharged chelates with divalent metal ions (e.g., Cu) in strongly alkaline medium (pH ~ 12) (61).

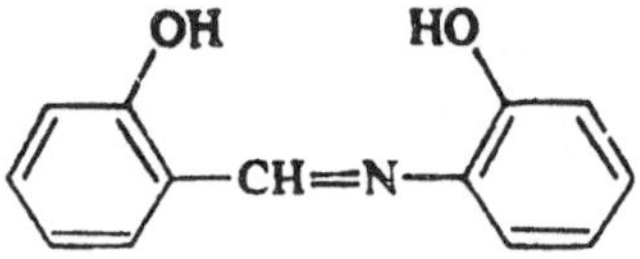

(60) 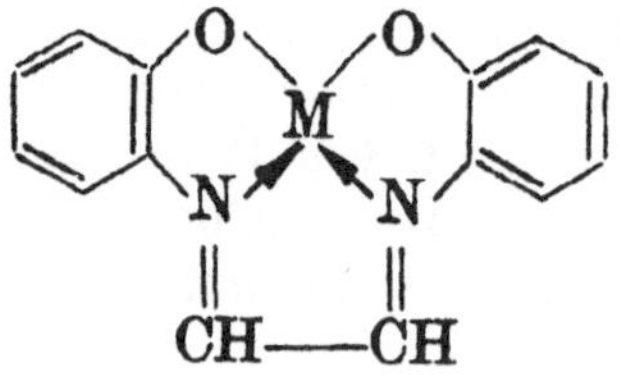   (61)

Regardless of the central metal ion, the absorption maxima of these chelates usually lie around 550 nm and correspond to the π–π^* transition. Two benzene rings are coplanar with the five-membered chelate ring, and thus an extensive π-electronic conjugation occurs.

Again, electroneutral 1 : 1 chelates are barely extractable into inert organic solvents. In the case of calcium and cadmium, it has been found that in the presence of zephiramine easily extractable chelates are formed [142].

Ligand coordination upon metals through O, N, and S atoms has been predicted for 2-(o-hydroxyphenyl)benzthiazoline complexes with Sn(II) and Pb(II) (62) [144].

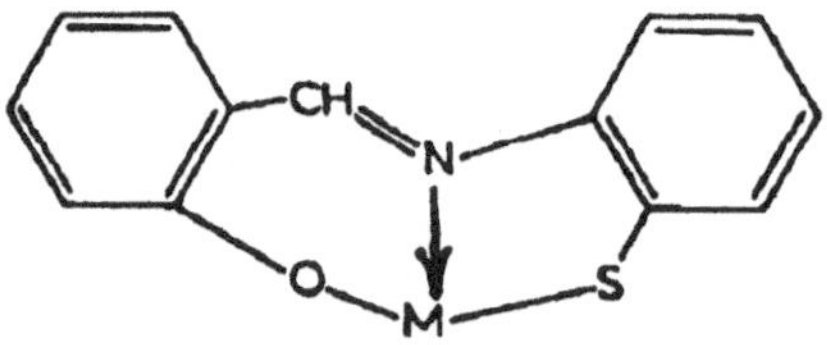

(62)

Aroylhydrazones.   These are compounds of the type (63). An example is salicylaldehydehydrazone (64).

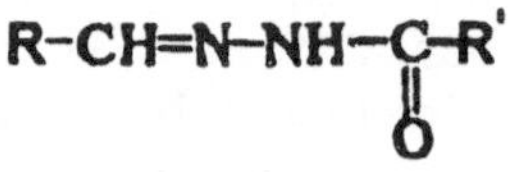

(63) 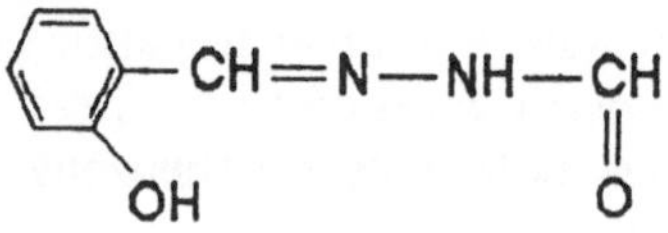   (64)

Reviews of analytical applications of hydrazones are provided in Refs. [145,146].

Bis-aroylhydrazones of α-diketones have been studied as reagents for spectrophotometric determinations of Ca(II), Cd(II), Bi(III), and La(III) [147]. An example is glyoxal bis(4-hydroxybenzoylhydrazone). It is assumed that the end form of this quadridentate ligand forms a chelate **(65)** with calcium ions in 0.01 mol L^{-1} NaOH.

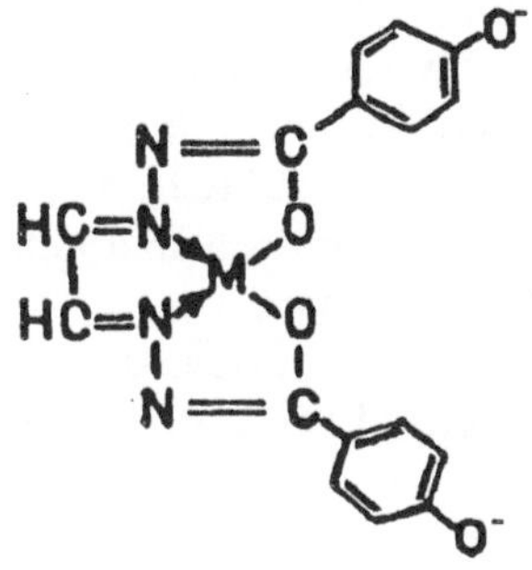

(65)

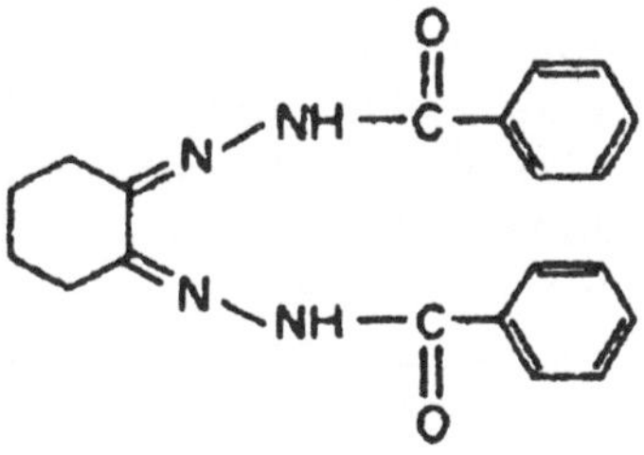

(66)

1,2-Cyclohexanedione bis benzoylhydrazones **(66)** and 1,2- and 1,3-cyclohexanedione bis (2-hydroxy)benzoylhydrazones have also been prepared [148,149]. They react with a number of divalent, trivalent, and tetravalent metals (λ_{max} ~ 450 nm). Mo(VI), Ti(IV), and Sn(II) react with the 1,2 derivative at a much lower pH than other ions, and so determinations in mixtures (primarily of Sn(II) and Sn(IV)) can be performed. The reaction with Ti(IV) is employed in the analysis of geochemical and metallurgical samples [150].

Carbazides (carbohydrazides) (67), carbazones (68), and semicarbazones (69). These compounds form chelates with oxygen donor atom, and with *-labelled nitrogen.

R—N*H—NH—C(=O)—NH—NH—R

(67)

R—N*H—NH—C(=O)—N=N—R

(68)

$(R_1)(R_2)$C=N*—NH—C(=O)—N$(R_3)(R_4)$

(69)

The most frequently used reagent of this group is 1,5-diphenylcarbazide, which still belongs to the most common as well as the most sensitive reagents for spectrophotometric determination of $Cr_2O_7^{2-}$. The reaction takes place in mineral acid medium, and the resulting product's λ_{max} = 540 nm. A number of authors who have studied this reaction mechanism have concluded that a redox reaction occurs to form Cr(III) and diphenylcarbazone, which then form a complex [151]. Diphenylcarbazide is also used for the determinations of Os, Re, Te, Pt, and Pd [152].

1,5-Diphenylcarbazone is used for highly sensitive determinations of Cu(II), with which it provides an intensely pink complex that is insoluble in water and extractable into benzene, isoamylalcohol, $CHCl_3$, and CCl_4. The reaction takes place in alkaline medium. This reagent is also applied for the determinations of Hg, Ga, In, and Pb, among others.

Analytical applications of semicarbazones are reviewed in Ref. [153]. Asymmetric derivatives have also been synthetized and proposed for analytical utilization [154].

Aminopolycarboxylic acids. The most common compounds of this type are the so-called complexons, used for chelatometric titrations: complexone 1 (nitrilotriacetic acid, NTA) (70), complexone 2 (ethylenediaminotetraacetic acid, EDTA) (71), and complexone 3 (disodium salt of EDTA).

$$
\begin{array}{lll}
CH_2-COOH & HOOC-CH_2 & CH_2-COOH \\
\diagup & \diagdown & \diagup \\
N-CH_2-COOH & N-CH_2-CH_2-N & \\
\diagdown & \diagup & \diagdown \\
CH_2-COOH & HOOC-CH_2 & CH_2-COOH \\
\textbf{(70)} & & \textbf{(71)}
\end{array}
$$

ethyleneglycol-bis-(2-aminoethylether)-N,N'-tetraacetic acid (EGTA), N-hydroxyethylethylenediaminetriacetic acid (HEEDTA), ethylenediamine-N,N'-bis-o-hydroxyphenylacetic acid) (EDHPA), and diethylenetriaminepentaacetic acid (DTPA) have also been prepared. All of these compounds are multidentate ligands. In their chelates, nitrogen and at least one oxygen atom participate in the bond on metal. Structure **(72a)** has been proposed for NTA 1 : 1, structure **(72b)** for 1 : 2 chelates, and structure **(73)** for EDTA 1 : 1 chelate with Co(II) [1].

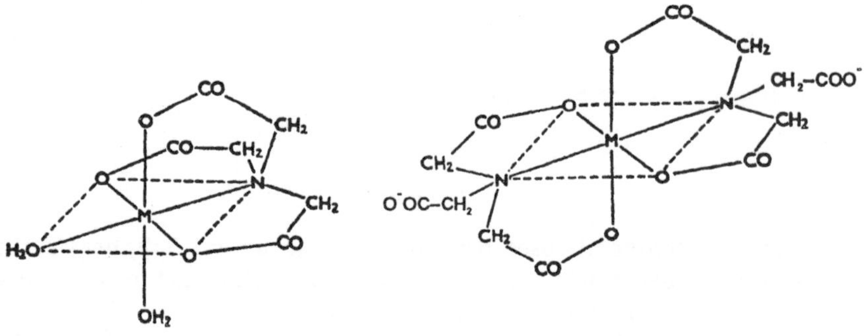

a   (72)   b

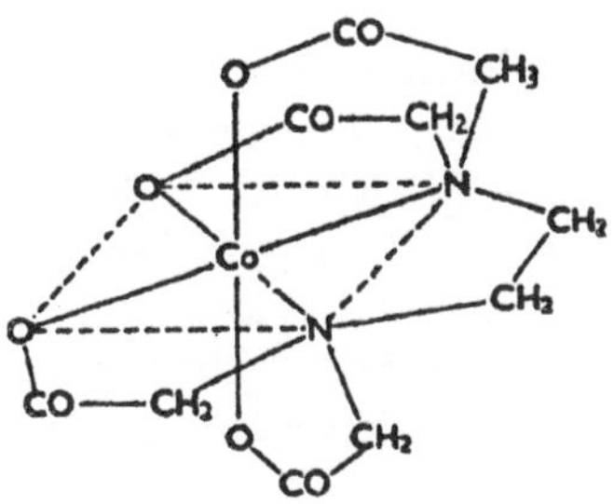

(73)

The EDTA chelates with most metals are colorless. However, bands in the ultraviolet region can be used for determinations (for Bi(III), for instance, λ_{max} = 263 nm). Colored chelates are formed with Cu(II), Co(II), and Ni(II). The Fe(III) chelate has λ_{max} = 260 nm in acidic medium; λ_{max} = 366 nm for pH = 5; and λ_{max} = 520 nm in ammonia (pH = 10–11) in the presence of H_2O_2.

Derivative spectrophotometry has been utilized for simultaneous determination of Fe(III) and Bi(III) with EDTA [155].

The high stability of metal chelates with EDTA is often utilized for the purposes of metal masking in spectrophotometric determinations.

Phenolphthalexone group dyes (74). These are derivatives of 4,4′-hydroxytriphenylmethane dyes (see p. 73) substituted in positions 5,5′ by the $-CH_2-N(CH_2-COOH)_2$ group. The basic compound is phenolphthalexone (phenolphthaleincomplexone) $(R_3=R_3{}'=R_6=R_6{}'=H, \quad R_2{}''= -COOH)$ or phenolphthalexone S $(R_2{}''=-SO_3H)$, respectively. A number of derivatives exist, distinguished by the substituent kind in the positions R_3, $R_3{}'$, and R_6, $R_6{}'$. Of them, the most frequently used is Xylenol Orange $(R_3=R_3{}'=CH_3, R_6=R_6{}'=H)$, and Methylthymol Blue $(R_3=R_3{}'=CH(CH_3)_2, R_6=R_6{}'=CH_3)$.

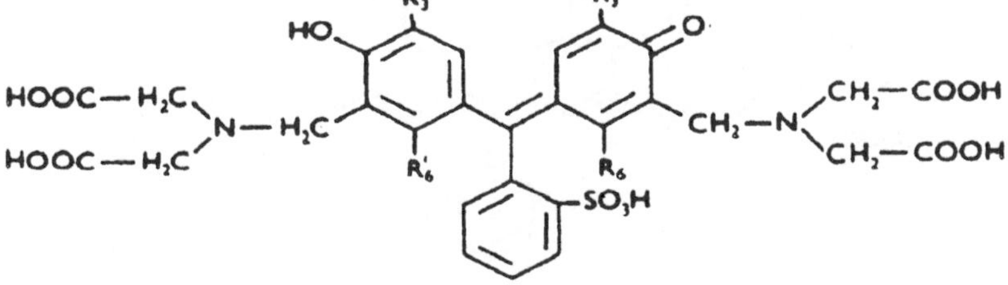

(74)

Semixylenol Orange (R_5=H, R_5'=CH_2—N(CH_2—COOH)$_2$, R_3=R_3' =CH_3, R_6=R_6'=H), Semithymol Blue (R_5=H, other substituents are the same as for Methylthymol Blue) and the dyes of Glycinecresol Red group (R_5=R_5'=CH_2—NH—COOH, R_3=R_3'=CH_3, R_6=R_6'=H; in the case of Glycinecresol Blue R_3=R_3'=H, R_6=R_6'=CH_3; for Glycinethymol Blue R_3=R_3'=CH(CH_3)$_2$, R_6=R_6'=CH_3) also belong to this structural group.

Fluorescein complexone **(75)** is a fluorescein derivative, which, under the name calcein, is used for Ca determination.

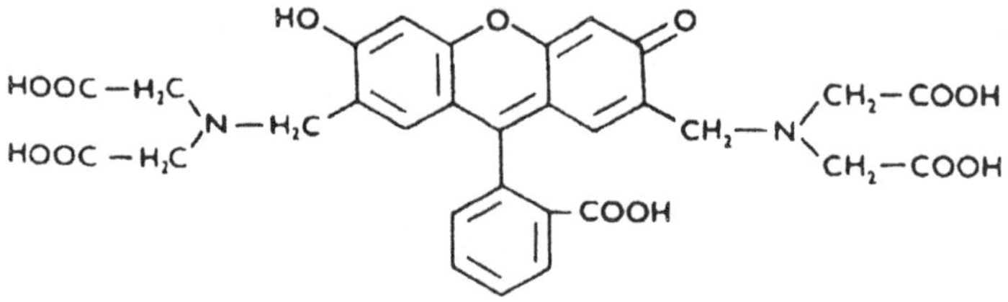

(75)

Phenolphthalexone-type dyes are used to determine Zr(IV), Hf(IV), Th(IV), Ti(IV), Fe(III), Cr(III), Al(III), Sc(III), lanthanoids, Ga(III), In(III), Pb(II), Zn(II), and so on, with which they form chelates in the stoichiometric ratios of 1 : 1 and 1 : 2. Their arrangement in the case of In(III) 1 : 2 complex with Xylenol Orange is shown in structure **(76)**.

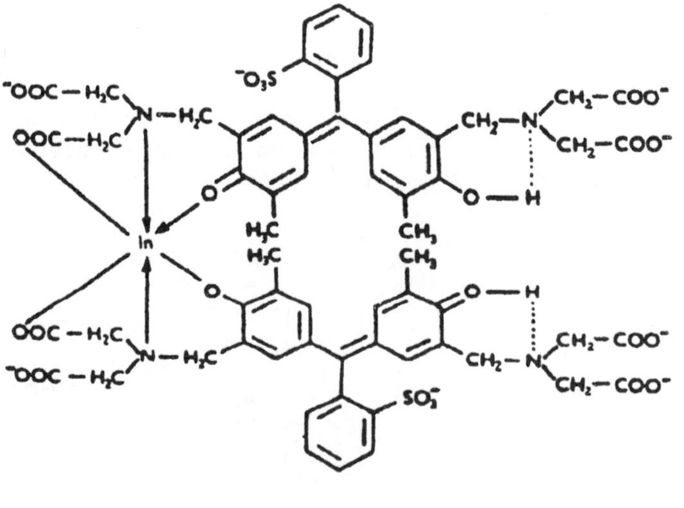

(76)

Structures of phenolphthalexone-type dyes, characteristics of their complexes (λ_{max}, pH range, stoichiometry, range of Lambert-Beer law validity), and examples of application for the analysis of practical samples are shown in Ref. [50]. The reaction of rare earths with Xylenol Orange, the sensitivity of which is enhanced by the presence of surfactant micelles [156], and the determination of Th(IV) with Semithymol Blue in soil samples [157] have been studied.

The decolorization of Zr(IV)-Xylenol Orange complex by fluorides can be used for fluoride determination. This reaction has been used in silicate analysis [158] and in HPLC postcolumn determination of fluorides [159].

Reactions of metals with dyes of this type have also been utilized for kinetic determination of substances (especially anions) that catalyze these reactions. For instance, the reaction of Cr(III) with Xylenol Orange is catalyzed by carbonates (and bicarbonates) [160], because they probably cause decomposition of aquo-complexes of Cr(III), which are the reason for Cr(III) kinetic inertness. The reaction of Zr(IV) with Xylenol Orange is catalyzed by fluoride ions [161]; these cause the depolymerization of zirconyl ions under the formation of ZrF^{3+} and ZrF_2^{2+} complex anions, which have a higher reaction rate. This approach permits the determination of fluorides in the concentration range of 5–50 ng mL^{-1}. The reaction is also catalyzed by SO_4^{2-}, PO_4^{3-}, AsO_4^{3-}, and organic anions (oxalates, citrates). The ions of SO_4^{2-} and F^- have a similar catalytic function in the reaction of Zr(IV) with Methylthymol Blue [162].

Nitrosophenols. Nitroso derivatives of phenols (e.g., of resorcinol), are known to form colored chelates with salts of heavy metals (Fe(II), Co(II), Cu(II), Ni(II), etc.) and procedures using in situ formation of dinitrosoresorcinol from resorcinol and nitrite have been recommended for the determination of those metals [163,164] or, conversely, for the detection and/or determination of nitrite [165] or resorcinol [164].

Nitrosonaphthols [1-nitroso-2-naphthol (77), 2-nitroso-1-naphthol].

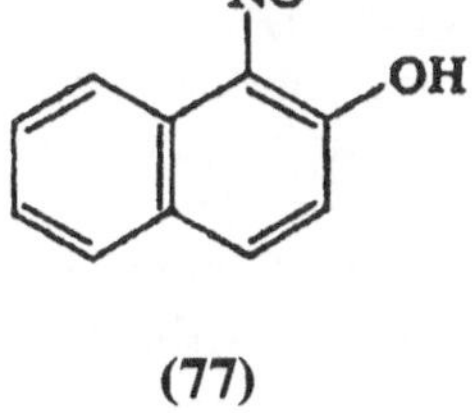

(77)

This reagent is most commonly used for the determination of Co(III), with which it forms a CoL$_3$ complex (λ_{max} = 520 nm). Less stable complexes are formed with Fe(III), Cr(III), Cu(II), and Pd(II).

The electrically neutral chelates formed are insoluble in water. Thus, 1-nitroso-2-naphthol-3,6-disulfonic acid (nitroso R-salt) is sometimes used instead [19].

Derivatives of anthraquinone with nitrogen substituent. Alizarin complexan (3-aminomethylalizarin-*N*,*N'*-diacetic acid) has been proposed for spectrophotometric determinations based on the formation of ternary complex with lanthanum or cerium and with a fluoride atom **(78)** [166].

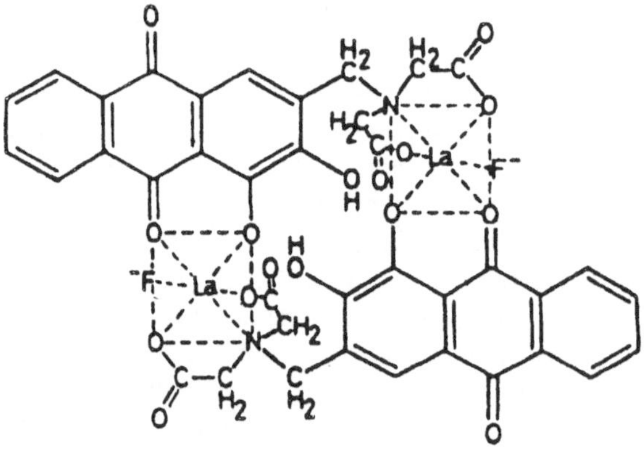

(78)

Azo dyes. Many reagents for inorganic photometric analysis are derived from different types of azo dyes. Based on quantum-chemical calculations, the nature of absorption bands and the electronic structure of the chromophore parts of the dye in the ground and excited states has been explained [167]. For a number of azo compounds, azo-hydrazo tautomerism has to be considered (see Chapter 2, Section C).

With respect to the number of donor atoms, the molecules of azo dyes behave as bi-(N,O; but also O,O), tri-(O,N,N; or O,O,N) or multidentate ligands, and chelation with metal ions can lead to the formation of five- or six-membered cycles. Their formation leads to the expansion of the conjugated system of the ligand double bonds, and a shift of absorption bands of π–π^* transitions occurs. Most significantly, their formation is manifested in ligands containing donor atoms in the *o*- or *o,o'*-position to the azo chromophore.

o-Hydroxyazobenzenes and metals provide a five-membered chelate ring with the metal bound through oxygen and nitrogen donor atoms **(79)**; *o,o'*-dihydroxyazo derivatives form chelates of the type **(80)**.

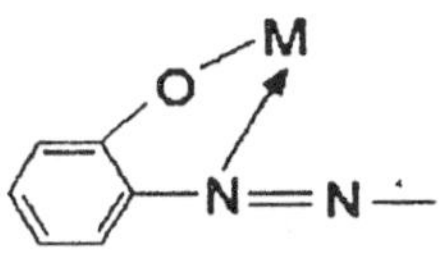

(79) 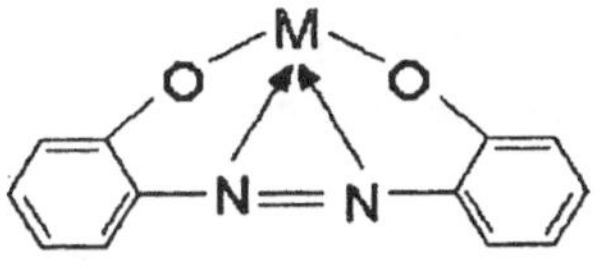   (80)

Of importance are, for instance, Tropaeolin O (2',4'-dihydroxyazo-benzene-4-sulfonic acid, Acid Orange 6) and Tropaeolin OOO (sodium salt of 1-(2-hydroxynaphthylazo)-benzene-4-sulfonic acid, Orange II) for the determination of palladium (λ = 550 or 570 nm, respectively). For determination of lanthanoids and magnesium, Eriochrome Black T (supra) (sodium salt of 1-(1-hydroxy-2-naphthylazo)-2-hydroxy-6-nitro-4-naphthalene sulfonic acid) is used (in the latter case, λ_{max} = 520 nm, ε = 2.2 × 10^4). Lumogallion (1-(2,4-dihydroxyphenylazo)-2-hydroxy-5-chlorobenzene-3-sulfonic acid, Mordant Red 72) is recommended for the determinations of tin and indium, as well as scandium and niobium. *o,o'*-Dihydroxyazo-benzene (DHAB) is suggested for determinations of Al(III), Ga(III), In (III) [169], and Ti(IV) [168] and has also been used in ion-pair reversed phase HPLC [170].

The application of the series of dyes Tropaeolin O, Tropaeolin OO, and Tropaeolin OOO for spectrophotometric determinations of Pd(II) has been evaluated [171]. Optimal conditions have been compared for the formation of chelates of sulfonated *o,o'*-dihydroxyazo dyes as HSN (i.e., 2-hydroxy-1-(2-hydroxy-4-sulfo-1-naphthylazo)-3-naphthoic acid) with Ca and Mg ions [172].

The *o*-hydroxyazo group is supplemented by the phenylarsonate group using the reagents Thorin (Thoron) I (Naftarson; APANS; which is a disodium salt of 2-(2-hydroxy-3,6-disulfo-1-naphthylazo)-benzenearsonic acid) and Thorin (Thoron) II (i.e., tetrasodium salt of 4,4'-diarsono-diphenyl-3,3'-bis(1-azo-2-hydroxynaphthalene)-3,6-disulfonic acid), which are suitable primarily for thorium determination.

Considerable attention has been paid to studying the effect of the surfactant CTMAC on the reaction of BDAS (i.e., sodium 2-bromo-4,5-dihydroxyazobenzene-4'-sulfonate) with Cu(II), Ti(IV), W(VI), and Mo(VI) ions. For instance, for titanium determination, λ_{max} = 515 nm, ε = 6.2 × 10^4 at a pH of 2.9 [173].

Other groups of azo dyes include mono- and bis azo derivatives of chromotropic acid. An overview of their applications is provided by Buděšín-ský [174], who divides these reagents into three groups with respect to the nature of the substituents.

In 3-arylazo and 3,6-bis(arylazo) derivatives, where the aryl group carries no donor group in the *o*-position with respect to the azo link, chelation of metal ions with this compound involves the peri-hydroxyl substituents of the naphthalene nucleus (**81**).

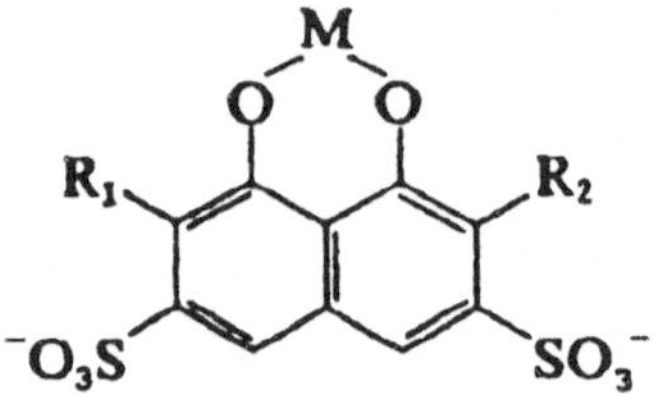

(**81**)

in which $R_1 = -N=N-Ar$, $R_2 = R_1$ or H, Ar = phenyl, naphthyl, and so on. One of the important representative of this group is SPADNS, which forms red complexes with Th and Zr. Other reagents from this group are di-SNADNS and Palladiazo and so on.

The second group contains 3-arylazo reagents, in which the donor group in the aryl substituent is in the ortho position with respect to the azo link. Ar can thus be ⟨ ⟩—, where Y is $-AsO(OH)_2$, $-PO(OH)_2$, $-COOH$, or it can be ⟨ ⟩—.

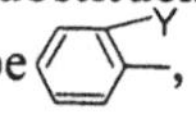

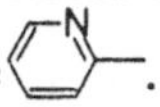

Chelation grouping allows the formation of two rings, one through the naphtholic oxygen and nitrogen of the azo group, and the second through the nitrogen of the azo group and oxygen of the arsonio, phosphonio, or carboxylic donor group or through the heterocyclic nitrogen of pyridine (**82**).

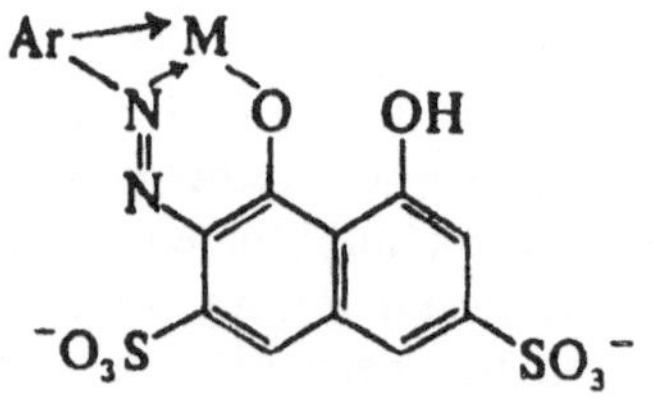

(**82**)

The selectivity of these reagents for certain metal ions (usually polyvalent with oxidation number (III) and (IV)) is based on the type of donor group in the ortho position of the aryl substituent. Nevertheless, some elements form chelates also with these reagents through the O,O bond of both naphtholic OH groups as was the case in the first group of reagents. To this group belong sensitive reagents for Th(IV), U(IV), (VI), lanthanoids, Y(III), and Sc(III).

3,6-Bis(arylazo) derivatives include the highest number of important and frequently used reagents. In these dyes, the aryl substituents carries a donor group ortho to the azo link. Chelation with the majority of metal ions occurs through the closure of several rings, in which two oxygen donor atoms from OH-groups and other donor atoms of o-groups of aryl substituents take part **(83)**.

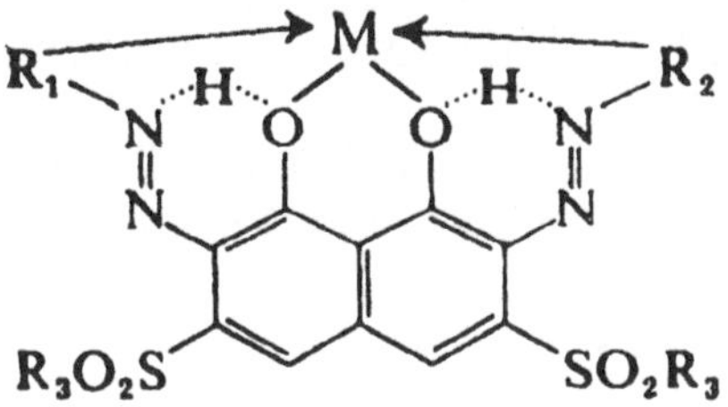

(83)

where R_1 is ⬡—Y or [N/S]— , and so on, and Y (donor group) = $-OH$, $-SH$, $-COOH$, $-SO_2(OH)$, $-PO(OH)_2$, $-AsO(OH)_2$, $-B(OH)_2$, $-CH_2N(CH_2COOH)_2$, and so on, $R_2 = R_1$ or phenyl or p-substituted phenyl, and $R_3 = -OH$, $-NH_2$ or $-NHC_6H_5$.

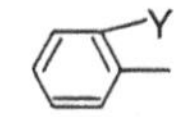

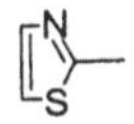

The chelation also depends on the acidity of the medium. Thus, chelates can have different structures with respect to the azo-hydrazo tautomery of the reagent under the formation of differently protonated complexes (e.g, BaL, BaHL, BaH$_2$L, BaH$_3$L). Antipyrylazo III reagent (or diantipyrylazo, i.e., 3,6-bis(4-antipyrylazo)-4,5-dihydroxy-2,7-naphthalenedisulfonic acid) provides with calcium a binuclear Ca$_2$HL complex of structure **(84)** [175].

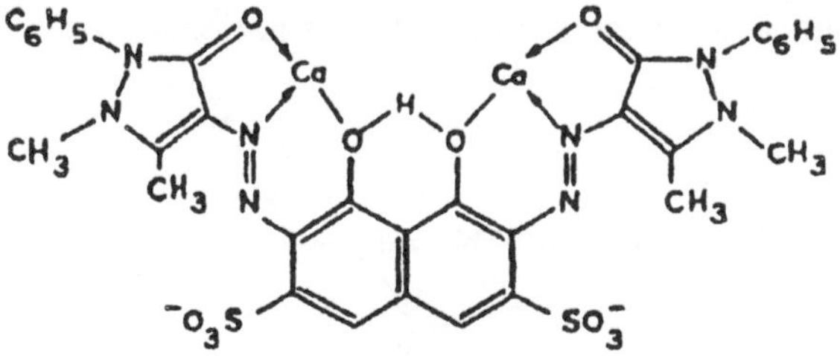

(84)

The nature of ortho donor groups in R_1 and R_2 substituents of (83) again predetermines the possible selectivity of the reagent. Arsenazo III is probably the most widespread reagent of the group [176].

The above reagents are often used in the presence of organic solvents to achieve a higher sensitivity. For instance, zirconium determination with chlorphosphonazo III in 2M HCl has $\varepsilon = 3.3 \times 10^4$, whereas the extraction of zirconium into the solution of the reagent with 3-methyl-1-butanol results in $\varepsilon = 2.1 \times 10^5$, $\lambda = 675$ nm [177]. In a number of cases, higher sensitivity is achieved by the presence of a surfactant (e.g., in the determination of Bi(III) [178].

Another large, important, and frequently used group of azo dyes consists of dyes with N, or N and S heteroatoms (85), where X is C or S.

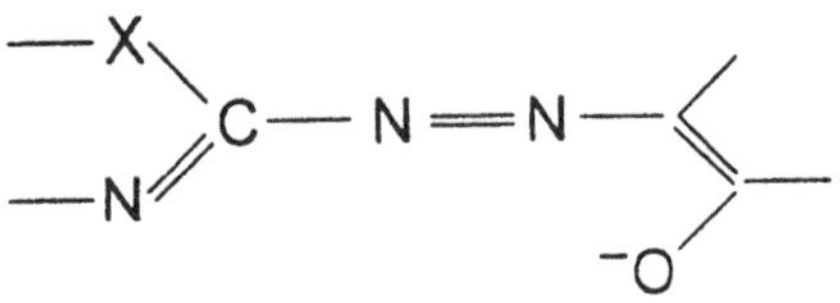

(85)

Compounds containing the 2-pyridylazo group have been summarized and evaluated [179]. The spectrophotometric applications of pyridylazo, thiazolylazo, and benzothiazolylazo compounds, and azo derivatives of 8-hydroxyquinoline have also been studied [180].

The reagents PAN (1-(2-pyridylazo)-2-naphthol; 2-PAN, β-PAN) and PAR (4-(2-pyridylazo)-2-resorcinol) are used for the determination and detection of more than 30 elements. In Ref. [179], 57 of the 2-pyridylazo derivatives with different substituents on naphthalene and benzene rings are listed. These differences account not only for different chromogenic properties of individual reagents but also for different complexation sites. PAN, for instance, is primarily a tridentate ligand, where metal is bound into the 1:1 chelates under the formation of two five-membered

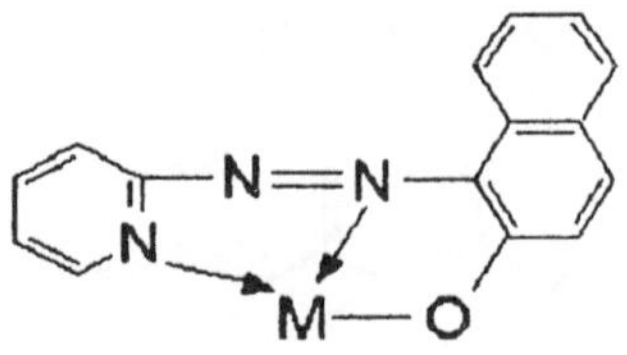

(86)

rings **(86)**. Sometimes, however, PAN behaves as a bidentate ligand. The PAR reagent is a tridentate ligand. Chromogenic properties of α-PAN (1-PAN, 2-(2-pyridylazo)-1-naphthol) are more pronounced relative to PAN. In *p*-PAN (4-(2-pyridylazo)-1-naphthol) complexes, the $-OH$ group does not take part in complexation, and chelation probably occurs through the heterocyclic nitrogen and the azo group nitrogen.

Complexes with PAN are mostly insoluble in water but can be extracted into $CHCl_3$, CCl_4, and benzene. They are soluble in various water-miscible solvents (acetone, ethanol, dioxane) and mostly are red (Fe(III), V(V)), with the exception of Pd(II) and Co(III) complexes, which are green. The stability of PAN complexes with some divalent transition metal ions in dioxane–water medium follows the sequence Cu > Ni > Co > Zn > Mn. The conditions for determination of different metal ions can be found in a number of monographs [179,180]. Complexation behavior of some pyridylazonaphthols (α-PAN) and pyridylazophenols (*o*-PAP and *p*-PAP) has also been studied [181,182]. Sulfonated derivatives have been prepared to increase the solubility in aqueous or water-miscible solutions (e.g., of α-PAN, denoted as α-PANS-*x*-S, where $x = 4$ to 8 (PANS)) [183,184].

Unlike PAN, PAR reacts with metal ions to form water-soluble complexes of intense red–to–red-violet color, with metal–ligand ratios of 1 : 1 and 1 : 2. The structure of chelate rings is analogous to the PAN reagent. The complex with palladium is green in acidic medium, and red in neutral medium. The reagent does not react with alkaline metals, Cr(VI), Sb(III), Mo(VI), W(VI), As(III) and As(V). The formation of some metal complexes depends on the medium acidity [185]: M(PAR)H complexes are formed in acidic medium, M(PAR)$_2$ complexes in alkaline medium. Depending on the medium and reagent concentration, copper, for instance, forms the following complexes: CuLH ($\lambda_{max} = 520$ nm, $\varepsilon = 1.8 \times 10^4$), CuL (510 nm, 3.9×10^4), CuL$_2$H (500 nm, 6×10^4), and CuL$_2$(500 nm, 7×10^4). The CuL$_2$H complex has not been confirmed. At a pH of 9.8–10.0, formation of the CuL$_2$ chelate ($\lambda_{max} = 500$ nm) is used for determination of the metal [185].

Determinations of Pd(II), Rh(III), and others can serve as examples of improved determinations of platinum elements with PAR in the presence of a surfactant or organic base [186,187] under ion association (see Chapter 4, Section D).

Apart from other *o*-hydroxy-2-pyridylazo derivatives, the halide analogs of 2-pyridylazoamino and of 2-pyridylazoaminophenol dyes, are of importance. Structure **(87)** is an example. PADAP ($R_1 = R_2 = H$), 5-Br-, and 5-Cl-PADAP ($R_1 = Br$, or Cl, $R_2 = H$), 3,5-di-Br-PADAP ($R_1 = R_2 = Br$) are the most often used.

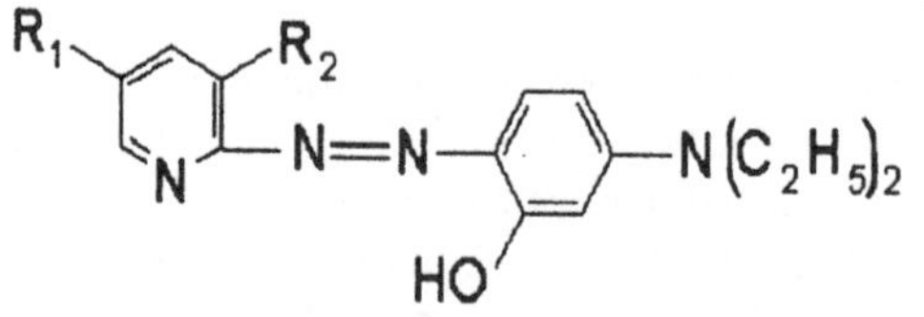

$$R_1 \text{---} \underset{N}{\overset{R_2}{\bigcirc}} \text{---} N{=}N \text{---} \bigcirc \text{---} N(C_2H_5)_2$$
$$HO$$

(87)

These reagents have the same chelation system as PAR. They form complexes with a number of metal ions, and relative to PAR, they exhibit a high sensitivity (mostly twofold; for bismuth up to fivefold). For transition metal complexes, ε values approach 1×10^5. Spectra of the complexes usually have two pronounced bands in the wide range of 520–600 nm and display a significant bathochromic shift relative to the single band of the reagent itself ($\Delta\lambda \sim 100$ nm). Examples of some complexes are listed in Table 3.

The determination of chromium and molybdenum with 5-Br-PADAP [194] is an example of how these reagents can be applied in modern separation techniques (reversed-phase liquid chromatography).

The reagents derived from thiazolylazo dyes or benzothiazolylazo dyes,

TABLE 3 Determination of Some Metals with PADAP Derivatives

Metal	Reagent	Medium (pH)	λ, (ε)[a]	Ref.
U(VI)	PADAP		(6.59×10^4)	188
U(VI)	5-Br-PADAP	6–8; F⁻ 45% (v/v) EtOH acetone	578 (7.4×10^4)	188
U(VI)	3,5-diBr-PADAP	9.3; SDS	576 (9.1×10^4)	189
Co(II)	5-Br-PADAP	7; TX-100 10% DMF	548 586	190
Gd(III)	5-Br-PADAP	9.2–11.6; TX-100, 5% (v/v) EtOH	540 580 (1.76×10^4)	191
Ni(II)	5-Br-PADAP	Tergitol NPx[b]	520; 560 (1.22×10^5)	192
Fe(II)	5-Br-PADAP	Tergitol NPx[b]	560 (8.2×10^4) 748 (3.35×10^4)	192
Ag(I)	3,5-diBr-PADAP	SDS	530; 570 (7.7×10^4)	193

[a]ε value shown for λ_{max} (L mol⁻¹ cm⁻¹)
[b]surfactant; pH 4.0–5.7

respectively, have the same chelate-forming as the 2-pyridylazo compounds above (e.g., TAR, 4-(2-(thiazolylazo)resorcinol; TAM, 2-(2-(thiazolylazo)-5-dimethylaminophenol; TAC, 2-(2-(thiazolylazo)-4-methylphenol; *p*-TAN, 4-(2-(thiazolylazo)-1-naphthol) and its 2-naphthol-sulfonic or disulfonic acid (TAN-6, TAN 3,6), and many others). For instance, 23 derivatives have been prepared from *o*-TAP (i.e., 2-(2-(thiazolylazo)-phenol) [180].

The sulfur atom in the thiazolyl ring does not participate directly in metal ion chelation; it merely affects the acid–base conditions of the chelate formation.

The above reagents are used for determinations of transition metals and uranium in aqueous medium, as well as after extraction into nonaqueous solvents. Examples include the determinations of zirconium and hafnium with TAN ($\lambda_{max} = 590$ nm, $\varepsilon = 3.36 \times 10^4$, respectively) [195], Ni(II) in aqueous ethanol with TAC ($\lambda_{max} = 580$ nm, $\varepsilon = 2.6 \times 10^4$) [196], and platinum elements with TAR in micellar medium of surfactant ($\lambda_{max} = 545$ nm, $\varepsilon = 3.75 \times 10^4$ for Rh(III), and 540 nm, $\varepsilon = 3.02 \times 10^4$ for Pd(II) in the presence of CPB) [197].

The effect of Ca(II) ions on the reaction rate of the ligand exchange reaction of Cu(II)-ethyleneglycol(2-aminoethylether) tetraacetic acid (EGTA) complex with the PAR reagent can be used to determine Ca(II) in the range of 0.4–40 μg mL^{-1}. The measurement is preformed at $\lambda = 515$ nm, where the absorption increase corresponds to the Cu(II)–PAR complex formation [198].

The ligand exchange in the reaction of Hg(II)-PAR with 1,2-cyclohexanediaminotetraacetic acid (CyDTA) is catalyzed by iodide ions. This effect can be utilized for iodide determination in the range of 0–10^{-7} M ($\lambda = 500$ nm) [199].

Oxygen derivatives of pyrimidine.   These belong to the large group of pyrimidines. Their analytical applications are reviewed in Ref. [200]. The most important derivative is barbituric acid **(88)**. Barbiturates form red-violet compounds **(89)** with Co(II) salts in nonaqueous media (methanol) in the presence of a base, ammonia, or an aliphatic (cycloaliphatic) amine.

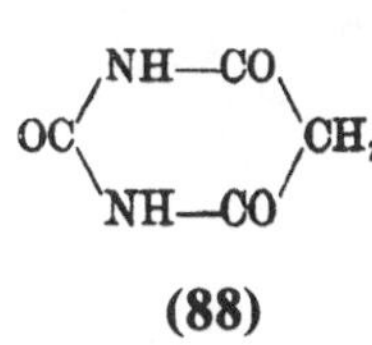

(88) 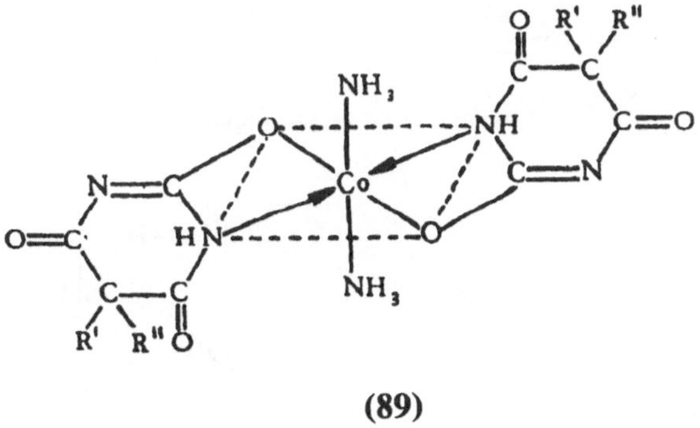   **(89)**

The composition of the complex formed depends on reaction conditions (ratio of individual components and the base used). Similar color reactions are given by theophylline, imidazole, 2-substituted imidazolines, 2-aminoisoxazole, 2-aminopyridine, 2-aminopyrimidine, hydantoins, and analogous thiocompounds (thiobarbiturates). All sulfonamids containing these heterocyclic compounds in their molecule also give a positive reaction. The presence of water causes the formation of unstable and less colored complexes. The procedure is used for the determination of barbiturates, and *N*-heterocyclic bases [26,27].

The addition of a solution of Hg(II) salts to aqueous solutions of barbiturates causes precipitation of water-insoluble complexes. These complexes can be dissolved in diluted acids or in chloroform, and the analyte content determined by the mercury determination with dithizone or diphenylcarbazone [26,27].

Uracil (2,4-dihydroxypyrimidine) derivatives form colored complexes with Pd, Pt(II), Au(III), Ru(III), and Os(VIII) and can be used for spectrophotometric determinations of these elements [201].

e. Ligands with oxygen and sulfur donor atoms

The ligands containing an $-SH$ group in the neighborhood of a hydroxyl, carbonyl, or carboxyl group belong to this group of substances. Representatives of the group include thioglycolic (mercaptoacetic) acid, thiosalicylic (mercaptobenzoic) acid, and *N*-(2-mercaptoacetyl)sulfanilic acid.

These compounds provide chelates with many metals. They are often applied in Mo(V), and Mo(VI) determinations [25].

f. Ligands with nitrogen and sulfur donor atoms

Thiocarbazones and thiosemicarbazones. Dithizone (diphenylthiocarbazone) **(90)** is the most well known reagent of the type used for metal determination. This compound is subject to keto-enol tautomerism:

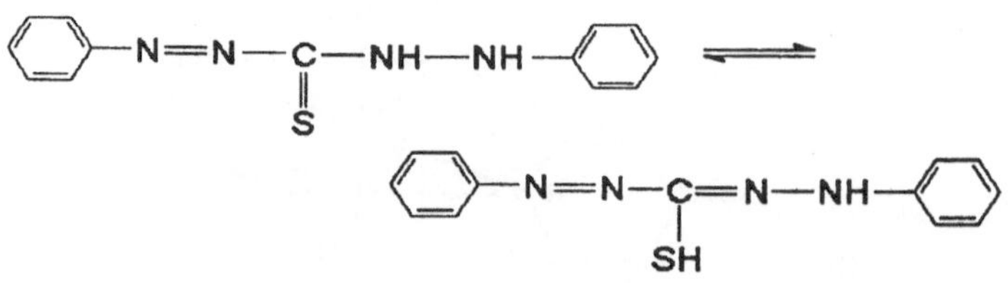

(90)

The keto form provides primary dithizonates of the ML_n composition (in the presence of excess ligand and low pH); the enol form provides secondary dithizonates (91) of $ML_{n/2}$ composition (in the presence of excess metal and higher pH values). The chelates are insoluble in water, and thus spectrophotometric determination is performed after extraction into organic solvent ($CHCl_3$ or CCl_4).

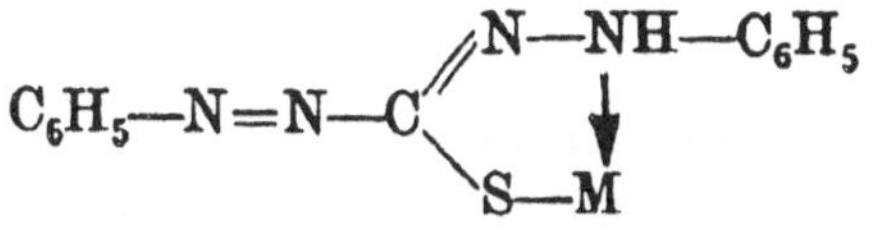

$$C_6H_5-N=N-C{\stackrel{\displaystyle \nearrow N-NH-C_6H_5}{\searrow}}_{S-M}$$

(91)

Ag(I), Au(III), Cd(II), Cu(II), Ni(II), and Pb(II) can be determined with dithizone by extraction spectrophotometry [4]. Hg(II) complex can be solubilized by Triton X-100 surfactant, as well as Zn(II) complex by sodium dodecylsulfate, and these determinations can be then performed in aqueous medium [202,203]. Pt(IV) ions can be determined after separation of their dithizonates [204,205]. The determination of Pd(II) and Pt(II) in mixture uses derivative spectrophotometry [206].

Structures of almost 40 derivatives of thiosemicarbazones, the characteristics of their complexes with metals (primarily with Fe(II, III), Co(II), Ni(II), Cu(II), Pd(II)), examples of spectrophotometric determinations of metals, and the utilization of N-phenylthiosemicarbazones in spectrophotometry have been reviewed in [207,208].

With respect to aldehydes or ketones used for the preparation of thiosemicarbazones, these compounds can act as unidentate (bond occurs through sulfur), bidentate, or multidentate ligands. They form complexes with metals in the metal–ligand ratio of 1 : 1 or 1 : 2.

The analytical use of 2-nitro-5,6-dimethyl-1,3-indanedionedithiosemicarbazone [209] has been studied, and palladium determination with glyoxal bis(4-phenyl-3-thiosemicarbazone) [210], phenanthraquinone monothiosemicarbazone [211], and o-hydroxyacetophenone thiosemicarbazone [212] has been reported. Cobalt and nickel determination with 2-hydroxybenzaldehyde thiosemicarbazone has been performed with FIA method (based on different complexation rates) [213].

Derivatives of thiocarbonohydrazide have also been prepared [214], for example, 1-[di(pyridyl)methane]salicylidene thiocarbonohydrazide (92), which has been proposed for the determination of trace amounts of Zn(II), with which it forms a chelate at the ratio of 1 : 2.

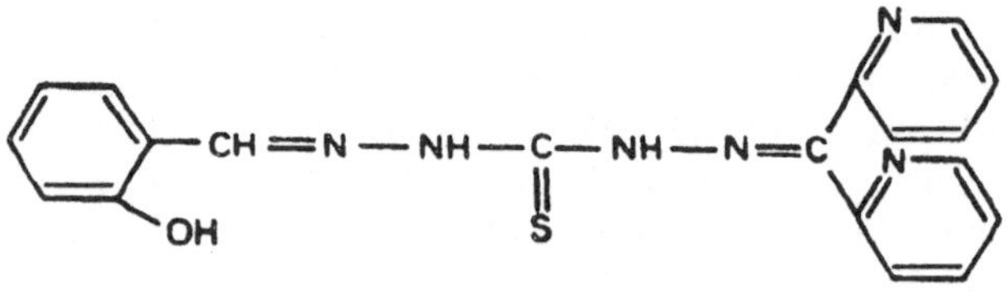

(92)

Rhodanine derivatives. These provide complexes with highly polarizable metal ions in acidic medium. In this medium, 5-dimethylaminobenzylidene rhodanine (93) is a selective reagent for Ag(I), Au(I), Pd(II), and also for Pt(II) and Cu(II). In neutral and in alkaline media, it reacts with other metals as well [19]. Sulfur from $>$C=S group and the adjacent nitrogen atom are the donor atoms in the complex.

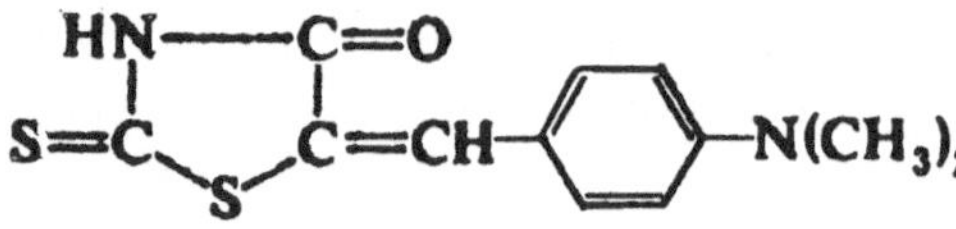

(93)

To identify suitable reagents for the determination of rare metals, 5-azo derivatives of rhodanine and their analogs (thiorhodanine, thiohydantoine, β-aminorhodanine, and N-substituted amino derivatives — a total of 42 compounds) have been studied [215].

Sulfochlorophenolazorhodanine was suggested for Pd determination using the FIA method [216].

Phenothiazine derivatives. These compounds of the basic structure (94) are utilized primarily as pharmaceuticals. However, a number of analytical methods have been proposed in which they are employed as reagents.

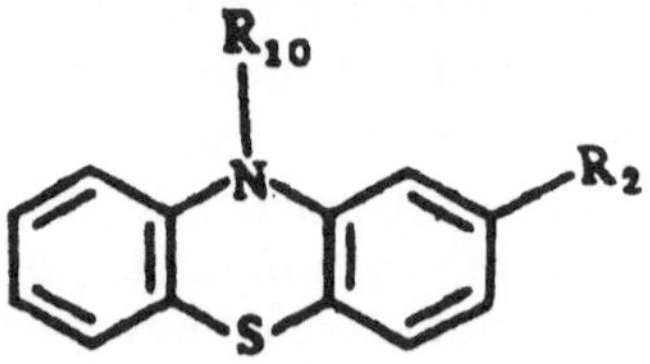

(94)

R_2 = H, $-Cl$, $-CF_3$, $-OCH_3$, $-CO-CH_3$

R_{10} = H, alkyl, $-(CH_2)_n-N(alkyl')_2$, $-(CH_2)_n-$piperidyl,
   ($-$piperazyl, respectively).

These methods take advantage of their redox properties (described in more detail in Chapter 5, Section D) and complex formation with metals. A combination of both these processes is applied in some cases.

Phenothiazine (R_{10} = H), and its derivatives with R_{10} = alkyl provide complexes especially with metals that have affinity towards sulfur atoms. In the case of Pd(II) and Pt(II), the formation of a red ML_2Cl_2 complex is assumed. The complex solubility and color intensity become enhanced in the presence of the anionic surfactant sodium dodecylsulfate [26].

In derivatives with the substituents R_{10} = $(CH_2)_nN(alkyl')_2$, (e.g., promethazine, R_{10} = dimethylaminopropyl), it has been confirmed by IR, NMR, and x-ray spectroscopy that in the reaction with Pd(II) (or even Pt(II)), the formation of 1 : 1 complex occurs. Here, promethazine is again coordinated through sulfur and the side chain is bent so that the quaternary nitrogen interacts with Pd(II) to form a "scorpion structure" [217,218].

The coordination of phenothiazine derivatives through the nitrogen atom of the side chain occurs in metals with a higher affinity for nitrogen atoms (e.g, Cu(II), Ru(III), Rh(III), Ir(III)). At the same time, the interaction between metal ions and heterocyclic nitrogen atoms occurs [219,220].

In some cases, where the direct metal complex formation with phenothiazine derivatives (PD) is anticipated (e.g., in the reaction with Pt(IV)), the redox reaction has to be considered first, followed by the complex formation [221].

Pt(IV) + PD = Pt(II) + PD$\cdot$ $^+$

Pt(II) + 2PD = Pt(II)(PD)$_2$

The complexes of phenothiazine derivatives have been proposed for the determinations of Pd(II) [222], Ru(III) [223], Rh(III) [224], and Hg [225]. On the other hand, the complexation reaction of phenothiazines can be used for their determination [26,27]. The reaction is performed mainly with $PdCl_2$ [226,227]. The determinations based on the formation of ionic associates, and acid–base and redox reactions are covered in chapters 4, 2, and 5, respectively.

Thiobarbiturates. As with barbiturates (and some *O*- and *N*-heterocycles), thiobarbiturates form complexes with Co(II) ions, like those discussed earlier. Their formation is used for the spectrophotometric determination of thiobarbiturates [26,27]. Thiobarbituric acid that forms complexes with metal ions has been used, for instance, for the spectrophotometric determination of Os(VIII) [228].

It is also possible to include thiourea and its derivatives in this group of compounds, although it is often considered to be a monodentate ligand. This reagent is especially used for determinations of Bi, but also of Sb, Sn, Pd, Ru, Os, and others [19].

B. MOLECULAR COMPLEXES

1. Charge-Transfer (CT) Complexes

The molecular complexes discussed in this section are formed between the reagent and analyte molecules, which are "bound" together only by weak electrostatic forces. These complexes have new characteristic physical properties, even though the original chemical properties of both components are retained.

Charge-transfer complexes or electron donor/acceptor (EDA) or π-complexes between electron donors having sufficiently low ionization potential and acceptors having sufficiently high electron affinity, represent a very attractive group of complexes [229]. This type of complexation is an important phenomenon in spectrophotometry. Many molecular complexes of this type are colored and give rise to new absorption bands in the electronic spectra. The intense coloration caused by this interaction is due to the partial transfer of one electron from the highest occupied molecular orbital of the donor molecule to the lowest unoccupied molecular orbital of the acceptor molecule. This transfer takes place when the energy required for it is considerably lower than that required for the transfer to the atom's own lowest unoccupied orbital.

The complexes are prepared simply by mixing solutions of the analytes (mostly electron donors, D) and the reagents (electron acceptors, A) in appropriate solvents. The solvents used are selected with respect to sufficient solubilizing power for both components and their possible influence on the properties of the expected complexes and while taking into consideration the possible interaction of the solvent with the π-acceptor. The color development is usually finished within 10 to 60 minutes, although it can be accelerated by carrying out the reaction at elevated temperatures (30–60°C). The color of the complexes can be strongly affected by the type of the solvent used. The formation of CT complexes is preferred in solvents of low polarity, whereas in solvents of high relative permitivity the formation of radical salts of the type D^+A^- and ion radicals ($D\cdot{}^+A\cdot{}^-$) can be observed, in which a total electron transfer from the donor to the acceptor is theorized also in the ground state. The resulting equilibrium can be expressed by Eq. (14).

$$D + A \rightarrow DA \rightleftharpoons D^+A^- \rightleftharpoons D\cdot{}^+ + A^- \qquad (14)$$

  CT complex   ionic radicals

Mostly 1 : 1 or 1 : 2 molar (donor–acceptor) complexes are formed. They can often be isolated in a solid state by evaporation of the solvent; however, their composition in a solid state can be different from that in the solution. As with metalloorganic complexes, each complex can be characterized by its constant of stability.

Depending on the nature of the donor–acceptor partners, the formation of redox or condensation products can also be observed. Every shift in the equilibrium and every further reaction are accompanied by color changes.

Donors fall into three categories: lone-pair donors (*n*-donors) such as amines, alkaloids, and alcohols; π-donors such as aromatics, particularly polycyclic systems; and σ-donors such as aliphatic and cyclic hydrocarbons [229,230]. Some compounds such as azaaromatics and aromatic amines may behave as *n*-donors towards some acceptors and π-donors towards others. Acceptors may be of σ- and of π-type. The former include iodine, the latter aromatic systems containing electron-withdrawing substituents such as nitro, cyano, or halogen groups. The following types of CT complexes can be distinguished according to the type of donors and acceptors: π–π, π–σ, σ–π, σ–σ, *n*–π, and *n*–σ complexes.

To enable the CT complex formation of electron acceptors with compounds that do not possess the properties of electron donors, derivatization of such compounds with appropriate reagents is used. Thus, corticosteroids, have been converted into their phenylhydrazones [231]; oximes [232] and hexoses into osazones [233].

This type of reaction offers simplicity of the procedure and is to be encouraged for the utility of these methods in routine analysis with a sensitivity that permits the determination of analytes in the microrange.

The most frequently used π-acceptors as spectrophotometric reagents are tetracyanoethylene, 2,3,5,6-tetrahalogenated 1,4-benzoquinones (fluoranil, chloranil, and bromanil), 2,3-dichloro-5,6-dicyano-1,4-benzoquinone, 7,7,8,8-tetracyanoquinodimethane, and 2,6-dimethoxy-1,4-benzoquinone.

a. Iodine

Iodine is used dissolved in chloroform (carbon tetrachloride, ethylenedichloride). When the chloroform solution is mixed with the chloroform solution of the electron donor, the violet color of iodine ($\lambda_{max} = 515$ nm) is changed immediately to yellowish purple or yellow. The absorption bands at 290 and 360 nm, those which are most often used for measurements, are attributed to the CT complex. Further shifts in the spectra are observed with prolonged reaction time, which leads to formation of triiodide ions by dissociation of the complex and reaction of iodide ions with free iodine. The situation is represented by Eqs. (15) and (16):

$$D + I_2 \rightarrow D\text{-}I_2 \rightleftharpoons D\text{-}I^+I^- \rightleftharpoons D\text{-}I^+ + I^- \tag{15}$$

$$I^- + I_2 \rightleftharpoons I_3^- \tag{16}$$

Charge-transfer complexes with iodine have been used for the determination of aliphatic sulfides [234], alkaloids [235–242], and a variety of drugs [243–250].

b. Tetracyanoethylene (TCNE, 1,1,2,2-ethenetetracarbonitrile)

TCNE is used in acetonitrile solution, whereas chloroform is used to dissolve the analytes. Acetonitrile promotes the formation of the TCNE·$^-$ radical anion (95), which has an absorption band of 473 nm and has a high molar absorption coefficient. In the presence of water, base–analyte hydrolysis can

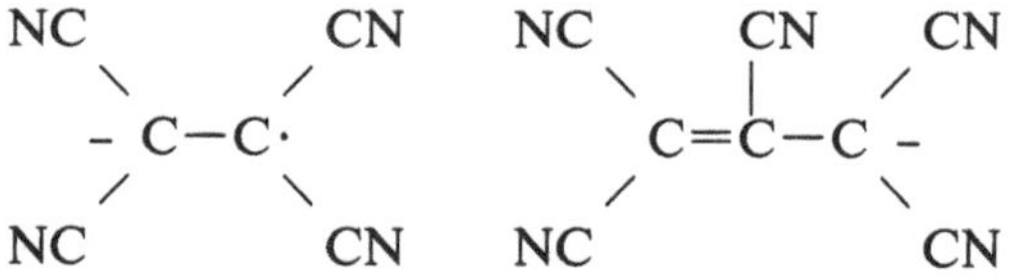

(95)   **(96)**

take place, and the 1,1,2,3,3-pentacyanopropenide anion PCNP$^-$ (96) is formed as a competitive reaction—in some cases, it can be the sole reaction. The doublet at 393 and 412 nm has been attributed to this anion. From the quantitative point of view, PCNP$^-$ is preferable to TCNE·$^-$ on the grounds of its higher ε_{max} value (2.26×10^4) as compared with TCNE·$^-$ (7.1×10^3) [238]. Primary and secondary aliphatic and aromatic amines react with TCNE to form a π-complex, which is rearranged to form a σ-complex and finally yield tricyanovinylamines (97) [251]. Complexes with TCNE have been used for the determination of a great variety of organic

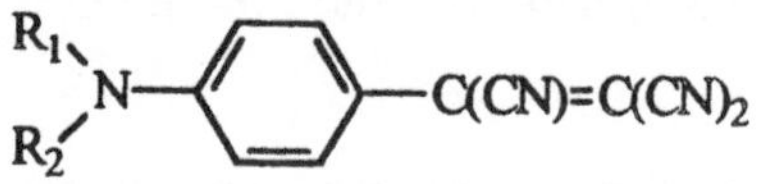

(97)

compounds, such as aromatic hydrocarbons [252], tertiary amines [251], aliphatic sulfides [253] and disulfides [254], organophosphorus compounds [255], alkaloids [238] and drugs [247,249,256–258].

The determinations can also be carried out by an indirect procedure. Thus anthracene and 1,3-dienes have been determined by measuring the decoloration of naphthalene-tetracyanoethylene complex after the analytes and TCNE have reacted via the Diels-Alder reaction. The decrease of absorbance is usually proportional to the concentration of the analytes present, though the reaction is not exactly stoichiometric [259,260].

c. 2,3,5,6-Tetrahalogenated 1,4-benzoquinones (fluoranil, chloranil, bromanil)

The often used reaction of aliphatic amines and amino acids with chloranil in the medium of ethanol and sodium borate buffer (pH 9) carried out at elevated temperatures proceeds via formation of the corresponding *N*-(dihydroxyphenyl)amines, which form CT complexes with the excess reagent exhibiting absorptions at 340–380 nm [27,261]. The reaction with some pharmaceutically important bases can also be carried out in chloroform or acetonitrile, and in that case the formation of CT complexes absorbing at 510–560 nm (or over 600 nm) in equilibrium with the radical anions absorbing at 440–450 nm is theorized [242–244,262–268]. Conversely, chlorinated benzoquinones can be estimated by the reaction with morpholine, thiomorpholine, or piperazine. The determination, however, is based on the reaction of yellow diaminobenzoquinone derivatives with a 50-fold excess of morpholine [269].

Bromanil has been used mostly in acetonitrile medium [248,250,262, 266,268]. Fluoranil has a higher electron affinity than chloranil and has been used in chloroform solution [270].

d. 2,3-Dichloro-5,6-dicyano-1,4-benzoquinone (DDQ)

The reaction with this acceptor is carried out in methanol. Mostly, the absorption band at 460 nm, attributed to the radical anion, is used for measurements. The bands at ~295 and ~395 nm characterize the CT complex. Bands at 576 and 545 nm in acetonitrile have been used for the determination of alkaloids, especially drugs [271–273].

e. 2,5-Dichloro-3,6-dihydroxy-1,4-benzoquinone (chloranilic acid)

This substance has been used as CT complexation agent with a characteristic absorption band at ~530 nm [273–276]. It must be emphasized, however, that at least in some cases the products formed are ion associates [277] (see Chapter 4, Section B).

f. 7,7,8,8-Tetracyanoquinodimethane (TCNQ)

TCNQ is used dissolved in acetonitrile, which promotes the dissociation of the CT complex and ensures the maximum yield of the TCNQ·⁻ anion radical [Eq. (17)].

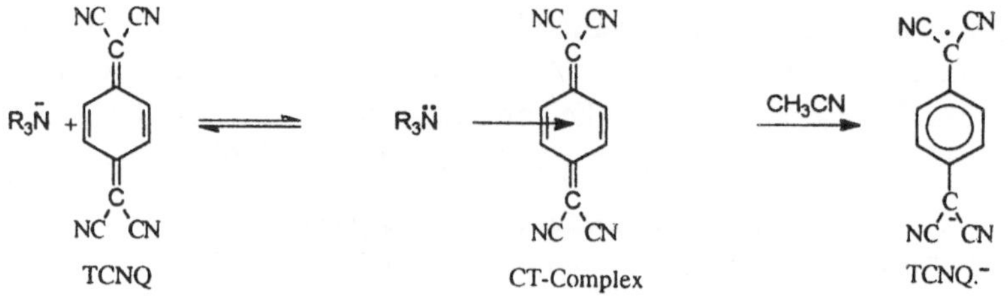

$$(17)$$

This anion exhibits three intense bands at 845, 752, and 685 nm. In comparative studies with other acceptors, TCNQ has been found to exhibit the most intense band and has therefore been recommended as superior to other reagents. The molar absorptivities of complexes of benzylpenicillin and atropine with individual acceptors are compared in Table 4. TCNQ in acetonitrile can be used to determine olefins in trichloroethylene solutions, with absorbance at 482–487 nm [278]. The reaction of TCNQ with dequalinium chloride in alkaline medium results in subsequent formation of condensation products according to Eq. (18) (λ_{max} = 488 nm) [279]. TCNQ is often used for determinations of drugs and related compounds [240,244, 245,247,257,266,276,280–283].

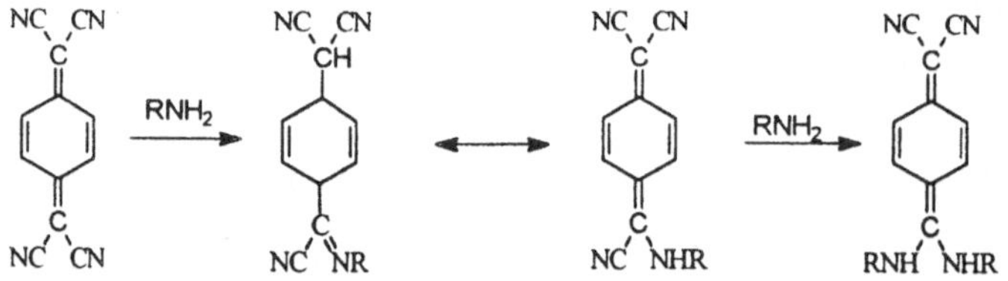

$$(18)$$

g. 2,6-Dimethoxy-1,4-benzoquinone (DMBQ)

DMBQ is used to determine isoniazid in its pharmaceutical formulations. A green complex is formed in the presence of sodium hydroxide in an aqueous-ethanolic solution. It can be shown that the reagent is reduced by the analyte to yield the corresponding hydroquinone, which forms a $1:1$ CT complex (**98**) with the reagent, analogous to quinhydrone [284].

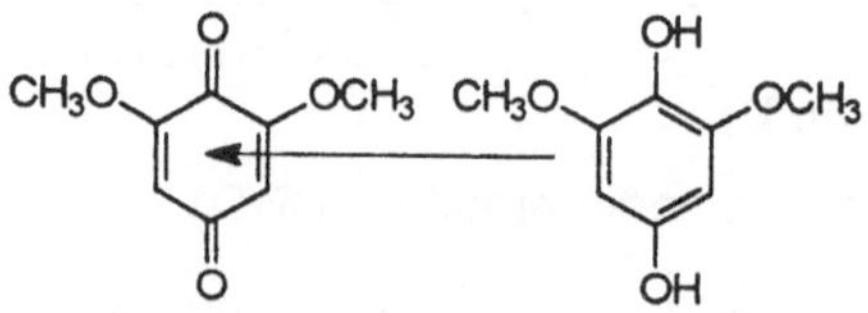

(98)

TABLE 4 Molar Absorptivities (ε_{max} in L mol^{-1} cm^{-1}) of Chromogens with Benzylpenicillin and Atropine [λ_{max} (nm) in parentheses]

| | Benzylpenicillin [257] in acetonitrile | Atropine [238] | |
		in acetonitrile	in 1,2 dichloroethane
TCNQ	3.04×10^4 (842)	3.86×10^4 (845)	2.95×10^4 (854)
DDQ[a]	2.96×10^3 (460)	1.03×10^4 (576)	7.58×10^4 (460)
TCNE	4.85×10^2 (393)	1.7×10^4 (393)	7.23×10^3 (398)
Chloranil	6.23×10^2 (540)	8.3×10 (442)	No reaction
Bromanil	3.20×10^2 (540)	—	—
Fluoranil	—	7.22×10^2 (538)	No reaction

[a]In methanol.

Besides the reagents discussed above, the following have also been used: 2,6-dichloro-4-nitro- or 2,4-dichloro-6-nitrophenol [250,262], dichlorophenolindophenol [273], alizarin and quinalizarin [285], 2,4,5,7-tetranitrofluorenone [244], 9,9-dicyanomethylene-2,4,7-trinitrofluorenone [244], and 1,3,5-trinitrobenzene [286].

2. Meisenheimer Complexes

Aromatic polynitrocompounds with nitro groups in the *m* position undergo characteristic color changes upon the action of alkalizing agents (alkali hydroxides or alkoxides, tetraalkylammonium hydroxides, etc.) and in polar organic solvents (dimethylsulfoxide, dimethylformamide, alcohols). The structure of species formed is strongly dependent on the reaction conditions, mainly on the concentration of the alkalizing agent, the presence of substituents on the aromatic nucleus, and the solvent polarity. The reaction is an aromatic nucleophilic substitution and occurs by electron transfer from the nucleophil into the antibonding orbital of the aromatic substrate to give a charge-transfer complex as an intermediate, which is transformed to Meisenheimer complexes (**99**) and (**100**). These can undergo further changes [287,288]. The position in which the reaction takes place is dependent on the type of substituent X.

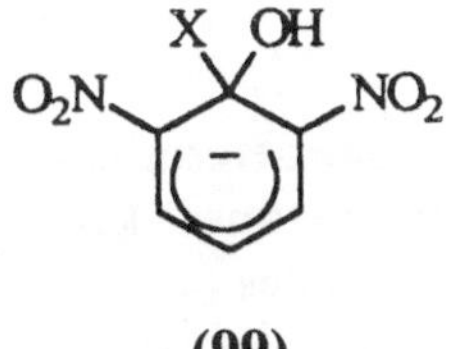

(99) 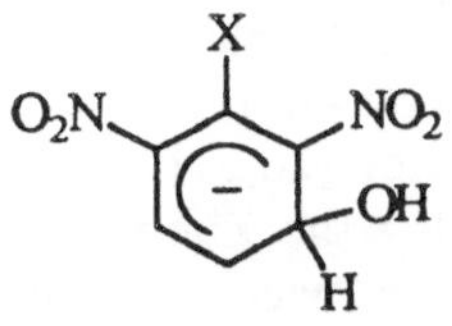　　　　　(100)

If the reaction is carried out in the presence of compounds containing active methylene groups (methylketones, nitroalkanes, ketosteroids, etc.), intense color changes occur. These are known as the Yanovsky or the Zimmermann reaction. These reactions differ in the structure of the chromogens formed since they are performed in a different mutual ratio of the reacting components: The former is carried out with an excess of acetone for the determination of *m*-polynitro compounds; the latter with an excess of *m*-dinitrobenzene for the determination of ketosteroids. Chromogens of the Yanovsky reaction are of the type (101) or (102), while chromogens of the Zimmermann reaction are of the type (103) as a result of the oxidative action of the excess *m*-dinitrocompound. The position of the substitution is affected by the substituents on the aromatic ring [289–291].

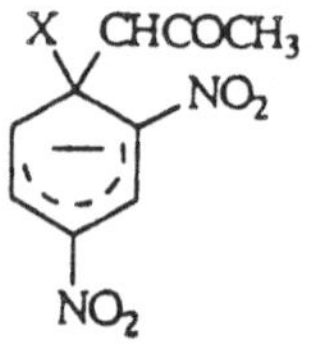

(101) 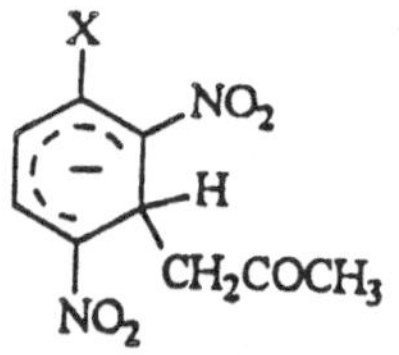   **(102)** 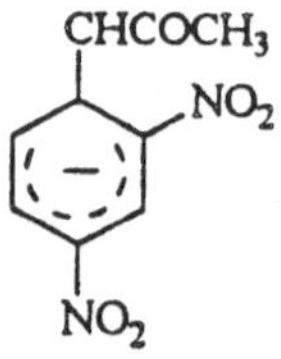   **(103)**

Acetone, methylethylketone, nitromethane, and dimedone are compounds with active methylene groups in their molecules, which are used as reagents for the determination of *m*-di- and *m*-polynitrocompounds. In this case, aqueous or alcoholic solutions of alkali hydroxides, ammonia or alkali alcoholates, tetraalkylammonium hydroxides, methylamine, and ethylenediamine are used for alkalization.

Compounds with an aromatic ring in their molecules which easily undergo nitration (HNO$_3$, H$_2$SO$_4$ + KNO$_3$) to yield *m*-dinitro derivatives can be determined using the Yanovsky reaction [292,293]. Compounds with an active hydrogen in their functional groups (such as −OH, −SH, −NH$_2$, −COOH, etc.), or with a reactive halogen atom, can be labeled by *m*-dinitro grouping by derivatization with a suitable reagent (e.g., 2,4-dinitrofluorobenzene, 3,5-dinitrobenzoylchloride, 3,5-dinitrobenzenesulfochloride, *p*-nitrophenacylbromide, or sodium 3,5-dinitrobenzoate), and then after separation of the products by extraction, they undergo the Yanovsky reaction [26,27].

By using the Zimmermann reaction for the determination of compounds with a reactive methylene group, methylketones, ketosteroids, nitromethane, dialkyl phosphites, and cardiac glycosides (α,β-unsaturated-χ-lactones) can be determined. *m*-Dinitrobenzene, dinitrodifenyl sulfone, 2,4,2′,4′-tetranitrobiphenyl, 3,5-dinitrobenzoic acid, and picric acid (2,4,6-

trinitrophenol) are the most frequently used reagents [26,27]. The very often applied determination of kreatinine using picric acid is known as the Jaffé reaction; the reaction with cardiac glycosides as the Baljet reaction.

The Zimmermann reaction can be used for the determination of organic bases using *m*-dinitrobenzene and nitromethane as the reagents [27].

REFERENCES

1. A. E. Martell and M. Calvin, *Chemistry of the Metal Chelate Compounds*, Prentice-Hall, New York, 1956.
2. D. D. Perrin, *Organic Complexing Reagents: Structure, Behavior and Application to Inorganic Analysis*, Interscience Publishers, J. Wiley, New York, 1964.
3. H. A. Flaschka and A. J. Barnard, Jr. (Eds.), *Chelates in Analytical Chemistry, Vols. 1 to 4*, Marcel Dekker, New York. Vol. 1: 1967, Vol. 2: 1969, Vol. 3: 1970, Vol. 4: 1972.
4. K. Burger, *Organic Reagents in Metal Analysis*, Akadémiai Kiadó, Budapest, 1973.
5. Z. Holzbecher, L. Diviš, M. Král, L. Šůcha, and F. Vláčil, *Handbook of Organic Reagents in Inorganic Chemistry*, Ellis Horwood, New York, 1976.
6. L. Sommer, *Analytical Absorption Spectrophotometry in the Visible and Ultraviolet: The Principles*, Akadémiai Kiadó, Budapest, 1989.
7. A. T. Pilipenko and L. I. Savransky, Selectivity and sensitivity of metal determination by coordination compounds, *Talanta 34*: 77 (1987).
8. M. T. Beck, *Chemistry of Complex Equilibria*, Publishing House of the Hungarian Academy of Sciences and Van Nostrand Ltd., Budapest, London, 1970.
9. A. Ringbom, *Complexation in Analytical Chemistry. A Guide for the Critical Selection of Analytical Methods Based on Complexation Reaction*, Interscience Publishers, J. Wiley, New York, 1963.
10. A. E. Martell and R. M. Smith, *Critical Stability Constants*, Vol. 3, Plenum Press, New York and London, 1977.
11. A. Ringbom and L. Harju, Determination of stability constants of chelate complexes, Part I. Theory, Anal. Chim. Acta 59: 33 (1972); Part II: Application, *Anal. Chim. Acta 59*: 49 (1972).
12. H. S. Dunsmore and D. Midgley, The simultaneous determination of the stability constants of protonated and unprotonated complexes in solution, *Anal. Chim. Acta 67*: 341 (1973).
13. E. Ohyoshi, Spectrophotometric determination of complex formation constants by estimation of free ligand concentration, *Anal. Chem. 55*: 2404 (1983).
14. F. A. Cotton and G. Wilkinson, *Advanced Inorganic Chemistry*, Fifth Ed., J. Wiley, New York, 1988.
15. F. Basolo and R. G. Pearson, *Mechanisms of Inorganic Reactions: A Study of Metal Complexes in Solution*, J. Wiley, New York, 1968.

16. M. Malàt, *Absorption Inorganic Photometry* (in Czech), Academia, Prague, 1973, p. 457.

17. K. Gorczyńska and A. Michalik, *UV/VIS Spectrophotometry in Chemical Analysis* (in Polish), Państwowe Wydawnictwo Naukove, Warszawa, 1988, p. 312.

18. A. K. Babko and A. T. Pilipenko, *Photometric Analysis, General Principles and Apparatus* (in Russian), Khimiya, Moskva 1968, p. 240.

19. Z. Marczenko, *Spectrophotometric Determination of Elements*, WNT and Ellis Horwood, Chichester, 1979.

20. A. Raychaudhuri, S. K. Roy, and A. K. Chakraburtty, The separation of W(V) from HCl-KSCN medium on polyurethane foam sorbents for its spectrophotometric determination in steels and silicates, *Talanta 39*: 1377 (1992).

21. R. Kuroda, I. Ida, and H. Kimura, Spectrophotometric determination of silicon in silicates by flow injection analysis, *Talanta 32*: 353 (1985).

22. cf. Ref. 6, p. 267.

23. L. Sommer, Über einige analytische Reaktionen der Polyphenole, *Z. Anal. Chem. 187*: 7 (1962).

24. L. Sommer, Analytische Reaktionen der Polyphenols. Der Nachweis von Eisen(III), Titan(IV), Vanadium(V), Niob(V), Uran(VI), Molybdän(VI), Cer(IV), *Z. Anal. Chem. 187*: 263 (1962).

25. R. Püschel and E. Lassner, Chelates and Chelating Agents in the Analytical Chemistry of Molybdenum and Tungsten, in H. A. Flaschka and A. J. Barnard, Jr. (Eds.), *Chelates in Analytical Chemistry*, Vol. 1, Marcel Dekker, New York, 1967.

26. Z. J. Vejdělek and B. Kakáč, *Color Reactions in Spectrophotometric Analysis of Organic Compounds*, Vol. II, Inorganic reagents (in German), VEB Fischer Verlag, Jena, 1973.

27. Z. J. Vejdělek and B. Kakáč, *Color Reactions in Spectrophotometric Analysis of Organic Compounds*, Supplementary Vol. II, (in German), VEB Fischer Verlag, Jena, 1982.

28. S. L. C. Ferreira, A. C. Spinola Costa, M. G. M. Andrade, B. F. Santos, and N. O. Leite, 2,3-dihydroxybenzoic acid as a reagent for the spectrophotometric determination of titanium, *Anal. Lett. 26*: 1001 (1993).

29. N. Iranpoor, N. Makeki, S. Razi, and A. Sapavi, Spectrophotometric determination of vanadium in different oxidation states with pyrogallol, *Talanta 39*: 281 (1992).

30. T. Taketatsu and N. Toriumi, Spectrophotometric study and analytical application of rare earth tiron complexes. I. Determination of neodymium, holmium and erbium, *Talanta, 17*: 465 (1970).

31. R. Purohit and S. Devi, Spectrophotometric determination of titanium(IV) using chromotropic acid and a flow injection manifold, *Analyst 117*: 1175 (1992).

32. T. Lussier, R. Gilbert, and J. Hubert, Determination of boron in light and heavy water samples by flow injection analysis with indirect UV-visible spectrophotometric detection, *Anal. Chem. 64*: 2201 (1992).

33. U. K. Gupta, P. G. Kulkarni, G. Thomas, N. Varadarajan, R. K. Singh, and M. K. T. Nair, Spectrophotometric determination of uranium using ascorbic acid as a chromogenic reagent, *Talanta 40*: 507 (1993).

34. O. M. Vilkova, V. M. Ivanov, and A. I. Busev, Catechol azo derivatives as reagents for zirconium (in Russian), *Zh. Anal. Khim. 33*: 716 (1978).

35. P. Solich, R. Karlíček and V. Jokl, The application of complex forming reactions in the quantitative analysis of drugs (in Czech), *Československ. Farm. 34*: 151 (1985).

36. M. Král, Electronic spectra of chelates important in analytical chemistry II (in Czech), *Chem. Listy 71*: 1 (1977).

37. J. Starý and E. Hladký, Systematic study of the solvent extraction of metal β diketonates, *Anal. Chim. Acta 28*: 227 (1963).

38. I. M. Korenman and M. I. Gryaznova, Spectrophotometric determination of acetylacetone in aqueous solutions (in Russian), *Trudy Khim. Khim. Tekhnol.* (2): 72 (1970); *Chem. Abstr. 76*: 135383s (1972).

39. A. D. Shendrikar, Substituted hydroxylamines as analytical reagents, *Talanta 16*: 51 (1969).

40. V. C. Bass and J. H. Yoe, Hydroxamic acids as colorimetric reagents, *Talanta 13*: 735 (1966).

41. A. R. Fields, B. M. Daye, and R. Christian Jr., Borate complexes of benzohydroxamic acid and some of its derivatives, *Talanta 13*: 929 (1966).

42. P. Solich, R. Karlíček, and V. Jokl, Spectrophotometric determination of cardioactive glycosides by hydroxamic reaction. I. Conditions, kinetics and mechanism of hydroxylaminohydrolysis (in Czech), *Československ. Farm. 36*: 327 (1987).

43. P. Solich, R. Karlíček, and V. Jokl, Spectrophotometric determination of cardioactive glycosides by hydroxamic reaction. II. Determination of lanatosides (in Czech), *Československ. Farm. 37*: 193 (1988).

44. R. Karlíček, and P. Solich, Hydroxylaminolysis of β-lactams and its use for the determination of penicillins by the method of flow-injection analysis (in Czech), *Československ. Farm. 39*: 77 (1990).

45. E. A. Krasnov and V. P. Fominykh, Photometric determination of tetracycline (in Russian), *Farmatsiya (Moscow) 31*: No. 1, 75 (1982).

46. M. Katyal, Flavones as analytical reagents: A review. *Talanta 15*: 95 (1968).

47. E. M. Nevskaya and V. A. Nazarenko, Oxyflavones in analytical chemistry (a review) (in Russian), *Zh. Anal. Khim. 27*: 1699 (1972).

48. K. Velasevic, Z. Radovic, and A. Stefanovic, Spectrophotometric study of a quercetin complex with copper(II), *Arh. Pharm. 31*: 175 (1981); *Chem. Abstr. 96*: 168813y (1982).

49. G. Stanic, B. Katusin-Razem, and J. Petricic, Spectrophotometric determination of flavonoids in Chamomillae flos and its extracts, *Pharm. Glas 44*: 179 (1988); *Chem. Abstr. 109*: 176415x (1988).

50. O. Valcl, I. Němcová, and V. Suk, *Handbook of Triarylmethane and Xanthene Dyes: Spectrophotometric Determination of Metals*, CRC Press, Boca Raton, 1985.

51. H. Shen, Z. Wang, and G. Xu, Spectrophotometric determination of trace amounts of germanium in minerals and ores with 9-(o-chlorophenyl)-2,6,7-trihydroxyxanthen-3-one in the presence of cetyltrimethylammonium bromide, *Analyst 112*: 887 (1987).

52. W. Quianfeng, Study of the titanium-phenylfluorone complex formed in the presence of Triton X-305 and emulsifieer OP, *Talanta 32*: 507 (1985).

53. M. Malát, Photometrische Zinnbestimmung mit Brenzcatechinviolett in Anwesenheit von Gelatine, *Z. Anal. Chem. 187*: 404 (1962).

54. M. C. Valencia, D. Gimedo, and L. F. Capitan Vallvey, Determination of tin in natural waters and fruit juices by solid phase spectrophotometry, *Anal. Lett. 26*: 1211 (1993).

55. N. L. Shestidesyatnaya, L. I. Kotelyanskaya, and M. I. Yanik, Extraction-photometric study of the tin(IV)-catechol violet diphenylguanidine system (in Russian), *Zh. Anal. Khim. 31*: 67 (1976).

56. L. N. Kharlamova, R. K. Chernova, and V. V. Belousova, Studies of reactions of p-, d-, and f-elements with some hydroxy derivatives of triarylmethane group in the presence of *o*-phenantroline (in Russian), *Zh. Anal. Khim. 32*: 1680 (1977).

57. D. Honová, I. Němcová, and V. Suk, Spectrophotometric determination of bismuth and EDTA by means of the reaction of bismuth with pyrocatechol violet in the presence of Septonex, *Talanta 35*: 803 (1988).

58. K. Hayashi and Y. Sasaki, Effect of nonionic surfactant as a dispersant on aluminium-aluminone lake (in Japanese), *Bunseki Kagaku 30*: T61 (1981).

59. L. Jančář, J. Havel, and L. Sommer, Spectrophotometric study of analytical reactions of triphenylmethane dyes with uranyl in the presence of cationic surfactants, *Coll. Czech. Chem. Commun. 53*: 1424 (1988).

60. J. Preisler, L. Jančář, and L. Sommer, The spectrophotometric determination of elements with chromazurol S in the presence of cetyltrimethylammonium bromide and Triton X-100, *Coll. Czech. Chem. Commun. 58*: 1495 (1993).

61. L. Jančář, J. Preisler, and L. Sommer, Multicomponent spectrophotometric determination of the sum of the rare earth elements, Al, Fe and the using multivariate calibration with PLS data evaluation, *Coll. Czech. Chem. Commun. 58*: 1509 (1993).

62. M. C. Valencia, S. Boundra, and J. M. Bosque-Sendra, Determination of trace amount of beryllium in water by solid phase spectrophotometry, *Analyst 118*: 1333 (1993).

63. J. M. Bosque-Sendra, M. C. Valencia and S. Boundra, Determination of trace amounts of aluminium in water by solid phase spectrophotometry, *Anal. Lett. 27*: 1579 (1994).

64. A. Y. El-Sayed and M. Abd-Elmottaleb, Determination of chromium(III) with eriochrome cyanine R by fourth-derivative spectrophotometry, *Anal. Lett. 27*: 1727 (1994).

65. R. M. Dagnall and T. S. West, A selective and sensitive colour reaction for silver, *Talanta 11*: 1533 (1964).

66. R. M. Dagnall, M. T. El-Ghamry, and T. S. West, Analytical applications of ternary complexes. V. Indirect spectrophotometric determination of cyanide, *Talanta 15*: 107 (1968).

67. R. Sukumar, T. P. Rao, and A. D. Damodaran, Determination of trace amounts of praseodymium by third-derivative molecular absorption spectrophotometry, *Analyst 113*: 1061 (1988).

68. M. S. El Ezaby, T. M. Salem, A. H. Zewail, and R. Issa, Spectral studies of some hydroxyderivatives of anthraquinones, *J. Chem. Soc. B*: 1293 (1970).

69. M. Roman Ceba, A. Fernandez-Qutierrez, and M. C. Mahedero, 1,8-dihydroxyanthraquinone-Mg(II) complex. II. Spectrophotometric study. Determination of Mg(II), *Microchem. J. 27*: 339 (1982).

70. J. A. Murello, J. M. Lemus, A. Muñoz de la Peña, and F. Salinas, Simultaneous determination of cobalt and nickel by first-derivative spectrophotometry, *Analyst, 113*: 1439 (1988).

71. N. K. Agnihotri, H. B. Singh, R. L. Sharma, and V. K. Singh, Simultaneous determination of beryllium and aluminum in mixtures using derivative spectrophotometry, *Talanta 40*: 415 (1993).

72. A. Arrebola Ramírez, D. Gazquez, I. M. de la Rosa, and F. Moreno, Spectrophotometric determination of lead by extraction of the mixed metal complex Pb(II)-Ca-purpurin (1,2,4-trihydroxyanthraquinone), *Anal. Lett. 27*: 1595 (1994).

73. K. A. Idriss, M. M. Seleim, M. S. Saleh, M. S. Abu-Bakr, and H. Sedaira, Spectrophotometric study of the complexation equilibria of zirconium(IV) with 1-amino-4-hydroxyanthraquinone and the determination of zirconium, *Analyst 113*: 1643 (1988).

74. T. Pal, N. R. Jana, and P. K. Das, Spectrophotometric determination of magnesium(II) with emodin (1,3,8-trihydroxy-6-methylanthraquinone, *Analyst 117*: 791 (1992).

75. J. Toei, Use of a stirred chamber to increase the efficiency of solid-liquid reactions in flow injection analysis, *Analyst 112*: 1067 (1987).

76. N. T. Son, J. Lasovský, and E. Růžička, Spectrophotometric study of reactions of sodium 6-hydroxy-5-dibenzo (a, j) phenoxazone-11,13-disulfonate with Al ions in the presence of cationoid tensides, *Coll. Czech. Chem. Commun. 45*: 1227 (1980).

77. M. T. M. Zaki and A. M. El-Didamony, Determination of gallium and indium with haematoxylin in a micellar medium, *Analyst, 113*: 1277 (1988).

78. S. M. Sultan, F.-E. O. Suliman, S. O. Duffuaa, and I. I. Abu-Abdoun, Complex-optimized and flow injection spectrophotometric assay of tetracycline antibiotics in drug formulations, *Analyst, 117*: 1179 (1992).

79. S. M. Galal, S. M. Blaih, and M. E. Abdel-Hamid, Comparative spectrophotometric analysis of rifampicin by chelate formation and charge-transfer complexation, *Anal. Lett. 25*: 725 (1992).

80. E. Blasius and K. P. Jansen, Analytical application of crown compounds and cryptands, *Top Current Chem. 98*: 163 (1981).

81. K. Datta and J. Das, Determination of silver(I) and gold(III) from the rate of catalysed substitution of α,α'-dipyridyl for cyanide in potassium ferrocyanide, *J. Indian Chem. Soc. 51*: 553 (1974).

82. V. Kubáň and D. B. Gladilovich, Simultaneous determination of iron and copper in waters by multiligand flow injection analysis, *Coll. Czech. Chem. Commun. 53*: 1461 (1988).

83. E. Tütem, R. Apak, and F. Baykut, Spectrophotometric determination of trace amounts of copper(I) and reducing agents with neocuproine in the presence of copper(II), *Analyst 116*: 89 (1991).

84. A. A. Schilt, T. A. Yang, J. Fu Wu, and D. M. Nitzki, New chromogens of

the ferroin type-VIII. Some di- and trisubstituted 1,2,4-triazines, 3-substi-tuted-9*H*-indeno[1,2-e]-1,2,4-triazin-9-ones, and di- and trisubstituted 1,2,4-triazolines, *Talanta 24*: 685 (1977).

85. A. A. Schilt and M. R. Di Tusa, Spectrophotometric determination of iron and reducing agents with PPTS, a new water-soluble ferroin-type chromogen of superior sensitivity, *Talanta 29*: 129 (1982).

86. G. L. Traister and A. A. Schilt, Water soluble sulfonate chromogenic re-agents of the ferroin type and determination of iron and copper in water, blood serum, and beer with the tetraammonium salt of 2,4-bis(5,6-diphenyl-1,2,3-triazin-3-yl)pyridinetetrasulfonic acid, *Anal. Chem. 48*: 1216 (1976).

87. M. A. Islam and W. I. Stephen, Evaluation of some as-triazines and re-evaluation of PDT and ferrozine as reagents for spectrophotometric determi-nation of ruthenium, *Talanta 39*: 1429 (1992).

88. M. K. Gadia and M. C. Mehra, Analytical reactions of substituted cyanofer-rates. 2. Pentacyanoamminoferrate(II) in catalytic spectrophotometric deter-mination of sub-parts million amounts of Ag, Au and Hg in solution, *Micro-chem. J. 23*: 278 (1978).

89. E. B. Buchanan Jr., D. Crichton and J. R. Bacon, Study of the factors which affect the reaction of the ferrous ion with tripyridyltriazene, *Talanta 13*: 903 (1966).

90. K. Burger, Selectivity and analytical application of dimethylglyoxime, in *Che-lates in Analytical Chemistry*, Vol. 2. Marcel Dekker, New York, 1969.

91. B. A. Jillot and R. J. P. Williams, Further complexes of ferrous dimethylgly-oxime, *J. Chem. Soc.*: 462 (1958).

92. E. W. Rice, Photometric determination of microgram quantities of palladium with beta-furfuraldoxime, *Anal. Chem. 24*: 1995 (1952).

93. D. G. Davis and E. A. Boudreaux, Nickel(IV) dimethylglyoxime, *J. Electro-anal. Chem. 8*: 434 (1964).

94. V. M. Peshkova and N. G. Ignatieva, 1,2-cyclopentanedione dioxime as a reagent for the gravimetric and extraction-photometric determination of nickel in the presence of copper (in Russian), *Zh. Anal. Khim. 17*: 1087 (1962).

95. H. Diehl and F. Smith, The iron reagents: Bathophenantroline, 2,4,6-tripyridyl-s-triazene, phenyl-2-pyridyl ketoxime, The F. Smith chemical com-pany, Columbus, Ohio, 1960.

96. A. J. Cameron, N. A. Gibson, and R. Roper, Analytical possibilities of a series of new aromatic-type chelate compounds, *Anal. Chim. Acta 29*: 70 (1963).

97. J. F. Geldard and F. Sions, Tridentate chelate compounds. VI. Copper(II) complexes derived from pyridine-2-aldehyde-2′-pyridylhydrazone, *Inorg. Chem. 4*: 414 (1965).

98. J. E. Going and C. Sykora, Photometric and potenciometric study of metal complexes of 2-(3′-sulfobenzoyl)pyridine-2-pyridyl hydrazone, *Anal. Chim. Acta 70*: 127 (1974).

99. H. Ishii, M. Yamaguchi, and T. Odashima, Studies of complexation equilib-

ria between water-soluble hydrazones which consist of 5-nitro-2-pyridylhydrazine and heterocyclic ketones and divalent metal ions, *Talanta 39*: 1181 (1992).

100. E. Kavlentis, Selective spectrophotometric determination of vanadium(II) in the presence of 35 common cations by means of di-2-pyridylketone-2-pyridylhydrazone, *Anal. Lett. 22*: 2083 (1989).

101. T. Takaoza, T. Taya, and M. Otsmo, Extractive-spectrophotometric determination of trace iron(II) with di-2-pyridylmethanone-2-(5-nitro)pyridylhydrazone, *Talanta 39*: 77 (1992).

102. G. Weber and G. Schwedt, Zur Analytik chemischer Bindungsformen von Nickelspuren in Kaffee, Tee und Rotwein mit chromatographischen und spektroskopischen Methoden, *Anal. Chim. Acta 134*: 81 (1982).

103. M. Katyal and Y. Dutt, Analytical applications of hydrazones, *Talanta 22*: 151 (1975).

104. C. Shiti and L. Xu, 1-(4-Nitropheny)-3-(2-quinolyl) triazene as an extremely sensitive reagent for spectrophotometric determination of mercury. *Talanta 39*: 1395 (1992).

105. S. Nai-kui, W. Fu-sheng, and Q. Qi-ping, Extraction and spectrophotometric determination of cadmium based on the formation of a ternary complex with cadion and 1,10-phenantroline, *Anal. Lett. 14*: 1565 (1981).

106. A. F. Danet and V. David, Complexation, extraction and determination of the H_3CHg^+ ion with Cadion [1-(*p*-nitrophenyl)-3-(*p'*-azobenzene)-triazene], *Talanta 39*: 1299 (1992).

107. W. Fu-sheng and Y. Fang, Spectrophotometric determination of silver with Cadion 2B and Triton X-100, *Talanta 30*: 190 (1983).

108. H. Ishii, K. Satoh, and H. Koh, Spectrophotometric and analogue derivative spectrophotometric determination of ultramicro amounts of cadmium with cationic porphyrins, *Talanta 29*: 545 (1982).

109. M. Tabata, Kinetic method for the determination of nanogram amounts of lead(II) using its catalytic effect on the reaction of manganese(II) with 5,10,15,20-tetrakis(4-sulphonatophenyl)porphine, *Analyst 112*: 141 (1987).

110. M. Tabata and M. Tanaka, A kinetic method for the determination of a nanogram amount of mercury(II) by its catalytic effect on the complex formation reaction of maganese with α, β, γ, δ-tetraphenyl porphine sulfonate, *Anal. Lett. 13*: 427 (1980).

111. M. Matsushita, T. Irino, Y. Muramoto, and A. Kushigemachi, Comparative study on biuret reactions with first transition metals (copper, nickel, and cobalt), *Rinsho Kagaku* (*Nippon Rinsho Kagakkai*) *20*: 91 (1991); *Chem. Abstr. 115*: 251368p (1991).

112. W. Reichardt and B. Eckert, On the estimation of the protein content of milk, cheese and meat by means of the biuret reaction (in German), *Nahrung 35*: 731 (1991).

113. Ch. Xie, Zh. Wu, L. Sun, and Zh. Wang, Analysis of biuret in slow-release urea, *Turang Tongbao, 23*: 88 (1992); *Chem. Abstr. 117*: 170174m (1992).

114. M. E. M. S. De Silva, A selective method for the determination of molybdenum using toluene-3,4-dithiol, *Analyst 100*: 517 (1975).

115. P. Bermejo-Barrera, J. F. Vazquez-Gonzalez, and F. Bermejo-Martinez, Extraction-spectrophotometric determination of molybdenum with toluene-3,4-dithiol, *Analyst 112*: 473 (1987).

116. A. A. Alwarthan, S. A. Fattah, and N. M. Zahran, Spectrophotometric determination of cephalexin in dosage forms with imidazole reagent, *Talanta 39*: 703 (1992).

117. R. Mendez, T. Alemany, A. Negro, and J. Martin-Villacorta, Spectrophotometric assay of a penem (SCH 29482) by reaction with imidazole, *Anal. Lett. 23*: 1335 (1990).

118. D. J. Halls, The properties of dithiocarbamates: A review, *Mikrochim. Acta 62*: (1969).

119. A. Hulanicki, Complexation reactions of dithiocarbamates, *Talanta 14*: 1371 (1967).

120. A. I. Busev, V. M. Byrko, and A. I. Dikusar, Disulphides of dithiocarbamates and possibilities of their use in analytical chemistry: A review (in Russian), *Zh. Anal. Khim. 27*: 1380 (1971).

121. M. C. Garcia, R. A. Ilvarez-Coque, R. M. Villanueva Camañas, M. C. Martinez Vaya, G. Ramis Ramos, and G. Mongay Fernandez, Spectrophotometric determination of mercury(II) and silver(I) with copper (II) and diethyldithiocarbamate in the presence of Triton X-100, *Talanta 33*: 697 (1986).

122. A. A. Khier, M. El-Sadek, and M. Baraka, Spectrophotometric method for the determination of flufenamic and mefenamic acids, *Analyst 112*: 1399 (1987).

123. B. C. Verma, B. G. Rao, P. Kumar, N. K. Sharma, D. K. Sharma, and N. Sharma, Spectrophotometric determination of amines, carbon disulfide and nickel(II) through their mutual reactions to form nickel(II)dithiocarbamates, *Natl. Acad. Sci. Lett. (India) 11*: 179 (1988); *Chem. Abstr. 110*: 224562n (1989).

124. M. N. El-Bolkiny, G. H. Ragab, and M. M. Ayad, Colorimetric determination of propranolol hydrochloride in formulations, *Egyptian J. Pharm. Sci. 33*: 741 (1992); *Chem. Abstr. 120*: 307601g (1994).

125. E. C. Hunt, W. A. McNally, and A. F. Smith, Modified field test for the determination of carbon disulfide vapor in air, *Analyst 98*: 585 (1973).

126. J. Jozwicka and P. Starostka, New spectrophotometric method to determine carbon disulfide vapor in air, *Wlokna Chem. 11*: 252 (1985); *Chem. Abstr. 104*: 154830y (1986).

127. S. P. Chattopadhyay and A. K. Das, Determination of trace amounts of carbon disulfide, *Indian J. Chem. Sect. A25A*: 1069 (1986).

128. V. Pindur and H. Witzel, Color reactions in drug analysis, *Deut. Apoth. Ztg. 128*: 2127 (1988).

129. S. Noriki and M. Nishimura, Effects of quaternary ammonium bases on valence-saturated but coordination-unstaturated chelates. Part II. Extraction of magnesium 8-hydroxyquinolinate, *Anal. Chim. Acta 72*: 339 (1974).

130. K. Goto, S. Taguchi, K. Miybe, and K. Hanuyama, Effect of cationic surfactant on the formation of ferron complexes, *Talanta 29*: 569 (1982).

131. J. I. Garcia Alonso, M. E. Diaz Garcia, and A. Sanz Medel, The surfactant-

sensitized analytical reaction of niobium with 8-hydroxyquinoline-5-sulphonic acid, *Talanta 31*: 361 (1984).

132. S. Takemoto, Q. Fernando, and H. Freiser, Structure and behaviour of organic analytical reagents: Some arylazo 8-quinolinols, *Anal. Chem. 37*: 1249 (1965).

133. K. Sriramam, G. S. N. Sastry, and T. Saratchandrababu, Extraction-spectrophotometric determination of 5-chloro-7-iodo-8-quinolinol (Quiniodochlor) in pharmaceutical preparations, *Indian Drugs 25*: 515 (1988).

133a. U. Saha, A. K. Sen, and T. K. Das, Spectrophotometric determination of clioquinol and diiodoquin in pharmaceutical preparations using uranyl acetate as chromogenic agent, *Analyst 113*: 1653 (1988).

134. A. K. Sen and T. K. Das, Spectrophotometric determination of diodoquin in multicomponent pharmaceutical formulations using iron(III) chloride as a chromogenic agent, *Indian Drugs 26*: 583 (1989).

135. F. M. Ashour, F. M. Salama, and M. A. E. Aziza, Colorimetric determination of diiodohydroxyquinoline, *J. Drug. Res. 19*: 317 (1990).

136. S. C. Mathur, P. D. Sethi, and C. L. Jain, Colorimetric determination of iodochlorohydroxyquinoline in formulations by transition metal complexation, *J. Indian Chem. Soc. 67*: 265 (1990).

137. K. Sriramam, L. Ramadevi, and J. Sreelakshmi, A simple, rapid spectrophotometric method for determining 5,7-diiodo-8-quinolinol and 5-chloro-7-iodo-8-quinolinol in pharmaceutical preparations using stannic chloride, *Analusis 19*: 248 (1991).

138. R. B. Singh, B. S. Garg, and R. P. Singh, Oximes as spectrophotometric reagents: A review, *Talanta 26*: 425 (1979).

139. R. K. Sharma, K. Shravah, and S. K. Sindhwani, Spectrophotometric determination of palladium after extraction with 3-hydroxy-2-methyl-1,4-naphthoquinone 4-oxime into molten naphthalene, *Analyst 112*: 175 (1987).

140. C. K. Rao, O. Babaiah, V. K. Reddy, and T. S. Reddy, Rapid and selective spectrophotometric determination of manganese in steels and alloys using resacetophenone oxime, *Talanta 39*: 1383 (1992).

141. E. Jungreis and S. Thabet, Analytical applications of Schiff bases, in *Chelates in Analytical Chemistry* Vol. 2, (H. A. Flaschka and A. J. Barnard, Jr., eds.), Marcel Dekker, New York, 1969.

142. M. Nishimura, S. Noriki, and S. Muramoto, Effect of quaternary ammonium bases on valence-saturated, but coordination-unsaturated chelates. I. Extraction of chelates of glyoxal bis(2-hydroxyanil) and *o*-(salicylidiamino)phenol, *Anal. Chim. Acta 70*: 121 (1974).

143. F. Lindstrom and C. W. Milligan, Derivatives of glyoxal bis(2-hydroxyanil) as direct calcium reagents, *Anal. Chem. 36*: 1334 (1964).

144. E. Uhlemann and V. Pohl, Extraction und photometrische Bestimmung von Zinn und Blei mit 2-(*o*-hydroxyphenyl)benzthiazolin, *Anal. Chim. Acta 65*: 319 (1973).

145. M. Katyal and Y. Dutt, Analytical applications of hydrazones, *Talanta 22*: 151 (1975).

146. R. B. Singh, P. Jain, and R. P. Singh, Hydrazones as analytical reagents: A review, *Talanta 29*: 77 (1982).
147. M. Lever, Bis-aroylhydrazones of α-diketones as reagents for colorimetric and fluorimetric determinations of calcium, cadmium and other cations, *Anal. Chim. Acta 65*: 311 (1973).
148. M. Garcia-Vargas, S. Trevilla, and M. Milla, Synthesis and characterization of 1,2-cyclohexanedione bis-benzoylhydrazone and its application to the determination of Ti in minerals and rocks, *Talanta 33*: 209 (1986).
149. M. Gallego, M. Silva, and M. Valcárcel, 1,2- and 1,3-cyclohexanedione bis(2-hydroxybenzoylhydrazone) S as analytical reagents, *Talanta 31*: 1075 (1984).
150. M. C. M. Andrey, M. D. Galinde Riañdo, J. A. Muñoz Leyva, and M. Garcia-Vargas, Spectrophotometric determination of trace amount of titanium in geochemical and metallurgical samples, *Anal. Lett. 26*: 2649 (1993).
151. H. Marchart, Über die Reaktion von Chrom mit diphenylcarbazid und diphenylcarbazon, *Anal. Chim. Acta 30*: 11 (1964).
152. K. Brajter, U. Kozicka, and J. Chmurska, Spectrophotometric determination of platinum and palladium with 1,5-diphenylcarbazide, *Fresenius' J. Anal. Chem. 328*: 598 (1993).
153. R. B. Singh, B. S. Garg, and R. P. Singh, Analytical applications of thiosemicarbazones and semicarbazones: A review. *Talanta 25*: 619 (1978).
154. D. Rosales, G. Gonzales, and J. L. Gomez Ariza, Asymmetric derivatives of carbohydrazide and thiocarbohydrazide as analytical reagents, *Talanta 32*: 467 (1985).
155. A. B. Bermejo-Barrera, P. Bermejo-Barrera, M. Guisasola-Escudero, and F. Bermejo-Martinez, Simultaneous derivative spectrophotometric determination of iron(III) and bismuth(III) with EDTA, *Analyst 112*: 481 (1987).
156. A. Hrdlička, J. Havel, C. Moreno, and M. Valiente, Micellar-enhanced highly sensitive reaction of rare earths with xylenol orange and surfactants. Study of reaction conditions and optimization of spectrophotometric method, *Anal. Sci. 7*: 925 (1991).
157. M. A. H. Hafez, I. M. M. Kenawy, and M. A. M. Ramadan, Comparative analytical determination of thorium in medium and low-grade minerals and ores using semimethylthymol blue, *Anal. Lett. 27*: 1383 (1994).
158. P. Sahu, J. P. Panda, and B. C. Sinha, A rapid spectrophotometric method for determination of fluoride in silicates with zirconyl-xylenol orange complex, *Talanta 39*: 541 (1992).
159. N. W. Barnett, P. J. Jones, and H. W. Handley, The use of zirconyl xylenol orange for the post column spectroscopic detection of fluoride in ion chromatography, *Anal. Lett. 26*: 2525 (1993).
160. R. P. Pantaler and A. M. Pulyaeva, Kinetic determination of small amounts of carbonates (in Russian) *Zh. Anal. Khim. 32*: 394 (1977).
161. L. Caballo-Tomas and T. S. West, Kinetochromic spectrophotometry. I. Determination of fluoride by catalysis of the zirconium-xylenol orange reaction, *Talanta 16*: 789 (1969).
162. R. V. Hems, G. F. Kirkbright, and T. S. West, Kinetochromic spectrophotometry. II. Determination of sulphate by catalysis of the zirconium-

methylthymol blue reaction, *Talanta 16*: 789 (1969); III. Determination of fluoride by catalysis of the zirconium-methylthymol blue reaction, *Talanta 17*: 433 (1970).

163. V. M. Peshkova, T. V. Polenova, and Yu A. Barbalat, In situ formed reagents in photometric analysis (in Russian), *Zh. Anal. Khim. 32*: 471 (1977).

164. I. Iohannes and L. Melder, Nitrosoresorcinols and their cobalt complexes (in Russian), *Eesti NSV Teaduste Akad. Toimet. Keemia 37*: 112 (1988). *Chem. Abstr. 109*: 157908 (1988).

165. J. Gabbay, Y. Almog, M. Davidson, and A. E. Donagi, Rapid spectrophotometric microdetermination of nitrite in water, *Analyst 102*: 371 (1971).

166. F. J. Langmyhr, K. S. Klausen, and M. H. Nouri-Nekoui, Complex formation of lanthanum(III) or cerium(III) with 3-aminomethylalizarin-*N,N*-diacetic acid and fluoride, *Anal. Chim. Acta 57*: 341 (1971).

167. A. T. Pilipenko and L. I. Savransky, Properties and construction of azo-dye reagents for inorganic photometric analysis, *Talanta 25*: 451 (1978).

168. J. R. Kirkby, R. M. Milburn, and J. H. Saylor, A spectrophotometric study of *o,o'*-dihydroxyazobenzene and its chelates with aluminium, gallium and indium, *Anal. Chim. Acta 26*: 458 (1962).

169. K. Watanabe, N. Miyakawa, and K. Kawagaki, The solvent extraction spectrophotometric determination of Ti(IV) with 2,2-dihydroxyazobenzene (in Japanese), *Bunseki Kagaku 31*: 632 (1982).

170. H. Hoshino, K. Nakano, and T. Yotsuyanagi, 2,2'-Dihydroxyazobenzene derivatives as reagents for trace metal determination by ion-pair reversed-phase high-performance liquid chromatography with spectrophotometric detection, *Analyst 115*: 133 (1990).

171. K. K. Saxena and A. K. Dey, Specific colorimetric reagents for the determination of palladium(II), *Anal. Chem. 40*: 1280 (1968).

172. Y. Noda and H. Miyata, 1,1'-dihydroxy-2,2'-azonaphtalene-4,4'-disulfonic acid as a new reagent for magnesium, *Bull. Chem. Soc. Japan 44*: 2291 (1971).

173. Y. Wakamatsu, The spectrophotometric determination of Ti(IV) with sodium 2-bromo-4,5-dihydroxyazobenzene-4'-sulfonate in the presence of cetyltrimethylammonium chloride, *Anal. Chim. Acta 89*: 199 (1977).

174. B. W. Buděšínský, Monoarylazo and bis(arylazo) derivatives of chromotropic acid as photometric reagents, in *Chelates in Analytical Chemistry*, Vol. 2, H. A. Flaschka and A. J. Barnard, Jr., (Eds.), Marcel Dekker, New York, 1969.

175. B. W. Buděšínský, Photometric determination of calcium with antipyrylazo III, *Anal. Chim. Acta 71*: 343 (1974).

176. S. B. Savvin, Analytical use of arsenazo III. Detemination of thorium, zirconium, uranium and rare earth elements, *Talanta 8*: 673 (1961).

177. T. Yamamoto, H. Muto, and Y. Kato, Extraction-spectrophotometric determination of zirconium with chlorphosphonazo III (in Japanese), *Bunseki Kagaku 26*: 515 (1977).

178. M. Teruyoshi, Solvent extraction of bismuth(III)-chlorphosphonazo III [3,6-bis-(4-chloro-2-phosphonophenylazo)chromotropic acid] chelate with hexadecylpyridinium chloride, *Miyakonjo Koyo Koto Semmon Gakko Kenkyu Hokoku 73* (1980); *Anal. Abstr. 41*: 1B109 (1981).

179. S. Shibata, 2-Pyridylazo compounds in analytical chemistry, in *Chelates in Analytical Chemistry*, Vol. 4, H. A. Flaschka and A. J. Barnard, Jr., (Eds.), Marcel Dekker, New York, 1972.

180. R. G. Anderson and G. Nickless, Heterocyclic azo dyestuffs in analytical chemistry: A review, *Analyst 92*: 207 (1967).

181. D. Betteridge and D. John, Pyridylazonaphtols (PANs) and pyridylazophenols (PAPs) as analytical reagents. Part I. Synthesis and spectrophotometric examination of reagents and some chelates, *Analyst 98*: 377 (1973).

182. D. Betteridge and D. John, Pyridylazonaphtols (PANs) and pyridylazophenols (PAPs) as analytical reagents. II. Spectrophotometric and solvent-extraction studies of complex formations, *Analyst 98*: 390 (1973).

183. V. Koblížková, V. Kubáň, and L. Sommer, Spectrophotometric study of complexation equilibria and methods of determining Cu(II) and Ni(II) with 2-(2-pyridylazo)-1-naphthol-4-sulfonic acid, *Collect. Czech. Chem. Commun. 43*: 2711 (1978).

184. V. Koblížková, V. Kubáň, and L. Sommer, Spectrophotometric study of complexes Hg(II) with 2-(2-pyridylazo)-1-naphthol-4-sulfonic acid, *Chem. Papers 33*: 485 (1979).

185. L. Sommer, V. Kubáň, and M. Langová, Spectrophotometric investigation of complexes with organic reagents for the optimization of spectrophotometric determinations (uv + vis) of inorganic analytes, *Fresenius' J. Anal. Chem. 310*: 51 (1982).

186. O. Coufalová and L. Čermáková, Spectrophotometric determination of the platinum metals. VIII. Highly sensitive extraction determination of palladium with 4-(2-pyridylazo)resorcinol in the presence of cetylpyridinium bromide, *Chem. Papers 39*: 83 (1985).

187. A. Alte and L. Čermáková, Spectrophotometric determination of the platinum metals. IX. Highly sensitive extraction determination of rhodium with 4-(2-pyridylazo)resorcinol in the presence of micellar hexadecylpyridium bromide, *Chem. Papers 42*: 483 (1988).

188. D. A. Johnson and T. M. Florence, A study of some pyridylazo dyestuffs as chromogenic reagents and the elucidation of the nature of their metal complex spectra, *Talanta 22*: 253 (1975).

189. S.-C. Hung, C.-L Qu, and S.-S. Wu, Spectrophotometric determination of uranium(VI) with 2-(3,5-dibromo-2-pyridylazo)-5-diethylaminophenol in the presence of anionic surfactant, *Talanta 29*: 629 (1982).

190. J. Zbíral and L. Sommer, Spektralphotometrische Bestimmung von Kobalt mit 2-(5-Brom-2-pyridylazo)-5-diethylaminophenol, *Fresenius' J. Anal. Chem. 306*: 129 (1981).

191. L. D. Martinez, E. Perino, E. J. Marchevsky, and R. A. Olsina, Spectrophotometric determination of gadolinium(III) with 2-(5-bromo-2-pyridylazo)-5-diethylaminophenol (5-Br-PADAP), *Talanta 40*: 385 (1993).

192. A. C. Spinola Costa, Sérgio L. C. Ferreira, M. G. M. Andrade, and I. P. Lobo, Simultaneous spectrophotometric determination of nickel and iron in copper-base alloy with bromo-PADAP, *Talanta 40*: 1267 (1993).

193. S.-C Hung, C.-L. Qu, and S.-S. Wu, Spectrophotometric determination of

silver with 2-(3,5-dibromo-2-pyridylazo)-5-diethylaminophenol in the presence of anionic surfactant, *Talanta 29*: 85 (1982).

194. C. Lin and X. Zhang, Determination of chromium and molybdenum with 2-(5-bromopyridylazo)-5-diethylaminophenol by reversed-phase liquid chromatography, *Analyst 112*: 1659 (1987).

195. S. Kalyanaraman, A. Sugiyama, and T. Fukasawa, Extraction-spectrophotometric determination of zirconium or hafnium with 1-(2-thioazolylazo)-2-naphthol, *Analyst 110*: 213 (1985).

196. S. L. C. Ferreira, Spectrophotometric determination of nickel in copper-base alloy with 2-(2-thiazolylazo)-*p*-cresol, *Talanta 35*: 485 (1988).

197. A. Müllerová and L. Čermáková, Spectrophotometric determination of the platinum metals. VI. Determination of rhodium and palladium with 4-(2-thiazolylazo)resorcinol in the presence of cation-active tensides, *Chem. Papers 35*: 651 (1981).

198. S. Funahashi, S. Yamada, and M. Tanaka, A kinetic method of determination of calcium in the presence of magnesium, *Anal. Chim. Acta 56*: 371 (1971).

199. S. Funahashi, M. Tabata, and M. Tanaka, Determination of submicrogram amounts of iodide by its catalytic effect on the substitution reaction of mercury(II)-PAR complex with CyDTA, *Anal. Chim. Acta 57*: 311 (1971).

200. A. K. Singh, B. Mukherjee, R. P. Singh, and M. Katyal, Analytic reactions of substituted pyrimidines, *Talanta 29*: 95 (1982).

201. D. A. Marshall and C. E. Meloan, A spectrophotometry study of the gold(III)-5-(*p*-ethoxyanilino)-5,6-dihydrouracil chelate, *Anal. Lett. 2*: 595 (1969).

202. K. Ueno, K. Shiraishi, T. Tōgō, T. Yano, I. Yoshida, and H. Kobayashi, Dual-wavelength spectrophotometric determination of traces of mercury(II) with solubilized dithizone: An approach to simplified analytical procedures for environmental pollutants, *Anal. Chim. Acta 105*: 289 (1979).

203. M. S. Abdel-Latif, Direct determination of zinc using dithizone in micellar medium, *Anal. Lett. 27*: 2341 (1994).

204. Z. Marczenko and S. Kus, Spectrophotometric determination of traces of platinum in palladium with dithizone after matrix precipitation as a compound with ammonia and iodide, *Anal. Chim. Acta 196*: 317 (1987).

205. D. Kundu and S. K. Roy, Spectrophotometric determination of platinum in glass after extraction with polyurthethane foam, *Talanta 39*: 415 (1992).

206. S. Kus and Z. Marczenko, Simultaneous determination of palladium and platinum as dithizonates by fifth-derivative spectrophotometry, *Analyst 112*: 1503 (1987).

207. R. B. Singh, B. S. Garg, and R. P. Singh, Analytical applications of thiosemicarbazones and semicarbazones: A review, *Talanta 25*: 619 (1978).

208. E. Cristofol, F. Sanchez Rojas, and J. M. Cano-Pavon, Evalutation of various *N*-phenyl thiosemicarbazones as chromogenic reagents in spectral analysis, *Talanta 38*: 445 (1991).

209. Y. Lingappa, K. H. Reddy, and D. V. Reddy, Analytical properties of 2-nitro-5,6-dimethyl-1,3-indanedione dithiosemicarbazone, *Talanta 34*: 739 (1987).

210. A. Asuero, A. M. Jimenez, and M. A. Herrador, Spectrophotometric deter-

mination of palladium in catalysts with glyoxal bis(4-phenyl-3-thiosemicarbazone), *Analyst 111*: 747 (1986).

211. K. Shravah, P. P. Sinha, and S. K. Sindhwani, Selective spectrophotometric determination of palladium(II) with phenanthraquinone monothiosemicarbazone, *Analyst 111*: 1339 (1986).

212. G. V. R. Murthy and T. S. Reddy, o-Hydroxyacetophenone thiosemicarbazone as a reagent for the rapid spectrophotometric determination of palladium, *Talanta 39*: 697 (1992).

213. A. Fernandez, M. D. Lugue de Castro, and M. Valcárcel, Flow-injection system for kinetic determination based on the use of two serial injection valves, *Analyst 112*: 803 (1987).

214. M. T. Morales, M. T. Montaña, G. Galán, and J. L. G. Ariza, Spectrophotometric determination of zinc with 1-[di(2-pyridyl)methylene]-5-salicylidenethiocarbonohydrazide, *Analyst 112*: 467 (1987).

215. S. B. Savvin and R. F. Gur'eva, 5-Azoderivatives of rhodanine and its analogues in the analytical chemistry of the noble metals, *Talanta 34*: 87 (1987).

216. P. M. Shiundu, P. D. Wentzell, and A. P. Wade, Spectrophotometric determination of palladium with sulfochlorophenolazorhodanine by flow injection, *Talanta 37*: 329 (1990).

217. W. J. Geary, N. J. Mason, I. W. Nowell, and L. A. Nixon, Interaction between phenothiazine drugs and metal ions. Part I. Palladium(II) and platinum(II) complexes. Crystal and molecular structure of protonated trichloro [10-(2'-dimethylaminopropyl)phenothiazine S]palladium(II), *J. Chem. Soc. Dalt. Trans. 6*: 1103 (1982).

218. B. Keshavan and R. Janardhan, Platinum(II) complexes of N-alkylphenothiazine, *Transition Met. Chem. 10*: 106 (1985).

219. G. M. S. Jayarama, Magnetic, spectral and thermal studies of copper(II) phenothiazine complexes, *J. Inorg. Nucl. Chem. 43*: 2329 (1981).

220. N. M. Gowda, M. M. Kuyi, and L. Zhang, Synthesis and characterization of iridium chloride complexes. I. Use of N-alkyl phenothiazine drugs as ligands, *Transition Met. Chem. 18*: 518 (1993).

221. I. Němcová, M. Charvátová, and V. Suk, A spectrophotometric study of the reaction of platinum(IV) with chlorpromazine, *Microchem. J. 30*: 221 (1984).

222. H. Puzanowska-Tarasiewicz, M. Tarasiewicz, and H. Basińska, Colorimetric determination of palladium(II) by means of promazine hydrochloride, *Microchem. J. 19*: 353 (1974).

223. H. S. Gowda and B. Keshavan, Chloropromazine hydrochloride as a new reagent for the spectrophotometric determination of ruthenium(III), *Microchim. Acta 1*: 211 (1977).

224. B. Keshavan and P. Nagaraja, Propionyl promazine phosphate as a new reagent for the spectrophotometric determination of rhodium(III), *Microchim. Acta 3*: 283 (1984).

225. H. S. Gowda and J. B. Raj, Perazine as new sensitive reagent for rapid spectrophotometric determination of mercury, *Ind. J. Chem. 25A*: 408 (1986).

226. L. G. Overholser and J. H. Yoe, Colorimeric determination of phenothiazine with palladous chloride, *Ind. Eng. Chem., Anal. Ed. 14*: 646 (1942).

227. L. Cavatorta, Assay of promazine and its separation from chlorpromazine and promethazine, *J. Pharm. Pharmacol. 11*: 49 (1959).

228. B. Morelli, Thiobarbituric acid as a reagent for the spectrophotometric determination of osmium, *Analyst 112*: 1395 (1987).

229. R. Foster, *Organic Charge-transfer Complexes*, Academic Press, London, 1969.

230. K.-A. Kovar and W. Mayer, Electron-donor-acceptor-complexes, *Pharm. Unserer Zeit* (in German) *8*: 46 (1979).

231. M. M. Ayad, S. Belal, S. M. El Adl, and A. A. Al Kheir, Spectrophotometric determination of some corticosteroid drugs through charge-transfer complexation, *Analyst 109*: 1417 (1984).

232. A. A. El Kheir, S. F. Belal, M. M. Ayad, and S. M. El Adl, The use of charge transfer complexation in the spectrophotometric determination of some corticosteroid drugs through intermediate oxime formation, *Anal. Lett. 19*: 1019 (1986).

233. S. Belal, A. Abou el Kheir, M. Ayad, and S. El Adl, Spectrophotometric determination of glucose and fructose in injections using tetracyanoethylene reagent, *Microchem. J. 37*: 25 (1988).

234. S. H. Hastings and B. H. Johnson, Spectrophotometric determination of aliphatic sulfides, *Anal. Chem. 27*: 564 (1955).

235. A. M. Taha, A. K. S. Ahmad, C. S. Gomaa, and H. M. El-Fatatry, Charge-transfer complexes in alkaloid assay, *J. Pharm. Sci. 63*: 1853 (1974).

236. C. Gomaa and A. Taha, Unit-dose assay of tropine alkaloids and their synthetic analogs, *J. Pharm. Sci. 64*: 1398 (1975).

237. A. M. Taha and C. S. Gomaa, Analysis of alkaloid mixtures by charge-transfer complexation, *J. Pharm. Sci. 65*: 986 (1976).

238. A. Taha and G. Rücker, Utility of π-acceptors in alkaloid assay, *Arch. Pharm. 310*: 485 (1977).

239. H. S. I. Tan, E. D. Gerlach, and A. S. Dimattio, Sensitive assay procedure for ethambutol hydrochloride via charge-transfer complex formation, *J. Pharm. Sci. 66*: 766 (1977).

240. H. Abdine, M. A. Elsayed, and Y. M. Elsayed, Spectrophotometric determination of hydrochlorothiazide and reserpine in mixtures, *Analyst 103*: 354 (1978).

241. M. A. Elsayed, M. A. Abdel-Salam, N. A. Abdel-Salam, and Y. A. Mohammed, Spectrophotometric determination of emetine and lobeline by charge-transfer complexation, *Planta Medica 34*: 430 (1978).

242. S. Belal, M. A. Elsayed, M. E. Abdel-Hamid, and H. Abdine, Use of charge-transfer complexation in the spectrophotometric assay of antazoline and naphazoline in combination, *Analyst 105*: 774 (1980).

243. M. S. Rizk, M. I. Walash, and F. A. Ibrahim, Spectrophotometric determination of piperazine via charge-transfer complexes, *Analyst 106*: 1163 (1981).

244. K.-A. Kovar and M. Abdel-Hamid, Molecular complexes and radicals of drugs containing the imidazole ring (in German), *Arch. Pharm. 317*: 246 (1984).

245. N. Omar, G. Saleh, M. Neugebauer, and G. Rücker, Spectrophotometric analysis of penicillins by detection with iodine, *Anal. Lett. 21*: 1337 (1988).

246. M. E. Abdel-Hamid and M. A. Abuirjeie, Utility of iodine and 7,7,8,8-tetracyanoquinodimethane for determination of terfenadine, *Talanta 35*: 242 (1988).
247. D. Kottke, T. Beyrich, and U. Meincke, Charge-transfer-complexation between pholedrin and various acceptors (in German), *Pharmazie 45*: 837 (1990).
248. G. A. Saleh and H. F. Askal, Spectrophotometric analysis of cinnarizine via charge-transfer complexation reaction, *Pharmazie 45*: 220 (1990).
249. H. F. Askal and G. A. Saleh, Spectrophotometric methods for the determination of sulfathiourea, *J. Pharm. Biomed. Anal. 9*: 297 (1991).
250. S. V. Kamath, R. Shivram, G. P. Uza, and S. Vangani, Use of charge-transfer complexation in the spectrophotometric determination of diclofenac sodium, *Anal. Lett. 26*: 655 (1993).
251. R. Tawa, S. Shimizu, and S. Hirose, Determination of some tertiary amines as π-complexes with tetracyanoethylene, *Chem. Pharm. Bull. 28*: 541 (1980).
252. H-J. Petrowitz and M. Wagner, Photometric determination of aromatic hydrocarbons after reaction with tetracyanoethylene, *Z. Anal. Chem. 330*: 125 (1988).
253. S. I. Obtemperanskaya, M. M. Buzlanova, O. S. Zhukovskaya, and M. G. Zhikhareva, Photometric method for determination of aliphatic sulfides as charge-transfer complexes (in Russian), *Zh. Anal. Khim. 34*: 1618 (1979).
254. S. I. Obtemperanskaya, O. S. Zhukovskaya, and L. N. Bukhtenko, Photometric determination of disulfides as charge-transfer complexes (in Russian), *Zh. Anal. Khim. 37*: 491 (1982).
255. V. P. Bystryakov, S. I. Obtemperanskaya and L. N. Bukhtenko, Photometric determination of organic derivatives of phosphine containing a thiophosphoryl group as charge-transfer complexes (in Russian), *Zh. Anal. Khim. 36*: 1600 (1981).
256. M. I. Walash, M. Rizk, and A. El-Brashy, Colorimetric determination of thioxanthenes with tetracyanoethylene, *Pharm. Weekbl., Sci. Ed. 8*: 234 (1986).
257. H. F. Askal, G. A. Saleh, and N. M. Omar, Utility of certain π-acceptors for the spectrophotometric determination of some penicillins, *Analyst 116*: 387 (1991).
258. M. S. Mahrous, Spectrophotometric determination of perphenazine and chlorpromazine through the formation of charge transfer complex with tetracyanoethylene, *Bull. Fac. Pharm. (Cairo Univ.) 29*: 37 (1991).
259. G. H. Schenk and M. Ozolins, Tetracyanoethylene chemistry: Indirect photometric determination of anthracene in naphthalene, *Talanta 8*: 109 (1961).
260. D. A. Williams and G. H. Schenk, Tetracyanoethylene π-complex chemistry: Indirect spectrophotometric determination of Diels-Alder-active 1,3-dienes, *Talanta 20*: 1085 (1973).
261. T. S. Al-Ghabsha, S. A. Rahim, and A. Townshend, Spectrophotometric determination of microgram amounts of amino acids with chloranil, *Anal. Chim. Acta 85*: 189 (1976).
262. Y. Tashima, H. Hasegawa, H. Yuki, and K. Takiura, Colorimetric determination of aromatic primary amines with chloranil (in Japanese), *Bunseki Kagaku 19*: 43 (1970).

263. S. Belal, M. A. Elsayed, M. E. Abdel-Hamid, and H. Abdine, Utility of chloranil in assay of naphazoline, clemizole, penicillin G sodium, and piperazine, *J. Pharm. Sci. 70*: 127 (1981).

264. K.-A. Kovar and M. Abdel-Hamid, Molecular complexes and radicals of drugs containing the imidazoline ring, *Arch. Pharm. 317*: 246 (1984).

265. T. S. Al-Ghabsha, T. N. Al-Sabha, and G. A. Al-Iraqi, Spectrophotometric determination of microgram amounts of ampicillin with chloranil, *Microchem. J. 35*: 293 (1987).

266. H. Y. Hassan, A.-M. I. Mohamed, and F. A. Mohamed, Utility of certain π-acceptors for the spectrophotometric determination of isoniazid, *Anal. Lett. 23*: 617 (1990).

267. A. M. El-Brashy, Determination of some pharmaceutically important aminoquinoline antimalarials, via charge-transfer complexes, *Anal. Lett. 26*: 2595 (1993).

268. S. M. Galal, S. M. Blaih, and M. E. Abdel-Hamid, Comparative spectrophotometric analysis of rifampicin by chelate formation and charge-transfer complexation, *Anal. Lett. 25*: 725 (1992).

269. U. Muralikrishna, K. S. Babu, and M. Krishnamurthy, A simple spectrophotometric determination of some chlorobenzoquinones with morpholine, thiomorpholine and piperazine, *Talanta 37*: 353 (1990).

270. S. El-Adl, Utility of fluoranil for spectrophotometric determination of griseofulvin and dexamethasone, *Anal. Lett. 26*: 2161 (1993).

271. A. S. Issa, M. S. Mahrous, M. Abdel-Salam, and M. E. Abdel-Hamid, Spectrophotometric determination of some antimalarials using 2,3-dichloro-5,6-dicyano-*p*-benzoquinone, *J. Pharm. Belg. 40*: 339 (1985).

272. M. E. Abdel-Hamid, M. Abdel-Salam, M. S. Mahrous, and M. M. Abdel-Khalek, Utility of 2,3-dichloro-5,6-dicyano-*p*-benzoquinone in assay of codeine, emetine and pilocarpine, *Talanta 32*: 1002 (1985).

273. M. A. Elsayed, M. Barary, M. Abdel-Salam, and S. Mohamed, Spectrophotometric assay of certain cardiovascular drugs through charge-transfer reactions, *Anal. Lett. 22*: 1665 (1989).

274. M. A. Elsayed and S. P. Agarwal, Spectrophotometric determination of atropine, pilocarpine and strychnine with chloranilic acid, *Talanta 29*: 535 (1982).

275. C. S. P. Sastry, T. E. Divakar, and U. V. Prasad, Spectrophotometric determination of penicillins and cephalosporins with chloranilic acid, *Chem. Anal. (Warsaw) 32*: 301 (1987).

276. A. M. El-Brashy, Determination of some pharmaceutically important aminoquinoline antimalarials via charge-transfer complexes, *Anal. Lett. 26*: 2595 (1993).

277. M. A. Elsayed, M. E. Abdel-Hamid, M. A. Korany, M. H. Abdel-Hay, and S. M. Galal, Spectroscopic investigation of the antazoline-*p*-chloranilic acid reaction product, *Spectroscop. Lett. 17*: 803 (1984).

278. S. I. Obtemperanskaya, S. V. Kalugina, and L. N. Bukhtenko, Spectrophotometric determination of unsaturated compounds as charge-transfer complexes (in Russian), *Zh. Anal. Khim. 36*: 2199 (1981).

279. H. A. Mohamed, Spectrophotometric determination of dequalinium chloride. *Anal. Lett. 26*: 2421 (1993).
280. M. E. Abdel-Hamid, M. S. Mahrous, M. M. Abdel-Khalek, and M. A. Abdel-Salam, Utility of 7,7,8,8-tetracyanoquinodimethane and *p*-chloranilic acid in the qualitative and quantitative analysis of pentazocine, *Egypt J. Pharm. Sci. 25*: 291 (1984); *Chem. Abstr. 106*: 2196912s (1987); *J. Pharm. Belg. 40*: 237 (1985).
281. S. I. Obtemperanskaya and E. El Kafravi Azza, Tetracyanoquinodimethane as reagent for spectrophotometric determination of nitrous bases (in Russian), *Zh. Anal. Khim. 37*: 1894 (1982).
282. A. M. I. Mohamed, H. Y. Hassan, H. A. Mohamed, and S. A. Hussein, Use of 7,7,8,8-tetracyanoquinodimethane for spectrophotometric determination of certain local anesthetics and procainamide hydrochloride, *J. Pharm. Biomed. Anal. 9*: 525 (1991).
283. M. M. Abdel-Khalek, M. E. Abdel-Hamid, and M. S. Mahrous, Use of 7,7,8,8-tetracyanoquinodimethane in the spectrophotometric determination of some antihistamines, *J. Assoc. Off. Anal. Chem. 68*: 1057 (1985).
284. N. M. A. Mahrouz and K. M. Emara, Colorimetric determination of isoniazid and its pharmaceutical formulations, *Talanta 40*: 1023 (1993).
285. Y. M. Issa and A. S. Amin, Spectrophotometric microdetermination of sulfamethoxazole and trimethoprim using alizarin and quinalizarin, *Anal. Lett. 27*: 1147 (1994).
286. J. P. Sharam and R. D. Tiwari, Charge-transfer complexes and their applications. Distinction and determination of some aromatic amines, *Microchem. J. 17*: 151 (1972).
287. R. Foster and C. A. Fyfe, Meisenheimer and related compounds, *Rev. Pure Appl. Chem. 16*: 61 (1966).
288. R. Bacaloglu, A. Blaskó, C. Bunton, E. Dorwin, F. Ortega, and C. Zucco, Mechanism of reaction of hydroxide ion with dinitrochlorobenzenes, *J. Am. Chem. Soc. 113*: 238 (1991).
289. R. J. Pollitt and B. C. Saunders, The Janovsky reaction, *J. Chem. Soc.* 4615 (1965).
290. M. Kimura, N. Obi, and M. Kawazoi, Studies on the reaction between polynitrobenzene compounds and active methylene groups. IX. On the Janovsky complexes derived from some 1-substituted 2,4,6-trinitrobenzene derivatives, *Chem. Pharm. Bull. 20*: 452 (1972).
291. P. A. Lehmann F. and G. A. Ciurlizza, A study of the Janovsky reaction on 2,4-dinitroaryl ethers. 3. Kinetics and structure of the complexes, *Rev. Latinoamer. Quím. 5*: 143 (1974).
292. A. A. El Kheir, S. Belal, M. El Sadek, and A. El Shanwani, Spectrophotometric determination of acetaminophen, oxyphenbutazone and salicylamide by nitration and subsequent complexation reactions, *Analyst 111*: 319 (1986).
293. M. El Sadek, Dimedone as an analytical reagent for the colorimetric determination of paracetamol and oxyphenbutazone, *Anal. Lett. 19*: 479 (1986).

4

Ion Association (Ion Pairing)

A. INTRODUCTION

Ion association is based on the formation of associates composed of the colorless (or colored) analyte ion and the colored (or colorless) reagent ion of opposite charge (counterion). The absorption of these associates can be measured directly in the reaction solution, provided that the absorption maximum is different enough from the absorption maximum of the reagent. These associates can also be extracted from the aqueous solutions into an organic solvent immiscible with water (e.g., chloroform, dichloromethane, toluene). The concentration of the original analyte is then determined by spectrophotometry of the organic layer, since only the amount of the counterion equivalent to the amount of the analyte is transferred to the organic layer (by ion-pair extraction spectrophotometry). Some associates can exist in only the organic phase. In cases when the ion associates form precipitates, the absorption of the excess dye or of the dye released from the isolated precipitate after dissolution in an appropriate solvent can be measured.

The associates formed can be classified into two main types according to their composition:

1. Binary ion associates $\{Q^+, X^-\}$, in which one ion does not absorb at all and the other is colored or has a characteristic absorption in the UV region. If the analyte is colorless, a colored counterion is usually chosen; for example, for the determination of an

135

organic base, an acidic dye (anionic dye) is used, and for acidic analytes, a basic dye (cationic dye) is an appropriate counterion. Instead of acidic and basic dyes, an inorganic colored ion can be used. If the analyte is colored, a spectrophotometrically inactive counterion may be used to enable extraction into the organic phase.

2. Ternary associates of the type $\{Q^+, ML_n^-\}$ or $\{ML_n^{j+}, X^{j-}\}$, where M is a metal and L represents different types of ligands. Their complex ions are usually colored, and it is possible to use these ternary associates to determine the colorless organic ion $\{Q^+ \text{ or } X^-\}$ or the metal M. Ternary associates of the analyte ion and a charge-transfer complex ion as the reagent ion have also been described.

The process that takes place in the extraction of the associates from the aqueous phase can be represented by Eqs. (1)–(3):

$$Q^+{}_{aq} + X^-{}_{aq} \rightleftharpoons \{Q^+, X^-\}_{org} \tag{1}$$

$$Q^+{}_{aq} + ML_n^-{}_{aq} \rightleftharpoons \{Q^+, ML_n^-\}_{org} \tag{2}$$

$$ML_n^{j+}{}_{aq} + X^{j-}{}_{aq} \rightleftharpoons \{ML_n^{j+}, X^{j-}\}_{org} \tag{3}$$

where subscripts aq and org refer to the aqueous and organic phases. The process of extraction is carried out in shaking flasks or separating funnels, and flow-injection analysis (FIA) method coupled with a phase separator can be used with success. Reextraction of the colored counterion into the aqueous layer is sometimes found to be advantageous. (See Chapter 1, Section D recommendations concerning the extraction step.)

The established equilibrium in Eq. (1), for example, is quantitatively expressed [1,2] by the extraction constant K_{ex}, defined as (4):

$$K_{ex} = \frac{[Q^+, X^-]_{org}}{[Q^+]_{aq} [X^-]_{aq}} \tag{4}$$

If analytical concentrations c of both the analyte ion and the counterion (i.e., the base and the acid) are equal, an almost complete recovery by one-step extraction is achieved [3] when $K_{ex} \geq 10^6$. When samples with lower extraction constants are extracted, an excess of the counterion has to be used for optimal recovery.

The constants characterizing the chemical equilibria (1)–(3) are connected, at the same time, with the association constants of the formation of the corresponding associates (K_a) and can be applied to other equilibria. The value of K_a expressed by Eq. (5) [4], based on the electrostatic character of the mutually interacting ions,

$$K_a = 4\pi N_A/1000 \, (z^+ z^- e^2/\varepsilon_r k \, T)^3 \, Q(b) \tag{5}$$

where N_A is the Avogadro constant, z the ion charges, e the elementary charge, ε_r the relative permittivity of the medium (solvent), k the Boltzmann constant, T the thermodynamic temperature, and $Q(b)$ the function of the minimum distance to which the interacting ions can approach each other. It can be seen from Eq. (5) that the value of K_a is dependent first of all on the physical properties of both ions forming the associate, and second on the relative permitivity of the medium. It is theorized that forces of coulombic character exist between the two spherical and nonpolarizable ions of opposite charge, but it should be emphasized that no chemical bond of any kind is formed and that the ions have only a so-called outer-sphere interaction. This theory is usually satisfactory for the elucidation of the character of ion pairs and the formation of most of the analytically important ion associates, though the theory does not give a true picture of the possible change of ε_r in the immediate environment (microenvironment) of the ions.

The effect of ionic strength (the presence of strong electrolytes and buffer components) on ion association in the presence of surfactants is significant for $I > 0.1$. This effect depends above all on the composition of the associate and the dyestuff present. The influence of the interfering ion of the electrolyte is dependent on its concentration in the solution [5].

Originally, the association interaction was assumed to be a simple salt formation. However, concept of ion association is preferred, since the process that takes place in the two-phase system is often complicated by a series of side reactions (e.g., protolysis and extraction of the original ions, aggregation and association in the aqueous phase under the formation of further ion associates, dimerization or the dissociation of associates in the organic phase, and the hydrolysis of the ion associates). Factors affecting the liquid–liquid extraction of the ion pair are properties of (1) the pairing ions (size, hydrophobicity, dissociation constants, concentration), (2) the extraction phase (polarity, specific solvation, presence of adduct-forming agents, volume), (3) the aqueous phase (salt concentration, pH, volume), (4) the nature of the ion pair formed (size, hydrophobicity, specific solvation behavior, dissociation), and (5) temperature [6]. In some cases the participation of the solvent in the formation of the associate has been considered (e.g., in the $\{\,{}^+H(H_2O)S_3, FeCl_4{}^-\,\}$, where S is ether).

The formation of an ion pair requires that the presence of both the analyte and the reagent be in the form of ions. Strong electrolytes undergo interaction in the whole pH region, whereas the interaction of weak electrolytes is limited to the pH region in which they are dissociated.

Furthermore, a certain degree of lipophilicity of both the analyte and the counterion is required to obtain an ion associate having an appropriate extraction constant [1,2,7]. For example, sample log K_{ex} value for tetra-n-butylammonium associates (chloroform as extractant) are as follows: Cl^-

(-0.11), Br^- (1.29), I^- (3.01), ClO_4^- (3.48), picrate (5.91), Methyl Orange (5.47), Bromothymol Blue (8.00). Possible interactions caused by anions would therefore decrease in the series $ClO_4^- > I^- > Br^- > Cl^-$. For each homologous series of compounds, determination by a single extraction step is possible, beginning with the member of sufficient lipophilicity. For example, in the series of aliphatic amine picrates (chloroform as extractant), this is possible beginning with 1-dodecylamine, di-1-hexylamine, and tri-1-butylamine; in the series of tetraalkylammonium salts, it is possible beginning with tetra-n-butylammonium picrate. Lower members of these series require repeated extraction since they are extracted only poorly. The log K_{ex} values of a particular ion associate differ for individual extractants; for example, the values for tetra-n-butylammonium picrate are for carbon tetrachloride 1.94, benzene 3.59, chloroform 5.91, and dichloromethane 6.68.

The situation becomes far more complicated when one of the ions of the associate is a surfactant. In that case the formation of the ion associate is only one of the steps of the interaction. The associates formed by the interaction of tensides with organic dyestuffs or complex metal ions have a stoichiometric composition. They are usually poorly soluble in water and they can be extracted directly into the organic phase. The mechanism of the participation of surfactants in the formation of ion associates is dependent above all on their analytical concentration in the solution (c_s): with increasing c_s, the formation of premicellar forms followed by the formation of true micelles is theorized. When the concentration of surfactants present approximates or is greater than the actual critical micellar concentration (cmc), the associates can be solubilized and incorporated on the surface or inside the micelles [8–13]. The ion associate solubilization (usually $c_S > 80$ μmol L^{-1}) is accompanied by changes in both the optical and acid–base properties of the products. The absorption bands are shifted ($\Delta\lambda_{max} \sim 30$ nm) and the conditional dissociation constant pK_a' values altered ($\Delta pK_a' \leq 2$).

Many of the dyes used in extraction spectrophotometry are acid–base or metallochromic indicators. Their frequent use can be explained by the simple fact that they are widely available in laboratories. They are of some importance, but above all when used with surfactants. Metallochromic indicators permit the application of metal chelation.

B. DETERMINATION OF ORGANIC CATIONS

1. Organic Bases and Quaternary Ammonium Compounds

These determinations proceed according to Eqs. (1) or (2), respectively. Acid dyes of different types are used. The most often used acid azo dye is Methyl Orange (**1**). Since its sulfo group is dissociated also in weak acid medium, Methyl Orange is applicable in the pH range 3–12. Higher concentrations of H^+ ions (pH 1–3), however, cause precipitation of the dye in its protonized form ((1a) in Eq. (6)) such that it can no longer be used for extraction. This property is valid for most aminoazo dyes.

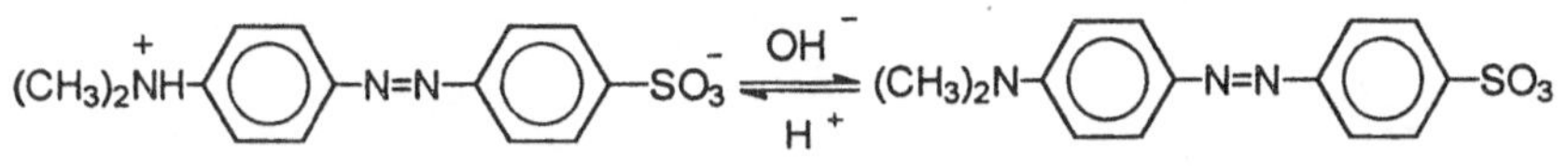

$$ (6) $$

(1a)   **(1b)**

The commercially available hydroxyazodye Orange II ((2), or TropaeolinOOO)is much more suitable. Its hydroxy group does not dissociate even in strong alkaline media, and therefore the dye is applicable over the whole region of pH. This means that the experimental conditions (pH) need to be adjusted only to the requirements of the analyte. The analogous dye having an extra dissociated sulfo group however, is not useful **(3)**, [14]. Some examples of the application of Orange II can be found in Refs. [15–23].

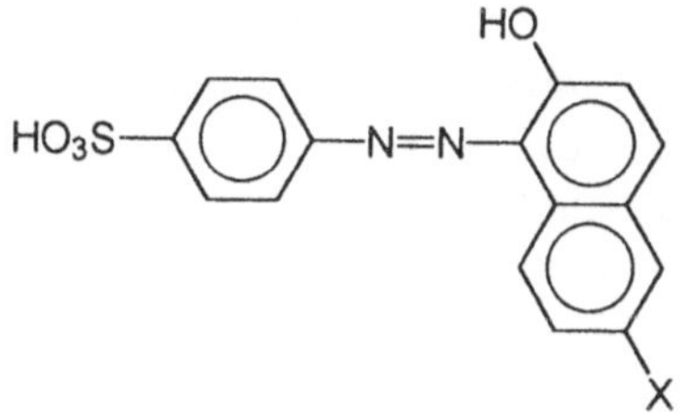

(2) X = H   **(3)** X = SO₃H

The great advantage of sulfonated azo dyes is a low blank, since the monosulfo anions are practically insoluble in the organic phase. According to the authors' experience, the determination with Methyl Orange and Orange II is best carried out using 10^4–10^5 molar solutions of the analyte, Britton-Robinson buffers of the appropriate pH required by the analyte, and a 10^{-3} molar solution of the dye and chloroform as the extractant. The time period required for shaking and establishing the extraction equilibrium is characteristic for each particular associate and must be estimated experimentally.

Azodyes of very similar structures have also been used with success (e.g., Croceine Orange G, Fast Red A, and Lithol Red [19,21,24]).

Bromothymol Blue is the most frequently used sulfophthalein. According to a comparative investigation of sulfophthaleins, Bromocresol Green, Bromocresol Purple, Bromophenol Blue, Thymol Blue, and Bromoxylenol Blue are also applicable [25]. Sulfophthaleins, however, undergo complex acid–base equilibria. These are shown schematically in a generalized and simplified form in Eq. (7) (the mesomeric forms of each dissociation form are not considered).

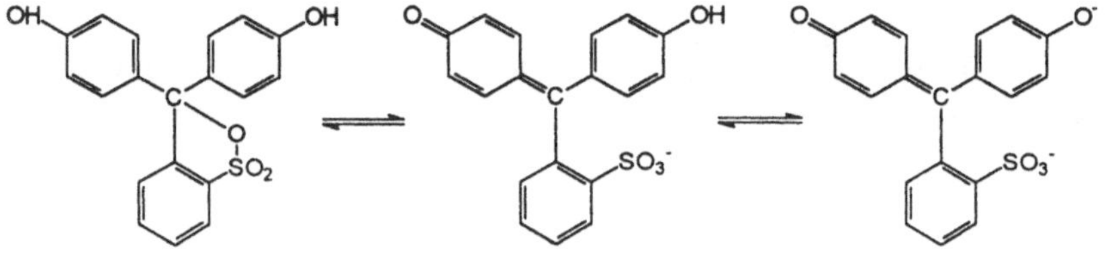

$$\qquad\qquad(7)$$

(4a)   **(4b)**   **(4c)**

Structure **(4b)** is the species suitable for ion pairing; **(4a)** does not form ion pairs and is also extracted into the organic phase (high blank); **(4c)** does not form extractable ion pairs. Individual sulfophthaleins differ considerably in the pH range in which the **b** forms exist (e.g., for Bromothymol Blue the range is pH < 7.2), which are dependent on their pK_a values (see Chapter 2, Section C).

Chromazurol S is the representative of triarylmethane dyes used to form ion associates with organic bases and cationic surfactants (acetate buffer pH 4, chloroform as extractant) [26,27]. Like sulfophthaleins, it undergoes also several acid–base equilibria (pK_{a1} = 2.45, pK_{a2} = 4.86, pK_{a3} = 11.47) [28].

Another metallochromic indicator, the azodye Eriochrome Red B, has also been recommended for the determination of drugs using acetate buffer (pH 3.6) and chloroform as extractant (λ_{max} = 475 nm) [29].

2,4,6-Trinitrophenol (picric acid) is a common nitrodye used especially in the photometric extraction of drugs [30,31]. The color produced by the interaction of basic drugs with p-chloranilic acid in dioxane is also attributed to ion-pair formation [32].

2. Cationic Surfactants

Orange II and Bromothymol Blue have been recommended for the determination of cationic surfactants [33]. Chloroform is used as a solvent for extraction (λ_{max} for the former is 485 nm, for the latter 420 nm). Disulphine Blue VN 150 is another suitable dye for the determination of cationic surfactants [34–36]. Chlorogallein methyl ester [37] and tetrabromophenolphthalein ethyl ester have also been used as ion-association reagents for organic bases and quaternary ammonium compounds, and for their determination in wastewaters. The wider optimal pH range and larger molar absorptivity (1×10^{-5}) of the latter have been found to be advantageous.

When determining cationic surfactants, interferences due to amine associates can be eliminated by measuring at 45 °C, the temperature at which the absorbance of amine associates approaches zero [38].

A procedure using Bromothymol Blue without extraction is based on the equivalent decrease of the absorbance at 610 nm [39]. In the same way, Bromocresol Purple buffered to pH 8 can be used [40].

Another procedure [41,42] based on the precipitation of surfactants with 2-bromo-5-chlorobenzene sulfonate is used for its indirect determination. The excess of the reagent is measured at 288.7 nm (the procedure is suitable for alkyltrimethylammonium chlorides with C_{12-18} alkylgroups). Picric acid may also be used as the precipitating agent: in this case the water-insoluble fraction of the tetraalkylammonium picrate is separated by filtration, and absorbance is measured after dissolution in chloroform at 365 nm ($\varepsilon = 1.5 \times 10^4$).

Two-phase or single-phase photometric titrations of cationic tensides based on the formation of associates with Bromophenol Blue, Methyl Orange, or Neutral Red make use of the same principle [43,44].

Ternary ion associates using bis-(2-[5-bromo-2-pyridylazo])-5-[N-propyl-N-sulfopropylamino-phenolato]-cobaltate or Pentachrome Azure Blue-metal(III) complexes as the counterions have been proposed as further possibilities to determine long-chain quaternary ammonium salts [45,46]. Ion-pair complexes with morin and quercetin [47] as well as molybdenum-(VI) and Pyrogallol Red [48] and of cetylpyridinium chloride in pharmaceuticals with quinidine and Bromocresol Green extractable with 1,2-dichloroethane [49] have also been used in analysis.

The use of pentanesulfonic acid as spectrophotometrically inactive counterion enables the extraction of analytically active (absorbing) analytes.

3. Nonionic Surfactants

The spectrophotometric determination of nonionic surfactants is another example of the application of ion association. The polyoxyethylene and polyether chains of these compounds can add ions of alkali metals (most often potassium) by weak interactions (solubilization) under the formation of "ionized cationic adducts" of surfactants, which subsequently form extractable products by ion association with different anionic counterions (e.g., with picric acid, thiocyanocobaltate or thiocyanozincate ions, some heteropolyacids, or derivatives of some dyestuffs). In some cases the chromophore of the counterion is utilized for the measurement (e.g., tetrabromophenolphthalein ethyl ester [50] at 620 nm, picrate [51,52] at 378 nm); in other cases the analysis is carried out indirectly by the determination of the associated chelating metal (in the neutral adduct of the tenside with $Zn(SCN)_4{}^{2-}$ the Zn content is determined in the 1,2-dichlorobenzene phase by the reaction with the azodye PAN). The Dragendorff reagent (potassium tetraiodobismuthate) and different other reagents have been used in a similar way [41].

4. Proteins

Whereas monosulfonated dyes are suitable counterions for organic bases, disulfonated dyes (e.g., Orange G, Ponceau S, Amidoblack 10B) are used as reagents for proteins. The ion associates formed are mostly insoluble in water. After filtration the excess dye is determined spectrophotometrically, or the dye is released from the isolated precipitate and determined spectrophotometrically [53–55].

The binding of the dye Coomassie Brilliant Blue G250 to protein causes a shift in the absorption maximum of the dye from 465 to 595 nm, and it is this increase in absorption at 595 nm that is monitored [56–59]. The method is suitable for the determination of proteins in biological fluids.

5. Other Groups

The methods that have been described in this section can be applied with success to several other groups of organic compounds (e.g., phosphonium and arsonium compounds, sulfonium and thiouronium salts, organotin cations, and steroids after conversion with the Girard reagent into derivatives).

C. DETERMINATION OF ORGANIC ANIONS

The determination of organic ions proceeds according to Eqs. (1) or (3). Basic (cationic) dyestuffs are used as the counterions for the determination of organic anionic compounds. Less research has been directed to the determination of simple acidic compounds such as sulfonic and carboxylic acids [60–62], nitrophenols [63,64], antibiotics [65–67], and acidic drugs [68–70].

Much of the literature has focused on the determination of anionic surfactants (tensides). Surfactants with one polar group form 1 : 1 associates with basic dyes; those containing two or more polar groups (alkanedisulfonates, for example) and those with a short alkyl chain with less than five carbon atoms do not form extractable ion associates since their molecules are too hydrophilic.

Methylene Blue, a thiazine dye, was used in traditional methods for anionic surfactant determination. These methods have involved into standard analytical procedures [71].

The batchwise Methylene Blue method was found to be troublesome, time consuming, and of very low sensitivity. It required repeating the extraction three times along with subsequent washing of the extract. Of the other dyes that have been tested, Ethyl Violet has been found to form ion associates with higher extraction constants using toluene as the nonpolar organic phase. Ethyl Violet also permits the determination of anionic surfactants using the batchwise procedure at the ppm level [72]. Acridine Orange is another dye that has been used for this purpose [41,73].

The formation of associates of anionic surfactants with Ethyl Violet or other similar dyestuffs in aqueous medium without extraction can be used for the determination of surfactants [74] in the concentration range 4 $\times$ 10^{-4} to 4 $\times$ 10^{-5} mol L^{-1}.

Further investigations led to the discovery of cationic azodyes forming ion associates of high extractibility into chloroform or toluene (e.g., 4-(4-*N*,*N*-dimethylaminophenylazo)-2-methylquinoline [75]). Of these compounds 1-(10-bromodecyl)-4-(4-aminonaphthylazo)-pyridinium bromide forms stable and soluble 1 : 1 associates with sodium dodecyl sulfate which show a sensitive color change in aqueous solution within the pH range 6–7.5 (phosphate buffer). The dye itself exhibits a maximum absorption at 595 nm, whereas the ion associates of dodecylsulfate and dodecylbenzene-sulfonate show maxima at 425 and 455 nm, respectively. Because of this fact, the extraction step becomes unnecessary. The associates with different surfactant types show different maxima, so that individual types of surfactants can be distinguished by measuring optical density at corresponding wavelengths. Therefore, the common determination is based on measuring the decrease in absorbance at 595 nm without extraction. Amounts less than 1 $\times$ 10^{-6} mol L^{-1} can be determined [76].

Furthermore, the determination of sulfonated and sulfated surfactants can be achieved using their association with a complex metal cation bonded as an azo dye chelate (Co-5-Cl-PADAP) [77,78]. On the other hand, optimization and innovations have been achieved by using FIA method [79–82]. Analogously, anionic dyes can be extracted from aqueous solutions using tetraalkylammonium cations [83].

To avoid the interference of low-molecular-weight compounds present in waters samples, solid-phase adsorption of all types of surfactants is used (e.g., on Amberlite XAD-2); their elution is achieved by methanol or 2-propanol [84,85].

D. DETERMINATION OF METALS

The determination of metal ions in the form of extracted ion associates can proceed according to Eqs. (2) or (3). These associates are classified by Sommer [86] as mixed-ligand ternary complexes, the so-called outer-sphere complexes. In contrast with inner-sphere complexes (see Chapter 3, Section A), their absorption spectra are similar to their components; their absorption bands, however, are more intense, and this is what brings about an increased sensitivity of the determination as well as improved selectivity

[87]. These associates can be classified into two groups according to whether the metal to be determined is part of the anionic or cationic part of the associate.

1. The Metal to be Determined Is a Constituent of the Anion

According to Eq. (2), in which M forms an anionic complex $ML_n{}^-$, use is made of metal halogenides complexes or metal anionic chelates (with 920- or triarylmethane dyes or gaffein derivative) and in some cases use of heteropolyacids. For the formation of ion associates it is possible to use the tetraalkylammonium salts (most often cationic tensides), other onium cations (e.g., tetraphenylarsonium, tetraphenylphosphonium) [88], other bases (e.g., diphenylguanidine [89], ephedrine), and chelating or nonchelating basic dyes (e.g., Brilliant Green, Crystal Violet, Rhodamines) as counterions. Amine extractions can also be used. The extraction of the complex anions is controlled by parameters that determine the conditions for the creation and existence of these complexes in aqueous medium as they do for simple binary associates.

Quaternary ammonium salts used for this purpose can usually be classified into two different types according to their solubility, surface activity, and ability of molecular association. Salts with more than 24 carbon atoms in their molecules dissolve in organic solvents only; those with fewer than 5 carbon atoms in their molecules are soluble in water. The presence of highly polar ion pairs of quaternary ammonium salts is assumed for nonionizable solvents. The low-molecular-weight types of quaternary ammonium salts show a high tendency to associate under the formation of inverse micelles.

Quaternary phosphonium and arsonium salts are significant not only for the extraction photometry of simple acid radicals or oxyanions of metals, but above all for the determination of anionic inorganic coordination complexes and anionic coordination complexes involving organic ligands. Anions (mostly Cl^-) originally associated with bulky organic tetraphenylarsonium or tetraphenylphosphonium cations participate competitively in the process of extraction, which can be expressed by the following equations:

$$Ph_4As^+ + Cl^- \underset{}{\overset{K'}{\rightleftharpoons}} \{Ph_4As^+,Cl^-\}_0 \tag{8}$$

$$Ph_4As^+ + MCl_4{}^- \underset{K''}{\overset{}{\rightleftharpoons}} \{Ph_4As^+,MCl_4{}^-\}_0 \tag{9}$$

where the values of K' and K'' are the measure of stability of the corresponding ion pairs and their distribution between two immiscible phases.

Examples of the application of anionic metal chelates for the purpose of association with basic dyestuffs have been compiled [90]. Associates of this type are also utilized for the determination of nonmetals. For example, the very sensitive determination of boron is based on the formation of the associate with the cation of Methylene Blue $\{MB^+, BF^{4-}\}$. For the determination of several metals in the form of their chelates with some azo dyes following association with cations of quaternary ammonium salts (usually tensides) or quaternary onium cations, see Table 1.

Complexes of 8-hydroxyquinoline-5-sulfonic acid (NQS) with different metals are used for ion-pair extraction with Zephiramine (TDBA$^+$) [98]. For instance, in the case of (TDBA$^+$,Co(NQS)$_3^-$), $\lambda_{max} = 530$ nm, $\varepsilon = 1.4 \times 10^4$, and (TDBA$^+$,Fe(NQS)$_3^-$), $\lambda_{max} = 695$ nm, $\varepsilon = 2.10 \times 10^4$, the formation of ion-associated charge-transfer complexes is assumed.

In the same way, derivatives of 8-hydroxyquinoline have been used for the determination of niobium [99], for which it has been found that the 7-sulfo derivative does not give any positive reaction. It is theorized that an ion-associate 1 : 3 : 3 complex (Nb : NQS : CPB) is formed with NQS in the presence of a cationic tenside CPB, with $\lambda_{max} = 383$ nm and $\varepsilon = 1.46 \times 10^4$ (pH 5.7, limit of determination < 6 μg Nb mL^{-1}). The ion pair of the anionic cobalt chelate of 2-nitroso-1-naphthol-4-sulfonic acid with the tetrabutylammonium ion ($\lambda_{max} = 307$ nm, $\varepsilon = 6.5 \times 10^4$) has been used for the determination of cobalt in nickel salts, iron, and steel [100].

TABLE 1 Application of $\{Q^+, ML_n^-\}$ for the Determination of M (extraction into chloroform; measurements carried out at λ_{max} of the organic layer)

M	Q$^+$	L	pH$_{opt}$	λ_{max} (nm)	ε (L mol^{-1} cm^{-1})	Ref.
Pd(II)	TDBA$^+$	PAR	6–11	540	3.29 × 10^4	91
U(VI)	TDEA$^+$	PAR	6.9–10.7	550	4.25 × 10^4	92
Cd(II)	CDBA$^+$	PAR	10	505	9.82 × 10^4	93
Co(II)	TDBA$^+$	TAR	6.7–10.2	550	—	94[a]
Ni(II)	TDBA$^+$	TAR	7.2–9.1	550	—	94[a]
Zn(II)	TDBA$^+$	TAR	8.2–10.1	550	—	94[a]
Cu(II)	TDBA$^+$	TAR	8.5–10.7	550	—	94[a]
Cu(II)	Ph$_4$As$^+$,Ph$_4$P$^+$	PAR	6–8	510	—	95[b]
Nb(V)	Ph$_4$As$^+$,Ph$_4$P$^+$	PAR	5.5	560	—	96
V(V)	Ph$_4$As$^+$,Ph$_4$P$^+$	TAR	3.8–4.0	555	2.55 × 10^4	97

TDBA$^+$ = tetradecyldimethylbenzylammonium (Zephiramine); CDBA$^+$ = cetyldimethylbenzylammonium; TDEA$^+$ = tridodecylethylammonium; PAR = 4-(2-pyridylazo)resorcinol; TAR = 4-(2-thiazolylazo)resorcinol.
[a]Extraction constants are increased in the series Ni < Cu < Zn ≪ Co.
[b]Extraction with mixture CHCl$_3$ + 5% EtOH (v/v).

The extraction determinations of $OsCl_6{}^{2-}$ with Septonex ($\lambda_{max} = 340$ or 376 nm, $\varepsilon_{340} = 1.4 \times 10^4$, limit of determination 1.5–18.6 μg Os mL^{-1}) [101] or $PdI_4{}^{2-}$ with cationic tensides can serve as examples of the determination of platinum metals based on the formation of ion-associated halogenide complexes with cationic tensides [102]. In the case of $PdI_4{}^{2-}$ and Zephiramine, $\lambda_{max} = 344$ nm and $\varepsilon = 2.5 \times 10^4$.

Gold(III) can be determined in the concentration range 2.4–10.2 ppm in the form of an associate of $AuCl_3{}^-$ with the cationic tenside Septonex directly in the aqueous solution (pH 0.1–1.0) at $\lambda_{max} = 332$ nm [103].

Another procedure for the determination of gold (and similarly for a very sensitive flotation spectrophotometric determination of platinum metals) is based on gold's simultaneous reduction to Au(I), which forms a simple associate with the basic dye Methylene Blue (MB), $\{MB^+, AuI_2{}^-\}$. In the given medium this is associated with the triiodide/dyestuff ion pair $\{MB^+, I_3{}^-\}$, and the product $\{MB^+, AuI_2{}^-\}, 3\{MB^+, I_3{}^-\}$ is formed. This associate (as with that for platinum metals) has a high value of molar absorptivity (up to 1.7 to 4.0 $\times$ 10^5, since each metal atom in the resulting product is associated with 2 to 5 molecules of the dyestuff). Flotation separation proceeds in the cyclohexane medium, and the resulting gold determination is carried out after the dissolution of the associate in a polar solvent (methanol; $\lambda_{max} = 655$ nm, $\varepsilon = 3.4 \times 10^5$), the procedure [104] obeying the Lambert–Beer law up to a gold concentration of 0.4 μg mL^{-1}.

The influence of different salting-out reagents (alkali metal and alkaline earth metal chlorides) has been studied for the extraction of associates of chlorocomplexes of indium with some basic triphenylmethane dyes (Ethyl Violet, Brilliant Green, Crystal Violet, Malachite Green, Methyl Violet), and for the extraction into a series of solvents of low polarity (benzene, toluene, xylenes). The optimal procedure has been applied to the determination of indium in semiconductor monocrystals [105].

In a series of α-hydroxyacids tested as chelating agents for extractive determination of titanium(IV) after association of its anionic complex with some basic dyestuffs, Malachite Green was found to give the best results for titanium determination in mild steels after extraction into chlorobenzene ($\lambda_{max} = 630$ nm, concentration range 0.05–1.44 μg mL^{-1}) [106].

Association reactions with phenothiazine derivatives as counterions are also recommended for the extraction determination of Mo(V), W(V), U(VI), Ti(IV) ions bonded in anionic complexes, usually with SCN$^-$. Titanium(IV), for example, can be extracted in the form of $Ti(SCN)_6{}^{2-}$ in the presence of chlorpromazine and 0.2–2.2 μg Ti(IV) mL^{-1} determined in the chloroform layer at 417 nm [107]. Similarly, the ion associate of the thiocyanate complex of cobalt(II) with ephedrine can be extracted ($\lambda = 620$ nm in methyl-iso-butyl ketone, $\varepsilon = 4.3 \times 10^3$; Co(II) can be determined in the range 0.5–19.5 μg mL^{-1}) [108].

Determinations have made use of the formation of ion associates of the metal-thiocyanate complex and some basic dyes in aqueous solution in the presence of a surfactant. Determinations of molybdenum(VI), zinc(II), cobalt(II), indium(III), selenium(IV), iron(II), and vanadium(V) are possible. The system $UO_2(SCN)_4^{2-}$ with some basic triphenylmethane dyestuffs (Crystal Violet, Malachite Green, Brilliant Green, Iodine Green, and Ethyl Violet under addition of gum arabic solution (as protective colloid) has been studied for the determination of uranium(VI)). Products of the type $\{Q_2^{2+}, UO_2(SCN)_4^{2-}\}$ are soluble in water and show high molar absorptivities (0.81×10^5 to 5.71×10^5). The associates with Crystal Violet and Malachite Green are the most suitable ones for the determination of uranium in waters and ores [109].

2. The Metal to be Determined Is a Constituent of the Cation

As with the preceeding examples, much attention is directed to the application of ion associates in which the metal to be determined is part of the cationic complex of the associate. The association proceeding in accordance with Eq. (3) makes the use of the formation of colorless, or even colored, cationic complexes of metals with 1,10-phenanthroline or 2,2′-bipyridyl (or their derivatives, in some cases also with derivatives of cupferron). The counterion is formed by different anionic chelating or nonchelating dyes, usually of the sulfophthaleine or gallein derivative type, the coloration of which and application for association is dependent on the pH of the medium (pH 6–9 is usually optimal). Structure (5) shows the simple associate (in which the ratio of the cationic and anionic component of the associate is 1 : 1) Ag(I)-1,10-phenanthroline-Bromophenol Blue:

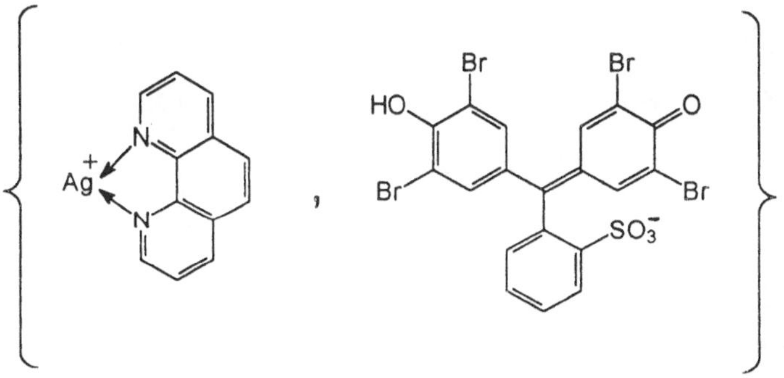

(5)

In other cases the dyestuff associates not only through its sulfogroup, but also through the OH- group of the carboxylic or hydroxygroup (after the loss of a proton). For this reason, different stoichiometries have been considered for scandium, yttrium, lanthanum, and lanthanoids associated with 1,10-phenanthroline and Chromazurol S under the formation of products of type $\{M(phen)_3^{3+}, dyestuff^{3-}\}$ [86]. Extractive determinations of metals are most often carried out in the chloroform layer, by extraction into polar aprotic solvents such as nitrobenzene, o-dichlorobenzene, or nitromethane is sometimes recommended [90,110–111].

Examples of the determination of metals in the form of cationic complexes with 1,10-phenanthroline after association with hydroxytriphenylmethane dyes or galleine derivatives [112] are given in Table 2.

The extractive determination of iron(II) in the form of cationic complexes with 2,2'-bipyridyl, 1,10-phenanthroline and bathophenanthroline after association with azodyestuffs (Methyl Orange, Benzyl Orange) has been studied. The molar absorptivities for chloroform extracts and Methyl

TABLE 2   Extraction-Photometric Determination of Metals in the Form of Associates $\{ML_n^{i+}, X^{i-}\}$ (L = 1,10-phenanthroline, pH_{opt} = pH of the aqueous phase, X^- = anion of the dye)

M	X^-	pH_{opt}	λ_{max} (nm)	$\varepsilon \times 10^{-4}$ (L mol^{-1} cm^{-1})	Determination range (μg mL^{-1})	Ref.
Fe(II)	PR	4.3	500[a]	0.8		112
Fe(II)	BP	4.0–8.0	485[a]	1.0		112
Fe(II)	BR	9.0	430[b]	2.45	up to 0.02	112
Cu(II)	BB		602[a]	1.0	0.06–0.6	112
Hg(II)	BB	6.0–8.0	610[a]	1.0	0.02–2.0	112
Fe(II)	BB	8.7–8.9	610[a]	5.9		112
Pb(II)	BB	5.7–6.1	606[a]	0.4	0.5–2.0	112
Fe(II)	BP	6.0	422[a]	1.6		112
Zn(II)	BG	6.2–7.2	416[c]		0.33–1.64	112
Mo(VI)	BPR	3.8–4.2	610	2.7	0.4–1.0	90
Ag(I)	BPR	7.0[d]	535[d];590[a]	5.1[d];3.2[e]	0.02–0.2	90

PR = Phenol Red, BR = Bromophenol Red, BB = Bromophenol Blue, BP = Bromocresol Purple, BG = Bromocresol Green, BPR = Bromopyrogallol Red.
[a]Chloroform.
[b]Nitromethane.
[c]1,2-Dichloroethane.
[d]Aqueous medium.
[e]Nitrobenzene.

Orange vary from 2.20 to 6.03 × 10^4 and for Benzyl Orange ∼ 5 × 10^4. The selectivity of the determination, however, is low [113].

REFERENCES

1. G. Schill, K. O. Borg, R. Modin, and B. A. Persson, Ion-pair extraction in the analysis of drugs and related compounds, *Progress in Drug Metabolism*, Vol. 2, J. W. Bridges and L. F. Chasseaud, eds., Wiley, London, 1977, p. 219.
2. G. Schill, K. O. Borg, R. Modin, and B. A. Persson, Ion-pair extraction in the analysis of drugs and related compounds, *Essays in Analytical Chemistry* E. Wanninen, ed., Pergamon Press, Oxford, New York, 1977, p. 379.
3. A. Bult and W. P. van Bennekom, Determinations based on two-phase ion pair and metal complex formation, *Trends Anal. Chem. 4*: 252 (1985).
4. Y. Marcus and A. S. Kertes, *Ion Exchange and Solvent Extraction of Metal Complexes*, Wiley-Interscience, New York, 1969.
5. S. B. Savvin, R. K. Chernova, and S. N. Shtykov, *Surface-active Substances*, Nauka, Moscow, 1991, p. 73.
6. E. Tomlinson, Ion-pair extraction and high-performance liquid chromatography in pharmaceutical and biomedical analysis, *J. Pharm. Biomed. Anal. 1*: 11 (1983).
7. K. Gustavii, Determination of amines and quaternary ammonium ions as complexes with picrate, *Acta Pharm. Suecica 4*: 233 (1967).
8. J. Havel, I. Burešová-Jančářová, and V. Kubáň, Quantitative description of the changes in acid–base properties of bromocresol green in the presence of submicelle and micelle concentrations of the cationic tenside-Septonex, *Collect. Czech. Chem. Commun. 48*: 1290 (1983).
9. E. Wyn-Jones and J. Gormally, eds., *Aggregation Processes in Solution*, Elsevier, Amsterdam, 1983.
10. M. E. Diaz Garcia and A. Sanz-Medel, Dye-surfactant interactions: A review, *Talanta 33*: 255 (1986).
11. V. Kubáň, I. Jančářová, J. Hedbávný, and M. Vrchlabský, Effect of tensides on the optical properties of sulfophthalein dyes, *Collect, Czech. Chem. Commun. 54*: 70 (1989).
12. V. Kubáň, J. Hedbávný, I. Jančářová, and M. Vrchlabský, Spectrophotometric investigation of interactions of sulfophthalein dyes with surfactants, *Collect. Czech. Chem. Commun. 54*: 622 (1989).
13. M. G. Neumann and M. H. Gehlen, The interaction of cationic dyes with anionic surfactants in the premicellar region. *J. Colloid Interface Sci. 135*: 209 (1990).
14. J. Gasparič, unpublished results.
15. Y. A. Beltagy, A. Issa, and S. M. Rida, Colorimetric determination of some organic bases using Tropaeolin OOO, *Pharmazie 31*: 484 (1976).
16. J. Manes Vinuesa and F. Bosch Serrat, Comparative investigation of different sulfonated azodyes for the analysis of alkaloids. I. Determination of

quinine and strychnine in biological fluids with Orange II, *An. Real. Acad. Farm. 48*: 525 (1982).

17. A. Balzsek-Bodó, I. Kiss, and J. Józsa, Study of the formation of lipophilic salts for benzodiazepine derivatives with azo dyes and their use in microdetermination of medazepam, *Farmacia (Bucharest) 33*(1): 15 (1985).
18. F. Bosch Serrat and G. Font Perez, Spectrophotometric determination of codeine and noscapine with Orange II, *An. Real. Acad. Farm. 52*: 637 (1986).
19. B. Pitarch, J. Manes, and F. Bosch, Extractive-colorimetric determination of mono-, di-, tri- and tetracyclic antidepressant using ion pairs, *An. Real. Acad. Farm. 52*: 279 (1986).
20. M. Y. Ebeid, B. A. Moussa, and A. A. Abdel Malak, Dye-salt formation method for the microanalysis of propantheline bromide, *Egypt. J. Pharm. Sci. 28*: 193 (1987).
21. J. Manes, J. Civera, G. Font, and F. Bosch, Spectrophotometric determination of benzodiazepines in pharmaceuticals after ion-pair extraction, *Cienc. Ind. Farm. 6*: 333 (1987).
22. K. Weclawska and A. Regosz, Colorimetric determination of pentoxiverine citrate in some pharmaceutical preparations, *Pharmazie 42*: 483 (1987).
23. C. S. P. Sastry, M. Aruna, M. N. Reddy, and D. G. Sankar, Extractive spectrophotometric determination of some anthelmintics using Fast Green FCF or Orange II, *Indian J. Pharm. Sci. 50*: 140 (1988).
24. F. Bosch Serrat, J. Manes Vinuesa, and M. Cerezo Llorca, Comparative study on different sulfonated azodyes for the analysis of alcaloids. II. Determination of quinine in tonic waters (in Spanish), *An. Real. Acad. Farm 49*: 263 (1983).
25. A. Sedmíková and J. Gasparič, Ion-pair extraction spectrophotometry of drugs using sulfonphthalein dyes as counter-ions (in Czech), *Československ. Farm.*, in press.
26. Z. Popelková-Malá and M. Malát, Extraction-spectrophotometric determination in pharmaceutical analysis. III. An extraction-spectrophotometric determination of some psychopharmaceuticals (in Czech), *Československ. Farm. 34*: 422 (1985).
27. J. Gasparič and M. Filipová, A contribution to the extraction-photometric determination of Septonex using Chromazurol S (in Czech), *Československ. Farm. 39*: 400 (1990).
28. M. Malát, Dissoziations konstanten von chromazurol S, *Anal. Chim. Acta 25*: 289 (1961).
29. M. Malát, Extraction spectrophotometric determination of organic bases with some metallochromic indicators, *Anal. Chim. Acta 109*: 191 (1979).
30. J. Gasparič, J. Šubert, and J. Čižmárik, Extraction-photometric determination of drugs using 2,4,6-trinitrophenol. I. Theoretical part (in Czech), *Československ. Farm. 39*: 458 (1990).
31. J. Šubert, J. Gasparič, and J. Čižmárik, Extraction-photometric determination of drugs using 2,4,6-trinitrophenol. II. Survey of applications (in Czech), *Československ. Farm. 41*: 69 (1992).
32. M. S. Mahrous, M. Abdel Salam, A. S. Issa, and M. Abdel-Hamid, Use of

p-chloranilic acid for the colorimetric determination of some antimalarials, *Talanta 33*: 185 (1986).

33. G. W. Scott, Spectrophotometric determination of cationic surfactants with Orange II, *Anal. Chem. 40*: 768 (1968).

34. H. K. Biswas and B. M. Mandal, Extraction of anions into chloroform by surfactant cations: Relevance to dye extraction method of analysis of long chain amines, *Anal. Chem. 44*: 1636 (1972).

35. H. K. Biswas, A. R. Das, and B. M. Mandal, Interaction of disulphine blue VN 150 dye with detergents in aqueous medium, *J. Surf. Sci. Technol. 1*: 113 (1985).

36. H. Hellman, Spectrophotometric determination of cation-active surfactants, *Z. Anal. Chem. 323*: 29 (1986).

37. I. Mori, Y. Fujita, and T. Enoki, Spectrophotometric determination of long chain quaternary ammonium salts with chlorogallein methyl ester, *Bunseki Kagaku 28*: 330 (1979).

38. T. Sakai, Spectrophotometric determination of quaternary ammonium salts by a flow injection method coupled with thermochromism of ion associates, *Analyst 117*: 211 (1992).

39. S. Z. El-Khateeb and E. M. Abdel-Moety, Determination of quaternary ammonium surfactants in pharmaceutical formulation by the hypochromic effect, *Talanta 35*: 813 (1988).

40. K. Yamamoto and S. Motomizu, Spectrophotometric method for the determination of ionic surfactants by flow-injection analysis with acidic dyes, *Anal. Chim. Acta 246*: 333 (1991).

41. R. A. Lienado and T. A. Neubecker, Surfactants, *Anal. Chem. 55*: 93R (1983).

42. G. F. Longman, *The Analysis of Detergents and Detergent Products*, J. Wiley and Sons, London, 1975, p. 262.

43. J. Jurasová and V. Kubáň, Determination of ionic tensides by photometric titration with equivalence point indicated by means of neutral red (in Czech), *Chem. Listy 81*: 215 (1987).

44. J. B. M. Hendry and H. Read, Photometric determination of cationic surfactants, *Analyst 113*: 1249 (1988).

45. I. Kasahara, M. Kanai, M. Taniguchi, A. Kakeba, N. Hata, S. Taguchi, and K. Geto, Bis[2-(5-bromo-2-pyridylazo)-5-(N-propyl-N-sulfopropylamino)-phenolate] cobaltate (III) as a counter ion for the extraction and spectrophotometric determination of long-chain quartery ammonium salts and tertiary alkylamines in the presence of each other, *Anal. Chim. Acta 219*: 239 (1989).

46. A. Seidel and E. Hoyer, Effect of cationic surfactant on the spectral behavior of the acid triphenylmethane dye pentachrome azure blue and its metal-(III)complexes: Micelle effect or ternary complex formation, (in German) *J. Prakt. Chem. 324*: 299 (1982).

47. T. A. Vasil'chuk, A. T. Pilipenko, and A. I. Volkova, Ionic associate of morin and quercetin with cationic surfactants and their application in analysis, (in Russian) *Ukr. Khim. Zhur. 51*: 278 (1985).

48. A. T. Pilipenko and N. G. Kulichenko, Evaluation of the sensitivity of spec-

trophotometric determination of hydrophobic organic cations by reaction with molybdenum(VI)–pyrogallol red complex in aqueous micellar solutions of non-ionic surfactants, (in Russian) *Dokl. Akad. Nauk SSSR 320*: 115 (1991).

49. T. Sakai, Y. Ohsugi, T. Kamoto, N. Ohno, and H. Sasaki, Spectrophotometric determination of cetylpyridinium chloride in pharmaceuticals by ion-pair extraction with quinidine and bromocresol green, *Bunseki Kagaku 37*: 174 (1988).

50. K. Toei, S. Motomizu, and T. Umano, Extractive spectrophotometric determination of non-ionic surfactants in water, *Talanta 29*: 103 (1982).

51. L. Favretto, B. Stancher, and F. Tunis, Determination of polyoxyethylene alkyl ether non-ionic surfactants in waters at trace level as potassium picrate active substances, *Analyst 105*: 833 (1980).

52. P. T. Crisp, J. M. Eckert, N. A. Gibson, and I. J. Webster, An extraction-spectrophotometric method for the determination of non-ionic surfactants, *Anal. Chim. Acta 123*: 355 (1981).

53. M. A. Pesce and C. S. Strande, New micromethod for determination of protein in cerebrospinal fluid and urine, *Clin. Chem. 19*: 1265 (1973).

54. W. Schaffner and C. Weissmann, A rapid, sensitive, and specific method for the determination of protein in dilute solution, *Anal. Biochem. 56*: 502 (1973).

55. E. Gracia and F. Fernandez-Belda, Ponceau S as a dye for quantitative protein assay: Its use in the presence of Triton X-100, *Biochem. Int. 27*: 725 (1992).

56. M. M. Bradford, A rapid and sensitive method for the quantitation of microgram quantities of protein utilizing the principle of protein–dye binding, *Anal. Biochem. 72*: 248 (1976).

57. J. J. Sedmak and S. E. Grossberg, A rapid, sensitive and versatile assay for protein using coomassie brilliant blue G250, *Anal. Biochem. 79*: 544 (1977).

58. T. Marshall and K. M. Williams, Bradford protein assay and the transition from an insoluble to a soluble dye complex: Effects of sodium dodecyl sulfate and other additives, *J. Biochem. Biophys. Methods 26*: 237 (1993).

59. L. P. Kirazov, L. G. Venkov, and E. P. Kirazov, Comparison of the Lowry and the Bradford protein assays as applied for protein estimation of membrane-containing fractions, *Anal. Biochem. 208*: 44 (1993).

60. V. G. Stavinchuk, V. Ya. Antipov, and I. M. Korenman, Basic dyes as reagents for extraction photometric determination of isomeric anthraquinone monosulfonic acids (in Russian), *Zh. Anal. Khim. 37*: 994 (1982).

61. Z. I. Chalaya and L. S. Mikhailova, Basic dyestuffs as reagents for the determination of chlorotoluenesulfonic acids (in Russian), *Zhur. Anal. Khim. 25*: 1829 (1970).

62. V. V. Kuznetsov and Yh. V. Glukchova, Extraction and photometric determination of naphthalenemonosulfonic acids in industrial mixtures of sulfur-containing compounds (in Russian), *Zavadsk. Lab. 31*: 1324 (1965), ref. *Z. Anal. Chem. 229*: 221 (1967).

63. I. M. Korenman, F. R. Sheyanova, and S. N. Maslennikova, Extraction-

photometric determination of nitrophenols (in Russian), *Zavodsk. Lab. 34*: 1300 (1968); *Probl. Anal. Khim. 1*: 259 (1970).

64. V. Stužka and M. Znojil, Extraction-photometric study of ion associate of fuchsine and picric acid, *Acta Univ. Palackianae Olomucensis Fac. Rer. Natur. Chemica xxvi, 88*: 85 (1987).

65. P. R. Bontchev and P. Papazova, Photometric extraction method for determination of oxacillin and its derivatives, *Mikrochim. Acta II*: 503 (1975).

66. N. Bergisadi, Spectrophotometric determination of rifamycin SV by ion-pair extraction technique, *Eczacilik Bul. 21*: 40 (1979).

67. A. N. Nayak, P. G. Ramappa, H. S. Yathirajan, and S. Manjappa, Spectro-photometric determination of microgram amounts of penicillins by solvent extraction with azure B, *Anal. Chim. Acta 134*: 411 (1982).

68. M. Abdel-Hady ElSayed, Saied F. Belal, Abdel-Fattah, M. Elwalily, and Hassan Abdine, Spectrophotometric determination of tolbutamide in tablets, *J. Assoc. Off. Anal. Chem. 62*: 533 (1979).

69. T. N. V. Prasad, B. S. Sastry, E. Venkata Rao, and C. S. P. Sastry, Extractive spectrophotometric determination of diuretics using basic dyes, *Pharmazie 42*: 135 (1987).

70. A. M. Wahbi, H. Abdine, M. A. Korany, and M. H. Abdel-Hay, Basic fuchsine as an ion-pairing reagent for some acidic drugs, *Analyst 103*: 876 (1978).

71. APHA-AWWA-WPCF, *Standard Methods for Examination of Water and Wastewater*, 14th, Ed., American Public Health Association, Washington, DC, 1975, p. 600.

72. S. Motomizu, S. Fujiwara, A. Fujiwara, and K. Tôei, Solvent extraction-spectrophotometric determination of anionic surfactants with ethyl violet, *Anal. Chem. 54*: 392 (1982).

73. A. M. Frigola Canoves and F. Bosch Serrat, Colorimetric determination of anionic surfactants with acridine orange by ion-pair extraction, *Afinidad* (in Spanish, *44*: 483 (1987).

74. M. Oshima, S. Motomizu, and H. Doi, Interaction of hydrophobic anions with cationic dyes and its application to the spectrophotometric determination of anionic surfactants, *Analyst 117*: 1643 (1992).

75. H. Kubota, M. Katsuki, and S. Motomizu, Batchwise and flow-injection methods for the spectrophotometric determination of anionic surfactants with 4-(4-*N,N-dimethylaminophenylazo)-2-methylquinoline*, *Anal. Sci. 6*: 705 (1990).

76. Y. Shimoishi and H. Miyata, Spectrophotometric determination of anionic surfactants in tap and river waters with 1-(10-bromodecyl)-4-(4-aminona-phthylazo)-pyridinium bromide, *Fresenius' J. Anal. Chem. 338*: 46 (1990).

77. S. Taguchi and K. Goto, Bis[2-(2-pyridylazo)-5-diethylaminophenolato]co-balt(III)chloride as a new extraction and spectrophotometric reagent for trace anions: Determination of sulphated and sulphonated surfactants, *Talanta 27*: 289 (1980).

78. S. Taguchi, I. Kasahara, Y. Fukushima, and K. Goto, An application of bis[2-(5-chloro-2-pyridylazo)-5-diethylaminophenolato]cobalt(III)chloride to the extraction and spectrophotometric determination of sulphonated and sulphated surfactants, *Talanta 28*: 616 (1981).

79. S. Motomizu, M. Oshima, and T. Kuroda, Spectrophotometric determination of anionic surfactants in water after solvent extraction coupled with flow injection, *Analyst 113*: 747 (1988).

80. S. Motomizu, Y. Hazaki, M. Oshima, and K. Tôei, Spectrophotometric determination of anionic surfactants in river water with cationic azo dye by solvent extraction-flow injection analysis, *Anal. Sci. 3*: 265 (1987).

81. V. Kubáň and Z. Vavrouch, Determination of trace concentrations of anionic tensides in water by extraction spectrophotometry (in Czech), *Chem. Listy 78*: 61 (1984).

82. Z. Vavrouch and V. Kubáň, Determination of anionic tensides in waters by using a two-phase titration method (in Czech), *Chem. Listy 78*: 561 (1984).

83. M. Reschke, W. Halwachs, and K. Schügerl, Ion-pair extraction of water soluble dyes, *Chem. Eng. Sci. 37*: 1529 (1982).

84. T. Saito, K. Hagiwara, and K. Higashi, Analysis of a trace of surfactants in water with polymer adsorbents, *Bunseki Kagaku 30*: 319 (1981).

85. T. Saito and K. Hagiwara, Analysis of trace of surfactants in water with anion-exchange resin and polymeric adsorbent, *Fresenius' Z. Anal. Chem. 312*: 533 (1982).

86. L. Sommer, *Analytical Absorption Spectrometry in the Visible and Ultraviolet: The Principles*, Studies in Analytical Chemistry 8, Akadémiai Kiadó, Budapest, 1989, p. 244.

87. Z. Marczenko, Sensitive spectrophotometric methods for elements determination based on some colour ternary systems (in Polish), *Chem. Anal. (Warsaw) 24*: 551 (1979).

88. A. J. Bowd, D. T. Burns, and A. G. Fogg, Analytical aspects of organo-P, As, Sb, S, Se, Te and Sn(IV) (onium) cations, *Talanta 16*: 719 (1969).

89. I. Rudzitis, L. Čermáková, K. Nedomová, and M. Malát, The extraction spectrophotometric determination of aluminum with bromopyrogallol red and diphenylguanidine, *Chem. Anal. (Warsaw) 26*: 1045 (1981).

90. V. Valcl, I. Němcová, and V. Suk, *Handbook of Triarylmethane and Xanthene Dyes: Spectrophotometric Determination of Metals*. CRC Press, Boca Raton, Florida, 1985.

91. T. Yotsuyanagi, H. Hoshino, and K. Aomura, Acid dissociation reaction of palladium-4-(2-pyridylazo)resorcinol complexes, *Anal. Chim. Acta 71*: 349 (1974).

92. Y. Shijo and K. Sakai, Analytical application of oleophilic quaternary ammonium salts. XV. Extraction of the uranium(VI)-4-(2-pyridylazo)resorcinol complex with tridodecylethylammonium bromide, *Bull. Chem. Soc. Japan 51*: 2574 (1978).

93. D. Nonova and S. Pavlova, Extraction-spectrophotometric determination of traces of cadmium with 4-(2-pyridylazo)resorcinol and a long-chain quaternary ammonium salt, *Anal. Chim. Acta 123*: 289 (1981).

94. K. Ueda, Spectrophotometric study on the extraction of 4-(2-thiazolylazo)-resorcinol chelates of cobalt(II), nickel(II), copper(II) and zinc(II) with Zephiramine, *Anal. Letters A11*: 1009 (1978).

95. Lj. Marić, M. Široki, and M. J. Herak, Extraction of copper(II)-4-(2-pyridylazo)resorcinol complexes by tetraphenylphosphonium and arsonium chloride, *J. Inorg. Nucl. Chem. 27*: 2309 (1975).

96. M. Široki, Lj. Marić, M. J. Herak, and C. Djordjević, Solvent extraction of niobium-4-(2-pyridylazo)resorcinol complex and spectrophotometric determination of niobium in oxalato solutions, *Anal. Chem. 48*: 55 (1976).

97. Lj. Marić, M. Široki, Z. Štefanac, and M. J. Herak, Extraction and spectrophotometric determination of vanadium with 4-(2-thiazolylazo)resorcinol and tetraphenylarsonium and phosphonium chlorides, *Microchem. J. 24*: 536 (1979).

98. T. Ozaki, Spectrophotometric study on the extraction of cobalt and iron chelates of 7-nitroso-8-hydroxyquinoline-5-sulfonic acid with Zephiramine, *Anal. Letters 15*: 595 (1982).

99. J. I. Garcia Alonso, M. E. Diaz Garcia, and A. Sanz Medel, The surfactant-sensitized analytical reaction of niobium with 8-hydroxyquinoline-5-sulfonic acid, *Talanta 31*: 361 (1984).

100. S. Motomizu and K. Toei, Selection of counter-cation in the solvent extraction of anionic chelates: Spectrophotometric determination of trace amounts of cobalt with 2-nitroso-1-naphthol-4-sulphonic acid and tetrabutylammonium ion, *Talanta 29*: 89 (1982).

101. V. Božkov, L. Čermáková, and M. Malát, Determination of platinum metals. III. An extraction spectrophotometric determination of tetravalent osmium, *Anal. Letters 12* (A12): 1259 (1979).

102. S. Motomizu, M. Oshima, and A. Arita, Liquid–liquid distribution of ion associates of tetraiodopalladate(II) with quaternary ammonium counter ions, *Anal. Chim. Acta 229*: 121 (1990).

103. E. Matoušková, I. Němcová, and V. Suk, Spectrophotometric determination of gold using Septonex, *Chem. Papers 35*: 501 (1981).

104. Z. Marczenko and K. Jankowski, Sensitive flotation-spectrophotometric determination of gold, based on the gold(I)-iodide-methylene blue system, *Talanta 32*: 291 (1985).

105. I. S. Balog, P. P. Kish, and V. V. Bagreev, Effect of salting-out agents on ion pair extraction of halide complexes of metals with basic dyes: Extraction of chlorocomplexes of indium with triphenylmethane dyes (in Russian), *Zh. Anal. Khim. 44*: 1213 (1989).

106. S. Sato and S. Uchikawa, Extraction and spectrophotometric determination of titanium(IV) with malachite green and *p*-chlormandelic acid, with application to mild steels, *Talanta 33*: 115 (1986).

107. H. Puzanowski-Tarasiewicz, M. Tarasiewicz, and W. Misiuk, Spectrophotometric determination of titanium(IV) with chlorpromazine hydrochloride, *Microchem. J. 29*: 341 (1984).

108. R. A. L. Ferreira, M. L. Enseñat y Berea, and F. Bermejo, Extractive spectrophotometric determination of cobalt(II) as ion association complex with thiocyanate and ephedrine, *Anal. Letters 22*: 1819 (1989).

109. L. Zhongfan and L. Shaopu, Highly sensitive spectrophotometric determina-

tion of trace amounts of uranium(VI) with the thiocyanate-basic triphenylmethane dyes-gum arabic system, *Analyst 116*: 95 (1991).

110. Y. Yamamoto, Determination of anions by solvent extraction with ferroin
and its derivatives, *Japan Analyst 21*: 418 (1972).

111. L. Sommer, *Analytical Absorption Spectrometry in the Visible and Ultraviolet: The Principles*, Studies in Analytical Chemistry 8, Akadémiai Kiadó,
Budapest, 1989, p. 269.

112. M. M. Tanannaiko and L. I. Gorenshtein, Reactions between complex cations of metals and bromophenol blue or bromopyrogallol red (in Russian),
Zhur. Anal. Khim. 37: 589 (1982).

113. A. Hulanicki and J. Nieniewska, Study of ion-association complexes used in
spectrophotometric determination of iron, *Talanta 21*: 896 (1974).

5

Redox Reactions

A. INTRODUCTION

Determinations based on redox reactions represent a broad group of spectrophotometric procedures. These reactions are employed to determine inorganic cations and anions, as well as organic substances. Redox reactions have also been often used as indicator reactions for kinetic catalytic methods, which are among the most sensitive determinations.

A simple redox reaction of substances 1 and 2 can be schematically expressed using the equation

$$n_2 Ox_1 + n_1 Red_2 \rightleftharpoons n_2 Red_1 + n_1 Ox_2 \tag{1}$$

where Ox_1 and Red_1 (Ox_2 and Red_2, respectively) are oxidized and reduced forms of substance 1 (substance 2) and n_1 (n_2) is the number of electrons exchanged during the reaction of substance 1 (substance 2). Substances 1 and 2 can be either organic or inorganic compounds or one of them may be an organic compound and the other an inorganic one. For each of these substances, the redox potential value can be expressed by the Peters equation:

$$E = E^f + \frac{RT}{nF} \ln \frac{[Ox]}{[Red]} \tag{2}$$

where [Ox] ([Red]) is the molar concentration of oxidized (reduced) form of the substance. Formal redox potential E^f is a measure of oxidizing or reducing ability of a substance under given conditions. In analytical chemistry, the formal redox potential is often used instead of the standard redox potential $E°$, which is defined based on activities of the reaction components.

The initial compounds represent the analyte and the reagent. The analyte may occur in a form capable of either oxidation or reduction, that is, it can be determined using the action of oxidizing or reducing reagents. Besides, a certain substance may be an analyte on one occasion and a reagent on another, as is true in complexation reactions. However, the substance applied as the reagent is always used in excess.

The relation between the redox reaction equilibrium constant and the formal redox potential of substances 1 and 2 is described by Eq. (3):

$$\ln K = \frac{(E_1^f - E_2^f)n_1 n_2}{RT/F} \tag{3}$$

The reaction is quantitative when the K value is greater than 10^6. With that corresponds a difference of more than 0.35 V between the formal redox potentials (under the exchange of one electron in the redox reaction).

In some cases, other substances also participate in the redox reaction. For instance, in the reduction of the Ox form, the following reaction might take place:

$$a\,Ox + b\,B + n\,e \rightleftharpoons c\,Red + d\,C \tag{4}$$

The reduction of MnO_4^- ions in acidic medium is probably the best-known example of this type of reaction:

$$MnO_4^- + 8\,H^+ + 5\,e \rightleftharpoons Mn^{2+} + 4\,H_2O \tag{5}$$

The Peters equation for redox potential of a substance concerned thus has the following form:

$$E = E^f + \frac{RT}{nF} \ln \frac{[Ox]^a [B]^b}{[Red]^c [C]^d} \tag{6}$$

In solution, the substances participating in redox reactions may also undergo side reactions, which then affect the value of the redox potential. Most frequently, the side reactions involved are complexation and acid-base reactions [1]. If, for instance, the oxidized, as well as the reduced form of a transition metal that takes part in the redox reaction provides complexes with ligand L, then the value of formal redox potential of this complex-bound metal E_L^f is given by the equation:

$$E_L^f = E_M^f - \frac{RT}{nF} \ln (\beta_{ox} - \beta_{red}) \tag{7}$$

where E_M^f is the formal redox potential of the free metal, and β_{ox} and β_{red} are overall stability constants of the complexes of oxidized and reduced forms of the metal (with the same number of ligands).

From the above equation it is apparent that if the oxidized form of the metal forms a more stable complex, the redox potential of the complex-bound metal decreases, while a more stable complex of the reduced metal form causes the redox potential to increase. For instance, for the reaction

$$Fe(III) + e \rightarrow Fe(II) \tag{8}$$

the value of $E^f = 0.77$ V (in water). In iron complexes with organic as well as inorganic anions, the complex Fe(III) has a higher stability constant, and thus, for instance, for the complex of iron with H_3PO_4, the value of E^f is lowered to 0.44 V (in 0.3 mol L^{-1} H_3PO_4). In reaction with neutral ligands, the formation of complex with Fe(II) is preferred [1], so that, for instance, for the Fe $(phen)_3^{3+}$/Fe $(phen)_3^{2+}$ system, $E^f = 1.04$ V (in 1 mol L^{-1} H_2SO_4).

In substances with a similar structure, the dependence of redox potential values of the complex-bound metal on pK_a of ligand has been observed. For instance, for the iron complexes with 5-substituted-1,10-phenanthrolines, the stability of Fe(II) complexes increases with increasing basicity, and therefore, their redox potential values also increase.

In series of ligands of this type, the effect of ligand substituents on the redox potential value has been found. The 5-nitro derivative of ferroin (i.e., of the Fe(II) complex with phenanthroline) has $E^f = 1.25$ V; for 5-methylferroin it is 1.02 V, and for 4,7-dimethylferroin it is 0.88 V (all in 1 mol L^{-1} H_2SO_4) [1]. Thus, the quantitative course of reaction between analyte and reagent can be affected using complex formation with a suitable ligand. This feature is often employed in practical analyses.

The combination of redox and complexation reactions is of special importance in kinetic catalytic determinations. Here the redox reaction serves as an indicator reaction. The analyte is the catalyst of reaction; usually it is a transition metal that is able to form coordination compounds with individual reagents of the indicator reaction. For instance, Ti(IV) ions catalyze primarily the redox reactions of oxygen-containing ligands, with which they readily form complexes (see Chapter 3); similarly Ag(I) ions catalyze primarily the reactions of nitrogen-containing ligands. Through the selection of a suitable indicator reaction, it can be achieved that only one of the metals of the analyzed mixture will catalyze the reaction, which means increased selectivity of metal catalytic determinations [2].

Organic compound can also be used as the catalyst if one of the reactants in the indicator reaction is a metal ion with which a complex formation of organic compound occurs.

In most catalyzed redox reactions, a change of catalyst oxidation number occurs in the course of reaction. If, for instance, the metal M^{n+} is a catalyst that reacts with the organic reagent and if one electron is exchanged in the course of reaction, the reaction could be described by the following equations:

$$Ox_1 + M^{n+} \quad \rightleftharpoons \quad Red_1 + M^{(n+1)+} \tag{9}$$

$$M^{(n+1)+} + Red_2 \rightleftharpoons Ox_2 + M^{n+} \tag{10}$$

The reaction is catalyzed if the formal redox potential of the couple $M^{(n+1)+}/M^{n+}$ (E_M^f) is lower under the reaction conditions than the formal redox potential of oxidizing agent Ox_1/Red_1 (E_{ox}^f), and higher than the potential of reducing agent Ox_2/Red_2 (E_{red}^f), i.e., $E_{ox}^f > E_M^f > E_{red}^f$. The same reaction applies if the catalyst reacts with the reducing agent.

In all cases, the direct reaction between Ox_1 and Red_1 is subject to kinetic hindrances, although the reaction must be possible according to the laws of thermodynamic whereas the catalyzed reaction M^{n+} with Ox_1 takes place fast.

As has been said in Chapter 1, Section B, the catalyst efficiency can be influenced by its reaction with a suitable substance. The efficiency of the resulting product is either greater than that of the free catalyst (i.e., the added substance acts as an activator), or it is lower (i.e., the substance acts as a reaction inhibitor). If metal ions are used as redox reaction catalysts, which is most frequently the case, organic compounds able to form complexes with these metals are then used as the activators or inhibitors. This way of affecting the reaction rate can be utilized to determine the concentrations of mentioned organic compounds [2,3].

From the above it is clear that in catalyzed redox reactions, and in reactions using an activator or inhibitor, the electron exchange reactions are closely linked with complexation reactions. With the knowledge of formation constant values of complexes formed, it is possible to predict the effect of ligands on metal ion catalysis.

B. CLASSIFICATION OF SPECTROPHOTOMETRIC REDOX REACTIONS

As with complexation reactions, redox reactions can be classified based on different perspectives. But the situation is more complicated due to the fact that while in complexation reactions the complex formed as the product of the reaction between analyte and reagent is used for determinations (with minor exceptions represented by indirect determinations), in redox reactions, the reaction products simultaneously include the oxidized (or reduced) form of analyte, as well as the reduced (or oxidized) form of reagent. Changes in absorbance of one of the reactants or products, induced by the reaction, can be employed in determinations.

In this chapter, spectrophotometric redox reactions are classified based on the type of change (oxidation or reduction) of the analyte in the course of reaction, which allows us to cover common features of these reactions. At the same time, the classification includes the spectral properties of the analyte, the reagent, and of their products.

The classification is based on the simplified redox equation

$$Ox_1 + Red_2 \rightleftharpoons Red_1 + Ox_2 \tag{11}$$

where 1 is used for the analyte, and 2 for the reagent. The metal ion, its complex, or the organic compound can be both the analyte and the reagent.

In equilibrium determinations, all redox reactions can be included within one of the following groups:

1. Reduction of analyte (Ox_1) by reagent (Red_2)
2. Oxidation of analyte (Red_1) by reagent (Ox_2)

In both cases, the redox reactions can be classified as follows:

a. The spectrophotometrically active analyte product is formed and evaluated
b. The spectrophotometrically active reagent product is formed and evaluated
c. The spectrophotometrically active reagent is used and its concentration (i.e., absorbance) decrease is evaluated
d. The reagent excess, or the reagent product, is determinated using other spectrophotometric reactions.

For some determinations, it is possible to perform spectrophotometric assessments of reactions by applying several of the above approaches (e.g., through the absorbance measurement of analyte reaction product, or through the reagent absorbance decrease).

In catalytic kinetic determinations, the analyte serves as the catalyst; thus it is not formally possible to include these reactions in one of the above groups. The indicator reaction in kinetic measurements, however, is again assessed using the above classification.

Reactions in which oxidation or reduction of the analyte serves only to transfer it into some other form suitable for another type of reaction (e.g., for complexation reaction) are not included in this chapter.

C. REDUCTION OF ANALYTE BY REAGENT

1. Spectrophotometrically Active Analyte Product Is Formed and Evaluated

This type of reaction is best represented by the reduction of metal ions to their metallic forms. An example is the well-known method of gold determination, which involves the use of ascorbic acid, $SnCl_2$, or Hg_2Cl_2 in the reduction of Au(III) ions to colloid solution of metallic gold (the so-called gold–tin purple, $\lambda_{max} = 440$ nm) [4]. Hg(I) provides a similar reaction. Selenites are reduced to metallic selenium through the action of hydroxylamine, hydrazine, sulfites, ascorbic acid, or tin(II) chloride. Selenium precipitation from solution is prevented by the addition of protective colloid. The reaction of Se(IV) with $SnCl_2$ is catalyzed by molybdenum ions, and thus it is the basis for the kinetic method for molybdenum determination ($\lambda_{max} = 390$ nm) [5]. The reduction of Ag(I) ions by Fe(II) is catalyzed by cysteine, which therefore is employed in its determination. The measurement is performed in the absorption maximum of metallic silver colloid solution ($\lambda_{max} = 530$ nm) [6].

Molybdates and their heteropolyacids (phosphomolybdic, silicomolybdic, arsenomolybdic, germanomolybdic acids), formed in acidic medium in the reaction of MoO_4^{2-} with appropriate anions (see Chapter 3), are reduced to so-called molybdenum blue through the action of reducing agents ($SnCl_2$, ascorbic acid, hydrazine sulfate). Molybdenum blue is a mixture of oxides with the medium oxidation state of molybdenum V–VI. The composition is usually described as Mo_9O_{26} or Mo_8O_{23}, or even $Mo_8O_{23} \cdot 8H_2O$. In the visible region, molybdenum blue provides a broad absorption band between 650 and 850 nm. The absorption maximum wavelength differs with respect to the reducing agent and the reaction conditions used [7].

The above reaction is frequently applied in the determination of PO_4^{3-}, SiO_3^{2-}, AsO_4^{3-}, Ge(IV), and As(III) in the presence of Ge(IV) [8]. It can also be employed to determine phosphorus-, arsenic-, or silicon-containing organic compounds, which first need to be appropriately converted to phosphates, arsenates, and silicates (by melting with sodium carbonate; combustion in oxygen atmosphere; or action of HCl, HNO_3, or $HClO_4$). After the addition of ammonium molybdate, heteropolyacids are formed, which are then reduced to molybdenum blue. Increased sensitivity is achieved by the extraction of molybdenum blue into an organic solvent (esters, alcohols, ketones, ethers, and their mixtures).

Many kinetic methods of metal determination are based on the catalytic influencing of the reduction of molybdate to molybdenum blue. They include, for example, determination of phosphorus in reduction with ascorbic acid [9] and determination of germanium, and phosphorus [10], and silicon [11] in reduction by iodide ions. The reduction by ascorbic acid is also catalyzed by Sb(III) ions. Absorbance measurement at 660 nm using the FIA method has been proposed for Sb(III) determination [12].

The determination of the herbicide Paraquat by reduction with ascorbic acid to a blue radical ion is an example of the determination of an organic analyte by an organic reagent [13].

2. Spectrophotometrically Active Reagent Product Is Formed and Evaluated

Variously substituted arylamines, phenols, suitable dyes, and some other compounds (e.g., phenothiazine derivatives) are often used as reagents to determine metal ions with oxidizing abilities.

Oxidation products of the reagents are colored compounds, and although they have not yet been identified in some cases, they are used in determinations. Based on current knowledge, several common features can be found in the redox reaction mechanisms of these structurally different compounds.

Variously stable radicals (their existence has been confirmed by EPR spectroscopy in a number of cases) are the product of one-electron oxidation of these reagents. Depending on their structure (primarily on the kind of substituents affecting the electron density in the conjugated system and the radical reactivity) and the properties of the medium (especially its acidity), either a further oxidation, polymerization, disproportionation, or hydrolysis of radicals occurs. Consecutive reactions of some compounds of this type are reviewed in Ref. [2]. Reactions of inorganic reagents are usually simpler, because the side reactions and consecutive reactions do not occur as often.

Some frequently used organic reducing agents and examples of their application are as follows.

a. Aromatic amines

Colorless benzidine (1) is oxidized by an analyte to form the colored oxidation product (2). This reaction has been used, for instance, to determine chlorine (after the reduction of chlorate). The determination is performed at $\lambda_{max} = 438$ nm [14].

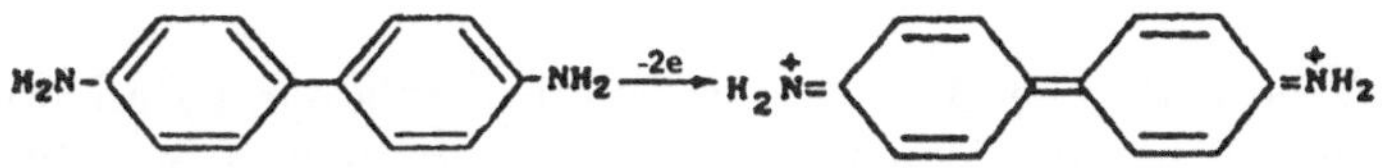

$$H_2N-\langle\!\rangle-\langle\!\rangle-NH_2 \xrightarrow{-2e} H_2\overset{+}{N}=\langle\!\rangle-\langle\!\rangle=\overset{+}{N}H_2 \qquad (12)$$

$$(1) \qquad\qquad\qquad (2)$$

Other commonly used derivatives react in the same way, for example, *o*-tolidine (3,3′-dimethylbenzidine), which is used for Au(III), VO_3^-, Cl_2, Br_2, I_2, and ClO_3^- determination [4], or *N,N,N′,N′*-tetramethyl-*o*-tolidine (tetron), which has been proposed for the determination of Au(III), Ce(IV), and BrO_3^- ($\lambda_{max} = 485$ nm) [15].

Analogously, organic peracids can be determined with benzidine as the reagent [16–19]: in the medium of ethanol and acetate buffer (pH 5–6), the reagent is oxidized to 4,4′-diaminoazobiphenyl, the absorbance of which is measured at 400–430 nm [18]. *o*-Tolidine can be applied in the same way using ethanol–phosphate buffer (pH 5.5) [18]. The determination of organic peroxides with *m*-phenylenediamine as the reagent in ethanolic solution and buffer (pH 2–3) at 50–60°C is supposed to be based on the formation of the Bandrowski base [18] (for its structure see Ref. [20]). The analogous reaction with *p*-phenylenediamine is catalyzed by aldehydes and enables their determination [18]. A further determination of organic peroxides is based on the oxidation of *N,N*-dimethyl-*p*-phenylenediamine at pH 7.5–9.1 [18].

Diphenylamine (**3**) (colorless) is oxidized by an analyte to benzidine violet (**4**) (the colorless diphenylbenzidine is an intermediate):

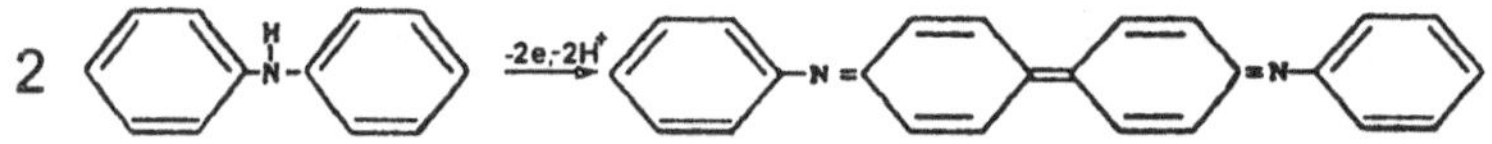

$$(13)$$

(3)   **(4)**

Diphenylamine is employed to determine Ce(IV), Au(III), VO_3^-, ClO_3^-, and Cl_2 [4].

Variamine Blue B (4-amino-4′-methoxydiphenylamine) (**5**) is the often-used derivative of diphenylamine, which on oxidation becomes an intensely blue-violet–colored imine (**6**) (for further reactions see [2]).

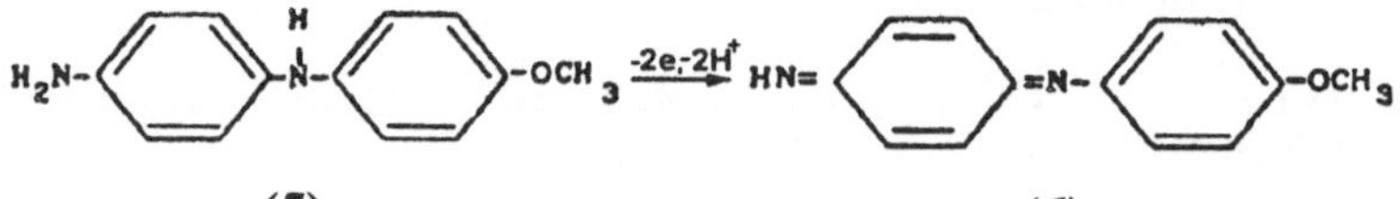

$$(14)$$

(5)   **(6)**

Variamine Blue B has been used to determine I_2, VO_3^-, ClO_4^-, and Au(III) [4].

Au(III) ions also catalyze the oxidation of this dye by ammonium persulfate, which has been used for kinetic determination of gold [21]. Variamine Blue oxidation by H_2O_2 is catalyzed by iodide ions, the concentration of which can be determined in this way [22].

Other aromatic amines and diamines are utilized especially for catalytic determinations. N,N'-Diethylaniline oxidation by KIO_4, when tetraethylbenzidine is formed, is catalyzed by Mn(II) (λ_{max} = 475 nm) [23]. Naphthylamine oxidation by $KBrO_3$ is catalyzed by molybdenum ions [24]. Fe(III) ions can be determined based on their catalytic effect on the reaction of p-phenetidine (4-ethoxyaniline) with KIO_4. The reaction is activated by 2,2'-bipyridyl [25], and inhibited by organic substances containing thiol or thione groups. The latter feature has been employed to prepare a method for the determination of thiocarbamide, dithiocarbamates, mercaptans, and some pesticides (a review on the determination of organic compounds by kinetic catalytic methods is given in Ref. [26]). p-Phenetidine oxidation by $KClO_3$, catalyzed by V(V) and activated by phenol, has been proposed for vanadium determination when the reaction product absorbance is measured at λ_{max} = 510 nm [27]. Sensitivity of the V(V) determination, which is based on the vanadium effect on the reaction of p-phenetidine with $KBrO_3$ in the presence of catechol, is enhanced by micellar catalysis (using cetylpyridinium bromide) by an order of magnitude [28].

The reaction of o-dianisidine (3,3'-dimethoxybenzidine) with H_2O_2 is catalyzed by the ions of Cr(VI) and Cr(III). The sensitivity of determination may be increased by a suitable activator. Using Cr(VI), the activation involves substances that accelerate its reduction to Cr(V); using Cr(III), it involves the substances that support its deaquation and the complex formation with the oxidant [29]. This reaction is also catalyzed by Mn(II) ions; their determination is performed at 440 nm [30].

The reaction of diaminophenol with H_2O_2 is catalyzed by Fe(III) ions. This catalytic action is inhibited by oxalates, citrates, and fluorides, which form complexes with Fe(III). The absorbance measurement of the reaction product at λ_{max} = 500 nm may be utilized for determination of these anions [31]. In oxidation by hydrogen peroxide, p-phenylenediamine provides a dark-colored imine of bis(2',5'-diaminophenyl)benzoquinone. The reaction is catalyzed by I^- ions and formaldehyde, which are used in their determination [32].

On oxidation by H_2O_2, 3,5-diaminobenzoic acid first provides a yellow intermediate, which later becomes a red-colored product (λ_{max} = 540 nm). The reaction is catalyzed by Fe(III) ions, which can be determined on this basis [33]. The sulfanilic acid oxidation by potassium periodate is catalyzed by Fe(III), here 1,10-phenanthroline acts as an activator by the complex formation with Fe(III). Thus, the reaction has been suggested for Fe(III) ion determination. Products of sulfanilic acid oxidation include o-benzoquinone-4-sulfonic acid (λ_{max} = 368 nm) and a small quantity of azobenzene-4-4'-disulfonic acid (λ_{max} = 446 nm). The Fe(III) determination is performed at λ = 368 nm, where the absorbance is higher for catalyzed as well as uncatalyzed reactions [34].

b. Aromatic hydroxy compounds

Hydroquinone is oxidized by hydrogen peroxide. The reaction is catalyzed by Cu(II) ions and thus has been employed in their determination. Heterocyclic amines act as the reaction activators, and their effect depends on electron donor properties of the activator, as well as of the solvent. Using pyridine, the catalytic effect of Cu(II) in solvents increases in the sequence acetonitrile < acetone < dimethylformamide < dimethylsulfoxide (measured at $\lambda_{max} = 453$ mn) [35]. Using 2,2′-bipyridyl, the most suitable medium is 30% (v/v) dimethylformamide-water ($\lambda_{max} = 480$ nm). This method has been applied for Cu(II) determination in blood serum after the separation of albumin [36].

In oxidation by H_2O_2, Tiron (for its structure see Chapter 4, Section A) provides a yellow-colored semiquinone ($\lambda_{max} = 440$ nm) that is stable for several hours. The reaction is catalyzed by Co(II) and Mn(II) ions and can be used for their determination [37]. A mathematical model has been developed for the determination of metal ions using this reaction.

In oxidation by H_2O_2, chromotropic acid (for its structure see Chapter 3, Section A) provides a yellow-colored product ($\lambda_{max} = 440$ nm). Fe(III) and Cu(II) ions catalyze the reaction, and it has been used for their determination [38].

c. Leucobases of basic dyes

These compounds are colorless reduced forms of dyes and may be used to determine analytes with oxidizing effects. Malachite Green leucobase **(7)**, for instance, turns into its blue-green–colored cationic form **(8)** during the course of the reaction ($\lambda_{max} = 620$ nm).

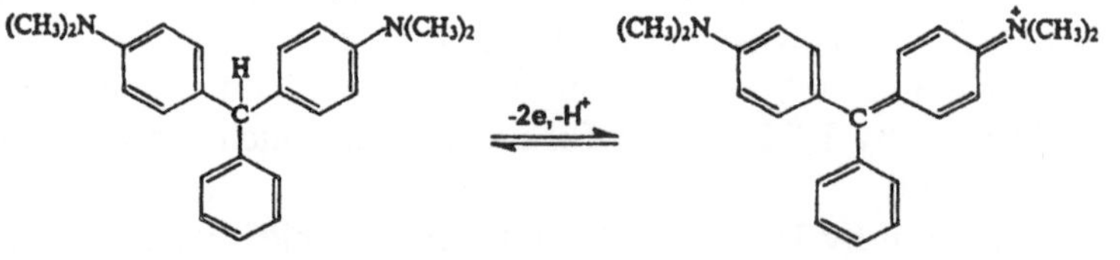

$$-2e, -H^+$$

(15)

(7)   (8)

Leuco Malachite Green has been used to determine Au(III) ions that oxidize it to Malachite Green [4]. It has also been used in a number of kinetic catalytic determinations. For instance, its oxidation by KIO_4 is catalyzed by Mn(II) ions. Thus, the reaction has been employed to determine manganese in seawater using the FIA method, with preconcentration into 8-hydroxyquinolinol immobilized onto a vinylpolymer gel [39].

Bindschendler's green leucobase **(9)** is oxidized to its colored form

(10) (λ_{max} = 730 nm) by H_2O_2. The reaction is catalyzed by Cu(II) ions and can be utilized for their determination (as the commercial dye provides a high blank value due to degradation during storage, it is better to prepare the leucobase by a simple reaction directly in the analyzed solution) [40]. The action of cationic and anionic surfactants in this redox reactions is an example of the positive effect of micellar catalysis, which accelerates the formation of colored product and enhances its stability and, at the same time, increases the dye absorbance, and thus the sensitivity of determination.

(9)   (10)   (16)

Methylene Blue leucobase (11) after oxidation turns into its colored cationic form (12) (λ_{max} = 660 nm). Yellow semiquinone, which is stable only in 10 mol L^{-1} H_2SO_4, forms an intermediate between these two forms.

(11)   (12)   (17)

Methylene Blue leucoform has been used to determine organic peroxides; the H_2O_2 that they liberate oxidizes it to the colored form (12) in acidic medium [41]. Methylene Blue reaction can also be employed in the kinetic determination of technetium, which catalyzes the reduction with $SnCl_2$ [42].

Leuco Thionine Blue (see structure (11) with $-NH_2$ groups instead of $-N(CH_3)_2$) has also been used as reagent in the determination of IO_4^-, CrO_4^{2-}, and other oxidants (λ = 670 nm) [43].

d. Phenothiazine derivatives

Considerable attention has been paid to the redox properties of this extensive group of compounds of the basic structure (13), R_2 = $-H$, $-Cl$, $-CF_3$, $-OCH_3$, $-CO-CH_3$, R_{10} = $-H$, $-alkyl$, $-(CH_2)_n-N(alkyl')_2$, $-(CH_2)_n$-piperidyl (piperazyl, respectively) (see Chapter 3), because it is assumed that oxidation processes play a decisive role in their pharmaceutical use, and the photochemical oxidation has been studied in relation to the construction of photoelectric cells. Some of these compounds have also been proposed as indicators in volumetric redox determinations [44].

Through the action of oxidizing agents (Ce(IV), H_2O_2, $Cr_2O_7^{2-}$, PbAc$_4$), these compounds are first subject to one-electron reversible oxidation [45], when the relatively stable red, violet, or even blue radical cations (14) are formed with λ_{max} = 500–600 nm. In the second one-electron step, the colorless sulfoxide (15) is irreversibly formed.

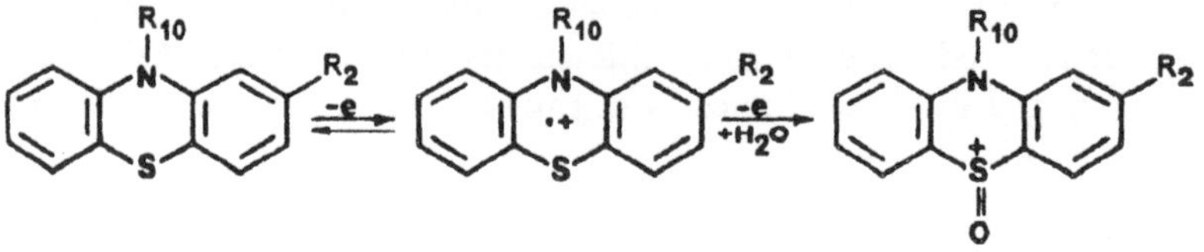

(18)

(13)   (14)   (15)

The kinetics of phenothiazine derivatives oxidations have also been studied, and reaction rate constants have been determined for oxidations by VO_3^-, BrO_3^-, $Fe(CN)_6^{3-}$, Co(III), Mn(III), and Np(IV) in aqueous medium, as well as for oxidations by Ce(IV) and Fe(III) in anhydrous acetonitrile [44].

The formation of intensely colored radical cations of variously substituted phenothiazine derivatives has been employed in the determinations of Fe(III), Au(III), Os(VIII), Os(VI), V(V), and Ce(IV) and in indirect determinations of As(III) and NO_2^- [44]. The determinations are performed at absorption maximum wavelengths of the formed cation radicals, which differ with respect to the phenothiazine derivative used (500–600 nm).

The cation radical remains sufficiently stable only in a strongly acidic medium. In media with a lower H^+ ion concentration, unsubstituted phenothiazine is subject to a number of consecutive and side reactions [45]. Under optimal conditions, intensely green dimer cation (λ_{max} = 460 and 635 nm) is formed. This reaction has been utilized in the determination of gold(III) which is reduced to Au(I) (16) [46].

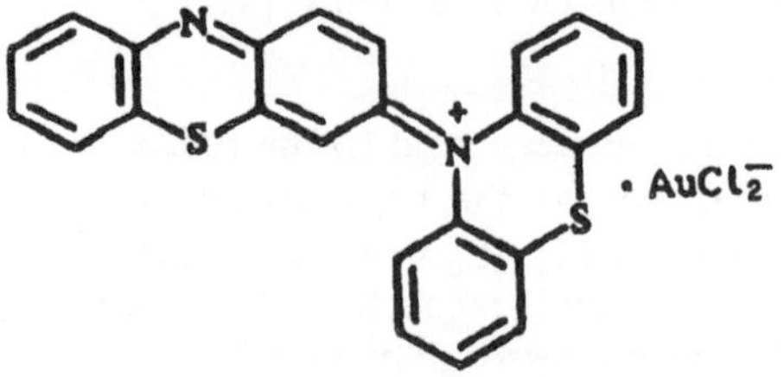

(16)

In aqueous medium, the radical cations of heterocyclic N-substituted phenothiazine derivatives ($PD^{+\cdot}$) are subject to reaction with nucleophilic anions (e.g., with anions of buffer salts) that gives colorless products [47]. The disproportionation reaction can also occur with the formation of the original derivative (PD), and its sulfoxide (PDO), both of which are also colorless.

$$2\,PD^{+\cdot} + H_2O \rightarrow PD + PDO + 2\,H^+ \tag{19}$$

Apart from the H^+ ion concentration and the kind and concentration of salts present [48], the $PD^{+\cdot}$ stability is also affected by the kind and concentration of the surfactants present [49]. This effect is based on the ability of micelles of the surfactant concerned to bind or repel the H^+ ions. The localization of $PD^{+\cdot}$ in surfactant micelles has also been studied [50].

3. Spectrophotometrically Active Reagent is Used and its Concentration Decrease is Evaluated

This way of determination is performed in the reduction of analyte by colored reagent which forms colorless oxidation products. As reducing agents, organic dyes are commonly used, as high values of their molar absorption coefficients allow very sensitive determinations. In their oxidation, the dye is completely degraded; the resulting products are not identified.

Pyrocatechol Violet (a chelation dye, see Chapter 3), for instance, provides colorless products in the oxidation by H_2O_2. The reaction is catalyzed by iodide ions and can be used for their kinetic determination [51]. It has been found that iodide determination is more sensitive when chloramine B is used as the oxidizing reagent [52]. The reaction of Pyrocatechol Violet with H_2O_2 is catalyzed by acetonitrile and has been proposed for its determination ($\lambda_{max} = 450$ nm) [53]. The same reaction catalyzed by Cu(II) has been used to determine a number of organic compounds [26], which have an inhibiting effect on this catalysis. This effect can be explained by the formation of a 1:1 complex of Cu(II) with these compounds, which is catalytically less active than Cu(II). Many amino acids, as well as thyroxine and 5-hydroxytryptophane, can be determined in this way. On the other hand, adrenaline, noradrenaline, and some phenothiazine antihistaminics form a complex with Cu(II) ions which exhibits greater catalytic effects than the free Cu(II) ions. The same applies for the anesthetic sodium thiobarbiturate [26]. In all cases, the absorbance decrease is measured at $\lambda = 480$ nm.

The indicator redox reaction of Pyrocatechol Violet with H_2O_2 is also catalyzed by Co(II) ions. This catalytic action is inhibited by histamine, some other compounds, and the vitamins of B group. These effects can serve as the basis for determinations of all the above compounds [26].

The oxidation of Pyrogallol Red by persulfate is catalyzed by Pb(II) ions, giving the colorless products. Hence, the reaction can be used for determination of Pb(II) ions [54].

Methyl Orange is oxidized by H_2O_2 under the catalytic action of Cr(VI). The reaction is activated by citric, salicylic, oxalic, hydroxybenzoic, and *p*-aminobenzoic acids and their salts, and it is utilized for Cr(VI) determinations ($\lambda_{max} = 490$ nm) [55]. Methyl Orange oxidation by bromate is inhibited by bromide ions in the first reaction phase; in the next phase it is catalyzed. This effect has been employed to determine Br^- ions (the absorbance decline of Methyl Orange has been measured at 510 nm) [56].

Malachite Green oxidation by KIO_4, accompanied by the absorbance decline at 620 nm, is catalyzed by Mn(II) ions and can be utilized for their determination [57]. Nitrilotriacetic acid (NTA) and EGTA, which forms a complex with Mn(II) promote the catalytic action. This effect can be used to determine the above acids [58]. Other aminopolycarboxylic acids can be determined in this way by so-called 'catalymetric titration' [59].

Methylene Green oxidation by periodate has been proposed for the catalytic determination of Mn(II). 1,10-Phenanthroline acts as the reaction activator; the determination is performed based on the dye's absorbance decrease at 620 nm [60].

The determination of Ag(I) is based on its catalytic effect on the oxidation of the dye Direct Green by persulfate. The determination sensitivity is increased considerably by the presence of the activators ethylene diamine and triethylenetetraamine [61]. Similarly, Ag(I) ions catalyze the oxidation of antipyrine-azo-8-hydroxyquinoline by persulfate. The rate-determining step is the formation of Ag(II). 2,2'-Dipyridyl acts as the activator of reaction (the formal redox potential of Ag(II)/Ag(I) couple in its presence decreases to 0.6 V). This increases the catalytic activity of silver by a few orders of magnitude [62]. Sulfur compounds (sulfides, thioacetamide, thiourea, thiosulfate) catalyze the reaction of indigocarmine with H_2O_2, and this reaction may be used in their determination. The catalytic effect is increased by the addition of Fe(II) or Al(III), which can be determined in this way. Because Al(III) is deactivated by F^-, this effect may also be used for F^- determination [63].

The inhibiting effect of tetracyclines on the azorubine S oxidation by H_2O_2, catalyzed by Mo(VI), may be exploited in kinetic determinations of tetracyclines [26]. The alizarine S oxidation by H_2O_2, catalyzed by Co(II), is inhibited by 8-hydroxyquinoline. The same reaction catalyzed by Mn(II) is inhibited by bacteriostatics and fungicides of 8-hydroxyquinoline type. All the above substances can be determined by this principle [26].

The oxidation of L-ascorbic acid by atmospheric oxygen is catalyzed by Cu(II) ions. The reaction leads to a decrease of the absorption band at $\lambda = 265$ nm. Complexing agents (cysteine, 2-aminoethanethiol, salicylic acid,

EDTA, 1,10-phenanthroline, ethylene diamine) inhibit the reaction. This phenomenon can be employed in determinations of these substances [64].

4. Reagent Excess, or Reagent Product, Is Determined Using Other Spectrophotometric Reactions

This relatively uncommon method may be illustrated by the determination of H_2O_2 and organic peroxides and hydroperoxides, as well as Ce(IV), through reduction by Fe(II). The addition of 1,10-phenanthroline or bathophenanthroline to the solution to be analyzed, gives, with the surplus Fe(II), an intensely red coloration, the absorbance of which is utilized to determine the concentration of H_2O_2 [65].

The quantity of Fe(III) ions produced in the reaction of H_2O_2 with Fe(II) ions is equivalent to the amount of H_2O_2. The Fe(III) ions can be determined by the reaction with potassium rhodanide or salicylic acid [66]. The determination of both aliphatic and aromatic nitro compounds is based on the same principle [19].

It is also possible to use reduction by Fe(II) in the determination of Cr(VI). A spectrophotometric method has been proposed in which the Fe(III) ions formed react with Tiron through the complex formation with $\lambda_{max} = 650$ nm [67].

D. OXIDATION OF ANALYTE BY REAGENT

1. Spectrophotometrically Active Analyte Product Is Formed and Evaluated

The determination of iodide ions by oxidizing agents can serve as an example of this type of reaction. In this reaction, free iodine is liberated, the concentration of which can be determined using one of the following ways:

Direct measurement at its absorption maxima (295, 365, or 400–410 nm)
Measurement after extraction into an organic solvent ($CHCl_3$, CCl_4, toluene, xylene)
Measurement of blue coloration of the iodine–starch complex ($\lambda_{max} = 600$ nm)
Utilization of the formation of mono- or diiodo derivatives of dyes, extractable into organic solvents (e.g., using Methyl Violet and extraction into toluene, $\lambda_{max} = 600$ nm)

These methods of spectrophotometric measurements are also applied in other modifications. For instance, the oxidation of I^- by bromine proceeds up to IO_3^-. After the evaporation of excess bromine and the addition of cadmium iodide, free iodine is formed, the concentration of which is measured after the addition of starch solution [68].

The iodide ion oxidation by H_2O_2 is catalyzed by the ions Mo(VI) and

W(VI). In the presence of citric acid, the catalytic effect of Mo(VI) is increased, while the catalytic effect of W(VI) disappears. This phenomenon has been used to determine molybdenum and tungsten in the presence of each other [69]. Conversely, organic peroxides and hydroperoxides can be determined by the reaction with KI or tetraalkylammonium iodides, where the free iodine that is liberated can be evaluated photometrically [19]; or after reaction with iodide in the cationic micellar medium provided by cetylpyridinium chloride (CPC), the triiodide ion produces associates with CPC micelles, giving rise to absorption at 500 nm [70].

The determination of Mn(II) ions is often performed by the oxidation to colored permanganate ions (using $K_2S_2O_8$, $NaBiO_3$, PbO_2, or KIO_4 in either H_2SO_4 or HNO_3). The measurement is performed at $\lambda_{max} = 525$ or 545 nm in the visible region; in the ultraviolet region at 310 nm. Leuco Malachite Green has been found to increase the sensitivity of Mn(II) ion determination by KIO_4 oxidation. The MnO_4^- ions formed by the analyte oxidation oxidize this colorless dye form to intensely colored Malachite Green ($\lambda_{max} = 620$ nm). The determination is 200 times more sensitive than the direct absorbance measurement of MnO_4^- ions [71].

Measurement of absorbance is also used in the determination of organic substances that are based on analyte oxidation to various colored products. Oxidation is performed by $KMnO_4$, $K_2Cr_2O_7$, $Ce(SO_4)_2$, $AgNO_3$, PbO_2, H_2O_2, KIO_3, KIO_4, $K_3Fe(CN)_6$, and $FeCl_3$ [17,19]. The following examples of determination can be given: of diphenylamine and its nitro derivatives by Ce(IV) sulfate resulting in the formation of diphenylamine blue ($\lambda_{max} = 570$ nm); using methods based on oxidation with $FeCl_3$($\lambda_{max} = 530$ nm) and $K_3Fe(CN)_6$ [72], or sodium periodate in sulfuric acid medium [73]; of diclofenac by different oxidizing agents [74]; of catecholamines by oxidation with H_2O_2, KIO_4, and $K_3Fe(CN)_6$ [17,19]; of *m*-aminophenol by oxidation of KIO_3; of ascorbic acid by oxidation with $K_3Fe(CN)_6$; and of benzidine and its derivatives, which are oxidized by hypochlorite in phosphate buffer (pH 3–4) to the corresponding halophenoquinones [17,19].

Oxidation to intensely colored radical cations has also been used for determinations of a number of phenothiazine derivatives by oxidizing agents [44,75]. These reagents include HNO_3 Fe(III) [19], H_2O_2 [76], KIO_4 [77], Cu(II) [78], 2-iodoxybenzoate [79], VO_3^- [80], $K_2Cr_2O_7$ [81], and Ce(IV) [82] and are used with the FIA method. The measurement of the absorbance of colored reaction products of analyzed organic compounds is performed even in cases when oxidation products have not been identified.

The conversion of organic analytes to colored products can also be achieved by organic oxidizing agents (e.g., with chloramines B and T [83–85]. *N*-Bromosuccinimide has also been used as the reagent in the determination of oxidizable organic compounds [86,87].

Oxidation of thiosemicarbazone derivatives (see Chapter 2, Section B), for example, autooxidation of 1,4-dihydroxyphthalimide dithiosemicarbazone [88] and oxidation by H_2O_2 of 4,4'-dihydroxybenzophenone thiosemicarbazone [89], is catalyzed by the ions of Mn(II) and Cu(II). Hence, these ions may be determined in this way. In oxidation, the elimination of electrons occurs probably on oxygen atoms of the above derivatives. The absorbance measurement is performed at $\lambda = 594$ nm for the former derivative, and at $\lambda = 415$ nm for the latter.

2. Spectrophotometrically Active Reagent Product Is Formed and Evaluated

Determinations based on this principle can be illustrated using the ferriin/ferroin system (Fe $(phen)_3^{3+}$/Fe $(phen)_3^{2+}$), which is also an example of the connection between redox and complexation reactions. Fe(II) complex with phenanthroline is more stable than the Fe(III) complex, hence the E^f value of this system is higher than that of the Fe(III)/Fe(II) system (i.e., Fe(III) phen$_3$ is a stronger oxidation agent than Fe(III)).

The ferriin/ferroin system is suitable primarily for the determination of substances that can be oxidized [90]. Their reaction with Fe(III) hydroxocomplex present in aqueous medium gives the red-colored, stable Fe(II)-phen$_3$ complex with a high value of ϵ at $\lambda = 510$ nm, which is suitable for determination:

$$Fe_2(III)(OH)_2phen_4^{4+} + 2\,phen + Red_1 \rightleftharpoons$$
$$2\,Fe(II)phen_3^{2+} + Ox_1 + 2\,OH^- \qquad (20)$$

(Red$_1$ and Ox$_1$ are the reduced and oxidized form of the analyte). The reaction is kinetically hindered, and therefore, in most cases, it must be performed at a higher than room temperature (60°C).

The above reaction has proved suitable primarily for the determination of phenols (with the exception of the derivatives that are substituted by groups with I-effect (i.e., nitro- and chloro- derivatives, salicylic acid, etc.), diphenols (also for dyes with aromatic -OH groups, e.g., triphenylmethane and azo dyes), naphthols, ascorbic acid, aliphatic enols (acetylacetone), aromatic amines, and thiols. Of the inorganic compounds, SO_3^{2-}, $S_2O_5^{2-}$, $S_2O_3^{2-}$, $S_2O_4^{2-}$, S^{2-}, $Fe(CN)_6^{4-}$, I^-, NO_2^-, NH_2OH, and N_2H_4 react at higher than room temperature [91].

The ferroin formed can be also extracted into organic solvents in the form of an associate with suitable anions (e.g., ClO_4^-, I^-), and in this way be used to determine these anions [90].

Another example of determinations based on the absorbance of the reaction product measurement is the group of reactions with molybdenum heteropolyacids used for the determination of reducing organic compounds. Molybdenum blue is the product such reactions, the absorbance of which is evaluated (see Section B of this chapter). The reagents used are known by the names Folin-Ciocalteu or Folin-Denis reagent. They are used to determine phenols, reducing steroids, reducing sugars, proteins, organic sulfur compounds, drugs, and so forth [17,19,92–99].

The application of the Folin-Ciocalteu reagent in the presence of Cu(II) salts is suitable for protein determinations (Lowry reaction). The presence of Cu(II) salts enhances the sensitivity; the selectivity of the determination, however, is low [19].

Another way of employing heteropolyacids in the determination of organic bases is based on the reaction in which they yield water-insoluble precipitates. Following the separation, the precipitates are reduced by $TiCl_3$ or $SnCl_2$ again into molybdenum blue [17].

A colored product of the reduced reagent is also measured in the oxidation of organic compounds by $K_2Cr_2O_7$, when the green-colored salt of Cr(III) is formed ($\lambda_{max} = 600$ nm) [17,19].

Different aromatic di- and polynitro compounds are reduced by reducing organic compounds in alkaline media to deeply colored quinonoid compounds — salts of the corresponding nitroso-nitroderivatives. Several methods for the determination of reducing sugars or hydrazines are based on this reaction [18].

5,5'-Dithiobis-(2-nitrobenzoic acid), the so-called Ellman reagent, is reductively split into the yellow 2-carboxy-4-nitrobenzenethiolate anion ($\lambda_{max} = 412$ nm) when reacting in alkaline media with thiols [18,100–103]. 2-Carboxy-4-nitrophenylalkyl(aryl)disulfide is formed as a by-product. Besides the determination of thiols, the method can be applied also to all compounds that can be transformed to thiols, for instance, to disulfides after their reduction. Similarly, bis-(4-nitrophenyl)disulfide [18] or 2,2'-dithiobis-(5-nitropyridine) [104] can be used as the reagent.

Tetrazolium salts (e.g., 1,3,5-triphenyltetrazolium chloride (17) form colorless solutions in ethanol. In alkaline media under mild conditions these solutions are easily reduced to intensively colored lipophilic and mostly insoluble formazans (e.g., triphenylformazan (18), Eq. (21)), which are dissolved by addition of pyridine + HCl.

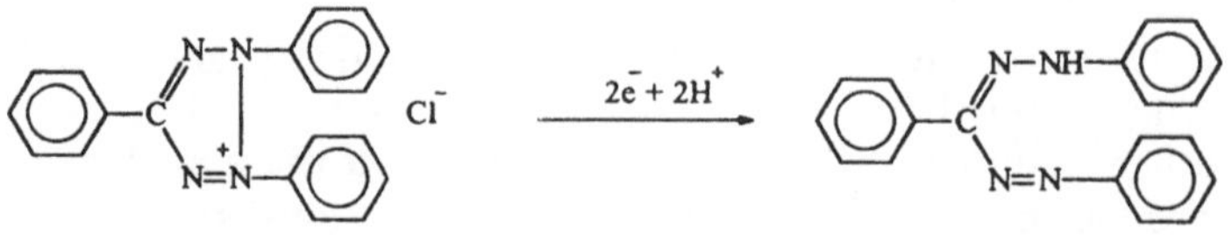

$$2\bar{e} + 2H^+ \qquad (21)$$

(17)   (18)

The quantity of the colored formazan is directly proportional to the concentration of the analyte. The molar absorptivity of triphenylformazan is in the range of 1.5×10^4 and determinations of different reducing compounds can be achieved, for example, of reducing sugars [16,18,105,106], ascorbic acid [16,18], reducing steroids [16,18,107–109], hydrazines and hydroxylamines [110–112], o-diphenols [113,114]. Several different tetrazolium salts are commercially available as reagents. Triphenyltetrazolium chloride (TTC) and tetrazolium blue (TB) are the most commonly used. The λ_{max} for the former is 485 nm, for the latter 525 nm. Bis(chlorobenzenesulfonic acid)tetrazolium salt T 32 has been introduced as a water-soluble reagent for steroids and ascorbic acid [115].

The reaction rate of individual reducing common hexoses and pentoses is different, so that the analysis of binary mixtures of certain sugars by a differential reaction rate technique is possible [116]. The reaction of corticosteroids with tetrazolium salts proceeds via a two-electron oxidation of the side chain [117]. Thiols are oxidized to disulfides (Eq. 22):

$$2\,R-SH \rightarrow R-S-S-R \tag{22}$$

while primary and secondary amines must be converted into dithiocarbaminates prior to the reaction with the reagent (Eq. 23):

$$R-NH-R' + CS_2 \rightarrow RR'N-CS-SH \tag{23}$$

and phenoxymethylpenicillin is nitrated prior to the reaction with TTC [118].

On the other hand, indirect determinations with tetrazolium salts are performed with organic peroxides and hydroperoxides. These compounds are treated with a known amount of excess Fe(II)sulfate and the unoxidized Fe(II)salt determined with the tetrazolium salt in the presence of sodium fluoride [18]. (This determination formally belongs into Section C.4 of this Chapter.)

3. Spectrophotometrically Active Reagent Is Used and Its Concentration Decrease Is Evaluated

This approach is applied in cases when the analyte oxidation is performed by colored reagent. The use of $KMnO_4$ to determine organic substances is such a case. With respect to medium acidity and the character of the substance being determined, MnO_4^- is reduced in this reaction to Mn(II) (in strongly acidic medium) or to MnO_2 (in slightly acidic, neutral, or slightly alkaline medium). The reaction in strongly alkaline medium that gives MnO_4^{2-} is not used for organic substances. The oxidation by potassium permanganate is employed in determinations of ethanol, adrenaline, noradrenaline, and amines [17,19].

When using $Cr_2O_7^{2-}$, the decrease of its absorbance in reaction with the analyte is often used. The measurement is performed in acidic medium at different wavelengths in the visible as well as the ultraviolet region [17,19].

A similar situation exists when the yellow-colored $Ce(SO_4)_2$ (or $(NH_4)_2Ce(SO_4)_3$) is used for the determination of alkaloids and steroids [17,19]. As(III) oxidation by Ce(IV), which is relatively slow, is catalyzed by ruthenium and therefore has been proposed for its determination [119]. The same reaction is also catalyzed by I$^-$ and Os(VIII) ions, and inhibited at Ag(I) and Hg(II) ions. All of these ions can be determined based on the absorbance change at $\lambda = 420$ nm [120]. The determination of Hg(II) [121] and Au(III) [122] ions can be also be done based on their inhibiting effect on As(III) oxidation by Ce(IV) catalyzed by I$^-$.

The effect of the micellar media of cationic surfactant dodecyltrimethylammonium bromide on the rate of As(III) oxidation by Ce(IV) is an example of how micellar catalysis can be used in chemical analysis. The reaction rate is enhanced in this media in both the presence and absence of iodide catalyst; the determination of Ce(IV) has a sensitivity and detection limit similar to the reaction catalyzed by iodide ions, but it achieves a higher selectivity. The selectivity of iodide determination is also improved [123]. The reaction of As(III) with Ce(IV) is also catalyzed by hormones containing bound iodine. Automated determination of these substances is performed at 365 nm [124].

The possibility of measuring the reagent absorbance decrease is also provided by $K_3Fe(CN)_6$, which is used to determine reducing sugars, phenols, and thiols. The determination is usually performed in a slightly alkaline medium [16,17]. The oxidation of CN$^-$ ions by this reagent has been applied to catalytically determine Cu(II). The measurement is performed at 422 nm [125].

The strong oxidative ability of Ag(II) ions (generated electrochemically in the flow system) has been used for the determination of inorganic as well as organic compounds. The decrease of the reagent absorbance is measured at 390 nm, where Ag(II) ions absorb strongly [126].

The colored organic reagents are also often used for analyte oxidation. 2,6-Dichlorophenolindophenol is an example of an organic dye that is reduced by ascorbic acid to its leucoform. The determination of ascorbic acid is based on measuring the absorbance decrease of the reagent [127]. For similar applications see Refs. [128,129].

Some colored organic radicals have also been used for the determination of organic compounds. For example, the absorbance decrease measurement of the colored chlorpromazine cation radical (prepared from chlorpromazine by oxidation with 2-iodoxybenzoic acid) due to the reduction to colorless chlorpromazine by the analyte has been used for the determination of isoniazid [130].

1,1-Diphenyl-2-picrylhydrazyl (19) is a free stable radical, which forms a red-violet solution in methanol (λ_{max} = 500–525 nm). The solution of this reagent is decolorized by phenols, thiols, and amines [131–136] in the medium of acetate buffer (pH 5). The acid hydrogen of the analyte is withdrawn by the reagent under formation of the yellow 1,1-diphenyl-2-picrylhydrazine (20), Eq. 24, and the corresponding phenoxy, thiol, and amine radicals. Finally, the thio radicals are converted into disulfides, Eq. 25:

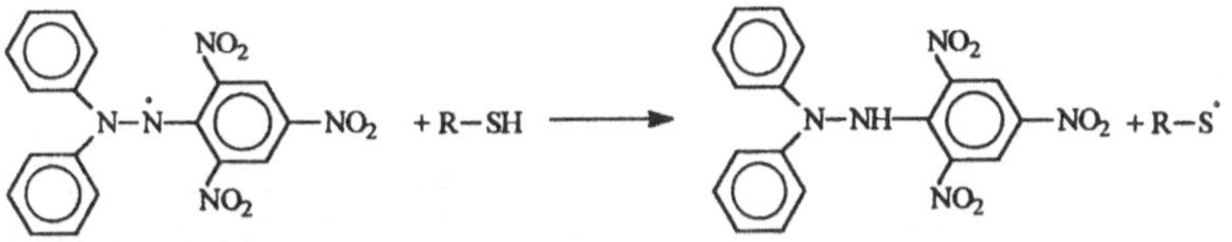

$$2\,R-S\cdot \rightarrow R-S-S-R \tag{25}$$

4. Reagent Excess, or Reagent Product, Is Determined Using Other Spectrophotometric Reactions

The oxidation of organic compounds (e.g., benzoic, ascorbic, citric, lactic, oxalic, and tartaric acids) by Ce(IV) ions belongs among the reactions that use this approach. The residual Ce(IV) can be determined from the oxidation of added ferroin (λ_{max} = 426 nm) [137]. Similarly, in the reactions of As(III) with Ce(IV), which are catalyzed by ruthenium, osmium, copper, and iodide ions, surplus Ce(IV) is determined from the reaction with the ferroin [138].

In the oxidation of U(IV) by Fe(III) ions in the presence of 1,10-phenanthroline, i.e. of ferriin, the Fe(II) ions formed give the red-colored ferroin, the concentration of which is proportional to U(IV) content [139]. Also many procedures for the determination of organic analytes are based on their oxidation with Fe(III) or Cu(II) ions in the presence of reagent yielding colored products with the resulting Fe(II) or Cu(I) ions (e.g., in the presence of 1,10-phenanthroline, 2,2'-dipyridyl, or neocuproine [140–143].

The oxidation of organic reducing substances (aldoses, ketoses, ascorbic acid, corticosteroids, diphenols, etc.) by $Fe(CN)_6^{3-}$ leads to the formation of $Fe(CN)_6^{4-}$. Its quantity, equivalent to the analyte quantity, may be determined by reaction with Fe(III) ions, when the complex Prussian Blue is formed with a broad absorption band between 600 and 700 nm [16,17]. The amount of $Fe(CN)_6^{4-}$ formed in the reaction of reducting sugars with $Fe(CN)_6^{3-}$ can also be determined (using the FIA method) by the reaction with the mixture of Fe(III) ions and 1,10-phenanthroline (i.e., with ferriine), the reduction of which gives the intensely colored ferroin [144]. Another option for determining $Fe(CN)_6^{4-}$ involves its reaction with arsenomolybdic acid, which yields molybdenum blue [17].

The organic analyte can also be oxidized by N-bromosuccinimide, and the excess of the oxidant determined, for example, by oxidative coupling using metol and isoniazid as the reagents [145].

REFERENCES

1. D. D. Perrin, *Organic Complexing Reagent: Structure, Behavior and Application to Inorganic Analysis*, Interscience, J. Wiley, New York, 1964.
2. P. R. Bontchev, Catalytic reactions. I. Mechanism, *Talanta 17*: 499 (1970).
3. P. R. Bontchev, Catalytic reactions. II. Activation, *Talanta 19*: 675 (1972).
4. Z. Marczenko, *Spectrophotometric Determination of Elements*, WNT and Ellis Horwood, Chichester, 1986.
5. G. D. Christian and G. J. Patriarche, Catalytic determination of molybdenum in blood and urine, *Anal. Lett. 12*: 11 (1979).
6. A. K. Babko, L. V. Markova, and T. S. Maksimenko, Photometric determination of subnanogram amounts of cystine by its catalytic effect on the reduction of silver ions (in Russian), *Zh. Anal. Khim. 23*: 1268 (1968).
7. B. E. Reznik and L. P. Tsyganok, The properties of heteropoly blue obtained on the reduction of phosphomolybdic acid with thiourea (in Russian), *Zh. Anal. Khim. 19*: 584 (1964).
8. S. Rosolowski, Determination of trace amounts of phosphorus by kinetic method, *Chem. Anal. 15*: 157 (1970).
9. V. S. S. Rao, S. C. S. Rajan, and N. Venkateswara Rao, Spectrophotometric determination of arsenic by molybdenum blue method in zinc-lead concentrates and related smelter products after chloroform extraction of iodide complex, *Talanta 40*: 653 (1993).
10. I. I. Alekseeva and I. Nemzer, Determination of germanium(IV) and phosphorus(V) by a kinetic method when they are present together in solution (in Russian), *Zh. Anal. Khim. 25*: 1118 (1970).
11. R. P. Morozova and L. V. Il'enko, A kinetic method for determination of microgram amounts of silicon in solution (in Russian), *Zh. Anal. Khim. 28*: 1835 (1973).
12. N. Lacy, G. D. Christian, and J. Ruzicka, Flow injection method for antimony(III) based on its enhancement of the molybdenum blue reaction, *Anal. Chim. Acta 224*: 373 (1989).
13. P. Shivhare and V. Gupta, Spectrophotometric method for the determination of paraquat in water, grain and plant materials, *Analyst 116*: 391 (1991).

14. E. A. Burns, Spectrophotometric determination of chlorate impurities in ammonium perchlorate. Determination of reducing and oxidizing impurities in hydrochloric acid, *Anal. Chem. 32*: 1800 (1960).
15. N. Jordanov and Ch. Daiev, *N,N,N',N'*-Tetramethyl-o-Tolidin (tetron) als Reagenz zur Bestimmung geringen Mengen von Oxydationsmitteln, *Talanta 10*: 163 (1963).
16. Z. J. Vejdělek and B. Kakáč, *Farbreaktionen in der spektrophotometrischen Analyse organischer Verbindungen. Band I. Organische Farbreagenzien*, Fischer Verlag, Jena, 1969.
17. Z. J. Vejdělek and B. Kakáč, *Farbreaktionen in der spektrophotometrischen Analyse organischer Verbindungen. Band II. Anorganische Farbreagenzien*, Fischer Verlag, Jena, 1973.
18. Z. J. Vejdělek and B. Kakáč, *Farbreagenzien in der spektrophotometrischen Analyse organischer Verbindungen, Organische Farbreagenzien*, Suppl. Vol. I, Fischer Verlag, Jena, 1980.
19. Z. J. Vejdělek and B. Kakáč, *Farbreaktionen in der spektrophotometrischen Analyse organischer Verbindungen, Anorganische Farbreagenzien*, Suppl. Vol. II. Fischer Verlag, Jena, 1982.
20. J. F. Corbett, Benzoquinoneimines III: The structure of Bandrowski's base, *J. Soc. Dyers Colour. 85*: 71 (1969).
21. O. A. Bilenko, N. B. Potekhina, and S. P. Mushtakova, Kinetic determination of gold by the oxidation of variamine blue with ammonium persulphate (in Russian), *Zh. Anal. Khim. 39*: 104 (1984).
22. S. U. Kreingol'd, L. I. Iosenkova, A. A. Pantheleimonova, and L. V. Lavrelasvili, Indicator reactions for kinetic determination of iodide ions in acid media (in Russian), *Zh. Anal. Khim. 33*: 2168 (1978).
23. I. Ya. Kolotyrkina, L. K. Shpigun, and Yu. A. Zolotov, A flow-injection system for catalytic spectrophotometric determination of manganese in sea water (in Russian), *Zh. Anal. Khim. 43*: 284 (1988).
24. M. Otto and H. Müller, Selektivitätssteigerung katalytischer Bestimmungsverfahren: Extractionskatalymetrische Molybdänbestimmung, *Talanta 24*: 15 (1977).
25. I. F. Dolmanova, V. I. Rychkova, and V. M. Peshkova, On the mechanism of catalytic action of iron in the oxidation of organic substrates with periodate (in Russian), *Zh. Anal. Khim. 32*: 1387 (1977).
26. G. A. Milovanovic, Determination of organic substances by kinetic-catalytic methods of analysis, *Microchem. J. 28*: 437 (1983).
27. P. R. Bontschev, Catalytic determination of trace amounts of vanadium, *Mikrochim. Acta*: 577 (1962).
28. M. Loreto Lunar, S. Rubio, and D. Pérez-Bendito, Combination of micellar and chemical catalysis as a means of enhancing the sensitivity of catalytic kinetic determinations, *Anal. Chim. Acta 237*: 207 (1990).
29. I. F. Dolmanova and T. N. Shekhovtsova, On the mechanism of activator action in reaction between *o*-dianisidine and hydrogenperoxide catalysed by chromium (in Russian), *Zh. Anal. Khim. 32*: 1154 (1977).

30. I. F. Dolmanova, G. A. Zolotova, and M. A. Ratina, Kinetic determination of manganese(II) using *o*-dianisidine oxidation with hydrogen peroxide in dimethylformamide-aqueous medium (in Russian), *Zh. Anal. Khim. 33*: 1356 (1978).

31. G. S. Vasilikiotis, C. Papadopoulos, D. G. Themelis, and M. C. Sofonions, Indirect kinetic determination of oxalate, citrate and fluoride ions, *Microchem. J. 28*: 431 (1983).

32. N. P. Evmiridis and M. I. Karayannis, Determination of formaldehyde using a kinetic-spectrophotometric method. Part I. Oxidation of *p*-phenylenediamine with hydrogen peroxide, *Analyst 112*: 831 (1987).

33. A. CH. Zotou and C. G. Papadopoulos, Indicator reaction for the kinetic-spectrophotometric determination of nanogram amounts of iron, *Analyst 112*: 787 (1987).

34. A. A. Alexiev and A. M. Stoyanova, A new catalytic reaction for the determination of nanomolar concentrations of iron(III), *Anal. Lett. 21*: 1515 (1988).

35. I. F. Dolmanova, Q. I. Melnikova, and T. N. Shekhovtsova, Kinetic determination of copper(II) by reaction of oxidizing hydroquinone with hydrogenperoxide in organo-aqueous media (in Russian), *Zh. Anal. Khim. 33*: 2096 (1978).

36. I. F. Dolmanova, O. I. Melnikova, G. I. Tsyzin, and T. N. Shekhovtsova, Effect of some organic solvents on catalytic activity of copper in oxidation of hydroquinone with hydrogenperoxide in presence of various activators (in Russian), *Zh. Anal. Khim. 35*: 728 (1980).

37. M. Otto and G. Werner, Mechanistic studies for modeling the metal ion-catalyzed Tiron-hydrogenperoxide indicator reaction, *Anal. Chim. Acta 147*: 255 (1983).

38. D. G. Themelis and G. S. Vasilikiotis, Kinetic determination of trace amounts of copper(II) using its catalytic effect on the oxidation of chromotropic acid by hydrogen peroxide, *Analyst 112*: 797 (1987).

39. J. A. Resing and M. J. Mottl, Determination of manganese in seawater using flow injection analysis with on-line preconcentration and spectrophotometric detection, *Anal. Chem. 64*: 2682 (1992).

40. L. Lunar, S. Rubio, and D. Pérez-Bendito, Micellar catalysis in kinetic methods of analysis: Improvement of spectrophotometric catalytic determination of copper, *Talanta 39*: 1163 (1992).

41. L. Dulog, Bestimmung organischer Peroxide, *Z. Anal. Chem. 202*: 192 (1964).

42. F. Grases and J. G. March, Determination of technetium by reduction of methylene blue with tin(II), *Anal. Chem. 57*: 1419 (1985).

43. C. Martinez-Lozano, T. Pérez-Ruiz, V. Tomás, and E. Yagüc, Flow injection determination of oxidants with leuco Thionine Blue, *Analyst 113*: 1057 (1988).

44. I. Němcová, N. Zimová-Šulcová, and I. Němec, Oxidation reduction properties of phenothiazine derivatives (in Czech), *Chem. Listy 76*: 142 (1982).

45. P. Hanson and R. O. C. Norman, Heterocyclic free radicals. IV. Some

reactions of phenothiazine, two derived radicals and phenothiazin-5-ium ion, *J. Chem. Soc. Perkin Trans. III*: 264 (1973).

46. I. Němcová, P. Rychlovský, and E. Kleszczewska, A spectrophotometric study of the determination of gold with phenothiazine, *Talanta 37*: 855 (1990).

47. H. Y. Cheng, P. H. Sackett, and R. L. McCreery, Kinetics of chlorpromazine cation radical decomposition in aqueous buffers, *J. Am. Chem. Soc. 100*: 962 (1978).

48. I. Jelínek, I. Němcová, and P. Rychlovský, Effects of salts on the stability of the cationic radical of phenothiazine derivatives, *Talanta 38*: 1309 (1991).

49. I. Němcová and I. Jelínek, The influence of some surfactants and inorganic salts on the stability of diethazine cation radical, *Chem. Papers 47*: 149 (1993).

50. I. Němcová, J. Novotný, and V. Horská, Spectrophotometric study of phenothiazine derivatives and their cation radicals in micelar media, *Microchem. J. 34*: 180 (1986).

51. K. Oiwa, T. Kimura, H. Makino, and L. Y. Kinoskita, Determination of microamounts of iodide ion by the catalytic reaction, *Japan Analyst 17*: 805 (1968).

52. E. I. Yasinskiene and Umbrazhkyunaite, Determination of iodides in soil by kinetic methods (in Russian), *Zh. Anal. Khim. 30*: 962 (1975).

53. G. A. Milovanovic and N. Božilovic, A catalytic method for the determination of acetonitrile, *Microchem. J. 27*: 345 (1982).

54. R. G. Anderson and B. G. Brown, The determination of lead in mosses by means of its catalytic effect on the persulphate oxidation of pyrogallol red, *Talanta 28*: 365 (1981).

55. E. I. Jasinskiene and E. B. Bilidiene, The use of carboxylic acids as activators for the determination of microamounts of chromium(VI) by a kinetic method (in Russian), *Zh. Anal. Khim. 23*: 143 (1968).

56. R. A. Hasty, E. J. Lima, and J. M. Ottaway, Bromate oxidation of methyl orange: Basis for the kinetic determination of bromide, *Analyst 102*: 313 (1977).

57. H. A. Mottola and C. R. Harrison, Sensitivity and detectability for manganese(II) determination in solution by kinetic methods of analysis, *Talanta 18*: 683 (1971).

58. H. A. Mottola and G. L. Heath, Detection and variable time kinetic determination of micro and submicrogram amounts of nitrilotriacetic acid, *Anal. Chem. 44*: 2322 (1972).

59. H. A. Mottola, Titration of microgram amounts of aminopolycarboxylic acids with manganese(II) and catalytic end-point indication, *Anal. Chem. 42*: 630 (1970).

60. H. Cordoba, P. Viñas, and C. Sanchez-Pedreño, Kinetic determination of traces of manganese in different materials by its catalytic effect on the methylene green-periodate reaction, *Talanta 33*: 135 (1986).

61. E. I. Yasinskiene and A. K. Rasevichute, The use of certain activators during

determination of microamounts of silver by a kinetic method (in Russian), *Zh. Anal. Khim. 25*: 458 (1970).

62. L. M. Matat, I. B. Mizetskaya, V. K. Pavlova, and A. T. Pilipenko, Kinetic determination of silver based on oxidation of antipyrine-azo-8-hydroxy-quinoline with potassium persulphate (in Russian), *Zh. Anal. Khim. 37*: 2165 (1982).

63. H. Weisz, S. Pantel, and G. Marquardt, Catalytic-kinetic absorptiostat technique with the indigocarmine-hydrogen peroxide reaction at the indicator reaction, *Anal. Chim. Acta 143*: 177 (1982).

64. A. Mottola, M. S. Haro, and H. Freiser, Use of metal ion catalysis in detection and determination of microamounts of complexing agents, *Anal. Chem. 40*: 1263 (1968).

65. B. Baily and D. F. Boltz, Differential spectrophotometric determination of hydrogen peroxide using 1,10-phenanthroline and bathophenanthroline, *Anal. Chem. 31*: 117 (1959).

66. M. Malát, *Absorption Inorganic Photometry* (in Czech), Academia, Prague, 1973.

67. K. A. Abdullah, Y. I. Hassan, and W. A. Bashir, Indirect spectrophotometric determination of chromium(VI), *Microchem. J. 27*: 319 (1982).

68. W. H. Crough, A spectrophotometric determination of iodine in silicate rocks, *Anal. Chem. 34*: 1698 (1962).

69. I. I. Alekseeva, L. P. Ruzinov, E. G. Khachaturyan, and L. M. Chernyshova, Kinetic determination of molybdenum(VI) and tungsten(VI) in the presence of each other (in Russian), *Zh. Anal. Khim. 35*: 60 (1980).

70. L. Lunar, S. Rubio, and D. Pérez-Bendito, Use of triiodide-cetylpyridinium chloride micellar system for the determination of benzoylperoxide in pharmaceutical preparations, *J. Pharm. Sci. 83*: 407 (1994).

71. S. H. Yuen, Determination of traces of manganese with leucomalachite green, *Analyst 83*: 350 (1958).

72. N. N. Mikhailova, N. P. Tolstopyatova, and L. K. Skripko, Photometric determination of *p*-hydroxydiarylamines (in Russian), *Zh. Anal. Khim. 37*: 477 (1982).

73. J. P. Rawat and P. Bhattacharji, Reactions of some nitrogen containing aromatic compounds with sodium periodate: Spectrophotometric determination of brucine, *p*-anisidine, benzidine, diphenylamine, *N*-phenyl-1-naphthylamine and indole, *Proc. Natl. Acad. Sci., India, Sect. A. 47*: 133 (1977); *Chem. Abstr. 89*: 156962p (1978).

74. B. V. Kamath and K. Shivram, Spectrophotometric determination of diclofenac sodium via oxidation reactions, *Anal. Lett. 26*: 903 (1993).

75. V. G. Belikov and G. F. Moiseeva, Pharmaceutical analysis of phenothiazines based on oxidation reactions (in Russian), *Farmatsiya (Moscow) 35*: 87 (1986).

76. H. Basińska, H. Puzanowska-Tarasiewicz, and M. Tarasiewicz, Colorimetric determination of perfenazine, perazine and promazine with H_2O_2, *Chem. Anal. 15*: 405 (1970).

77. L. Kužmicka, H. Puzanowska-Tarasiewicz, and M. Tarasiewicz, Spectrophotometric determination of phenothiazines with potassium periodate, *Pharmazie 43*: 288 (1988).

78. P. Rychlovský and I. Němcová, A novel spectrophotometric determination of phenothiazine by copper dichloride (in Czech), *Československ. Farm. 37*: 101 (1988).

79. S. M. Hassan, F. Belal, F. Ibrahim, and F. A. Aly, Colorimetric determination of some *N*-substituted phenothiazine derivatives using 2-iodoxybenzoate, *Anal. Lett. 22*: 1485 (1989).

80. S. M. Sultan, Flow injection method for the assay of phenothiazine neuroleptics in pharmaceutical preparations using ammonium metavanadate, *Analyst 116*: 177 (1991).

81. S. M. Sultan, Computer assisted optimization of a flow-injection method for the assay of promethazine, chlorpromazine and trimeprazine in drug formulation, *Talanta 40*: 681 (1993).

82. J. Martinez Calatayud and T. Garcia Sancho, Spectrophotometric determination of promethazine by flow injection analysis and oxidation by Ce(IV), *J. Pharm. Biomed. Anal. 10*: 37 (1994).

83. G. R. Rao, Y. P. Rao, and I. R. K. Raju, Spectrophotometric determination of amodiaquine hydrochloride in pharmaceutical dosage forms, *Analyst 107*: 776 (1982).

84. F. Bosch, J. Manes, G. Font, and J. Salmeron, Colorimetric determination of sulfonamides with chloramine T, *Cienc. Ind. Farm. 3*: 239 (1984); *Chem. Abstr. 103*: 183637a (1985).

85. M. A. Korany, D. Heber, and J. Schnekenburger, Colorimetric determination of *p*-aminophenol with 3-cyano-*N*-methoxypyridinium perchlorate, *Talanta 29*: 332 (1982).

86. V. N. Pathak and I. C. Shukla, Microgram determination of some antipyrines with *N*-bromosuccinimide, *J. Indian. Chem. Soc. 60*: 206 (1983).

87. T. Hassib Sonia, M. Safwat Hany, and I. El-Bagry Ramzeia, Spectrophotometric determination of some antiinflammatory agents using *N*-bromosuccinimide, *Analyst 111*: 45 (1986).

88. D. Pérez-Bendito, M. Valcarcel, M. Ternero, and F. Pino, Kinetic determination of traces of manganese(II) by its catalytic effect in the autooxidation of 1,4-dihydroxyphthalimide dithiosemicarbazone, *Anal. Chim. Acta 94*: 405 (1977).

89. J. L. Ferrer-Herranz and D. Pérez-Bendito, Kinetic determination of traces of copper(II) by its catalytic effect on the oxidation of 4,4'-dihydroxybenzophenone thiosemicarbazone by hydrogen peroxide, *Anal. Chim. Acta 132*: 157 (1981).

90. S. Koch and G. Ackermann, Application of redox reaction in spectrophotometry. I. The iron(III)/1,10-phenanthroline complex as a reagent for the determination of some anions and organic compounds, *Talanta 39*: 687 (1992).

91. S. Koch, G. Ackermann, and P. Lindner, Application of redox reaction in spectrophotometry. II. Detection and spectrophotometric determination of

phenolic compounds with the iron(III)/1,10-phenantroline complex, *Talanta* *39*: 693 (1992).

92. J. G. Jacangelo, V. P. Olivieri, and K. Kawata, Thiosulfate-dechlorination interference with the Folin-Ciocalteau reagent method for protein determination, *Water Res. 21*: 1143 (1987).

93. B. Guvener and N. O. Sahin, A spectrophotometric determination of levothyroxine sodium in tablets, *Acta Pharm. Turc. 30*: 93 (1988); *Chem. Abstr. 109*: 135060j (1988).

94. C. S. P. Sastry and A. R. M. Rao, Application of Folin-Ciocalteau reagent for the spectrophotometric determination of some nonsteroidal antiinflammatory agents, *J. Pharmacol. Methods 19*: 117 (1988).

95. J. Emmanuel and S. D. Haldankar, A simple and sensitive spectrocolorimetric method for the estimation of chlorquinaldol and its formulations, *Indian Drugs 25*: 346 (1988), *Chem. Abstr. 109*: 27673c (1988).

96. L. Bruno and S. K. John, New spectrophotometric methods for estimation of nifedipine, *Indian J. Pharm. Sci. 50*: 109 (1988); *Chem. Abstr. 109*: 197294x (1988).

97. C. S. P. Sastry, N. R. P. Singh, M. N. Reddy, and D. G. Sankar, Spectrophotometric determination of α-tocopherol, *J. Inst. Chem. (India) 60*: 53 (1988); *Chem. Abstr. 109*: 216087a (1988).

98. C. S. P. Sastry, T. T. Rao, A. Sailaja, and J. V. Rao, Spectrophotometric estimation of acebutolol hydrochloride and captopril in their dosage forms, *Indian Drugs 28*: 523 (1991); *Chem. Abstr. 115*: 189919p (1991).

99. C. S. P. Sastry, A. Sailaja, T. T. Rao, and D. M. Krishna, Three simple spectrophotometric methods for the determination of sulphinpyrazone, *Talanta 39*: 709 (1992).

100. M. A. Raggi, M. R. Cesaroni, and A. M. Di Pietra, Colorimetric determination of mercaptopropionylglycine in pharmaceutical formulations, *Farmaco, Ed. Prat. 38*: 312 (1983).

101. A. C. Hsu, A quick colorimetric method for the determination of total thiols in water and wastewater, *Proc. AWWA Water Qual. Technol. Conf. 13*: 571 (1986); *Chem. Abstr. 105*: 84803m (1986).

102. K. Y. Chan and B. P. Wasserman, Direct colorimetric assay of free thiol groups and disulfide bonds in suspensions of solubilized and particulate cereal proteins, *Cereal Chem. 70*: 22 (1993); *Chem. Abstr. 118*: 232431z (1993).

103. M. F. Guingamp, G. Humbert, and G. Linden, Determination of sulfhydryl groups in milk using Ellman's procedure and Clarifying Reagents, *J. Dairy Sci. 76*: 2152 (1993).

104. A. V. Anisimov, S. M. Panov, and E. A. Viktorova, Photometric determination of thiols using 2,2′-dithiobis(5-nitropyridine) (in Russian), *Zh. Anal. Khim. 39*: 2248 (1984).

105. S. Mashiba, K. Uchida, S. Okuda, and S. Tomita, Measurement of glycated albumin by the nitroblue tetrazolium colorimetric method, *Clin. Chim. Acta 212*: 3 (1992).

106. P. Gillery, C. Perier, A. Stahl, and J. Delattre, Recommendations for the

determination of glycoproteins using the fructosamine method (in French), *Ann. Biol. Clin. 50*: 603 (1992).

107. B. Heintz and R. Kalusa, α-Ketolsteroids of the pharmacopoeias and codexes: The assay of α-ketolsteroids, listed in the pharmacopoeias and codexes and assayed after reaction with tetrazolium blue by the method of the European Pharmacopoeia (in German), *Deut. Apoth. Ztg. 118*: 1000 (1978).

108. V. Nevrekar, Spectrophotometric estimation of spironolactone, *Indian Drugs 21*: 349 (1984); *Chem. Abstr. 101*: 216519c (1984).

109. W. Horsch, I. Finke, and B. Wolf, Contribution to extraction and assay of prednisolone: besides drugs and auxiliary ingredients of ointments and creams and the uniformity of prednisolone distribution in such preparations, *Pharmazie 42*: 261 (1987).

110. E. Kovacheva, P. Vasileva-Aleksandrova, and V. Ivanova, Spectrophotometric determination of isoniazid using 2,5-diphenyl-3-thiazolyltetrazolium chloride in pharmaceuticals, *Mikrochim. Acta* 1983 I, 483.

111. P. V. K. Rao, R. S. Rao, and C. Rambabu, Spectrophotometric determination of aroyl hydrazines using 2,3,5-triphenyl tetrazolium chloride, *Acta Cienc. Indica 4*: 13 (1978); *Chem. Abstr. 89*: 190558w (1978).

112. Z. Zhang, L. Chen, H. Yang, and T. Ma, Colorimetric reaction of tetrazolium salts with diethylhydroxylamine, *Lanzhou Daxue Xuebao, Ziran Kexueban 18*: 99 (1982); *Chem. Abstr. 99*: 47317t (1983).

113. N. A. El-Rabbat and N. M. Omar, Colorimetric determination of catecholamines by 2,4,5-triphenyltetrazolium chloride, *Egypt. J. Pharm. Sci. 18*: 35 (1979); *Chem. Abstr. 92*: 116489v (1980).

114. P. B. Issopoulos, Sensitive colorimetric assay of carbidopa and methyldopa using tetrazolium blue in pharmaceutical products, *Pharm. Weekbl., Sci. Ed. 11*: 213 (1989).

115. M. Grote and A. Kettrup, Reduction of a new water-soluble tetrazolium salt by steroids and ascorbic acid, *Anal. Chim. Acta 212*: 273 (1988).

116. H. B. Mark Jr., L. M. Backes, D. Pinkel, and L. J. Papa, 2,3,5-Triphenyl-2H-tetrazolium chloride as a reagent for the determination of sugar mixtures by differential reaction-rate technique, *Talanta 12*: 27 (1965).

117. S. Görög and P. Horváth, Analysis of steroids. Part XXXI. Mechanism of the tetrazolium reaction of corticosteroids, *Analyst 103*: 346 (1978).

118. V. K. Shormanov, Ye. P. Duritsyn, M. M. Inozemtseva, and M. Yu. Markelov, Extraction photometric determination of phenoxymethylpenicillin (in Russian), *Farmatsyia (Moscow) 41*: 29 (1992).

119. Ch. Surasati and E. B. Sandell, Determination of submicrogram quantities by catalysis of the cerium(IV)–arsenic(III) reaction, *Anal. Chim. Acta 22*: 261 (1960).

120. H. Weisz and H. Ludwig, Eine kinetische Analysenmethode unter Verwendung katalysierter Systeme im stationaren Zustand, *Anal. Chim. Acta 60*: 385 (1972).

121. P. J. Ke and R. J. Thibert, Kinetic method for the determination of mercury at the nanogram level using an iodide catalysed arsenite-cerium reaction, *Mikrochim. Acta*: 768 (1972).

122. T. I. Fedorova, K. B. Yatsimirski, L. V. Shvedova, and T. G. Ermolaeva, Cerium-arsenite reaction for determination of gold(III) microconcentrations by catalymetric titration (in Russian), *Zh. Anal. Khim. 30*: 62 (1975).
123. S. Rubio, D. Pérez-Bendito, Micellar media in kinetic determinations, *Anal. Chim. Acta 224*: 185 (1989).
124. G. Knapp and H. Leopold, Automatic digital system for quantitative kinetic analysis: Application to catalytic determination of thyroid hormones, *Anal. Chem. 46*: 719 (1974).
125. G. López-Cueto and J. A. Casado-Riobó, Catalytic effect of copper on the hexacyanoferrate(III)-cyanide redox reaction, *Talanta 26*: 151 (1979).
126. R. C. Schothorst and G. Den Boef, The application of strongly oxidizing agents in flow injection analysis. 1. Silver(II). *Anal. Chim. Acta 169*: 99 (1985).
127. S. H. R. Davies and S. J. Masten, Spectrophotometric method for ascorbic acid using dichlorophenolindophenol: Elimination of the interference due to iron, *Anal. Chim. Acta 248*: 225 (1991).
128. C. Lopez Erroz, M. Hernandez Cordoba, and C. Sanchez-Pedreno, Kinetic determination of ascorbic acid in fruit juices, *An. Quim. 87*: 683 (1991); *Chem. Abstr. 116*: 254146w (1992).
129. L. E. Leon and J. Catapano, Indirect flow-injection analysis of ascorbic acid by photochemical reduction of methylene blue, *Anal. Lett. 26*: 1741 (1993).
130. A. M. El-Brashy and S. M. El-Ashry, Colorimetric and titrimetric assay of isoniazid, *J. Pharm. Biomed. Anal. 10*: 421 (1992).
131. G. J. Papariello and M. A. M. Janish, Diphenylpicrylhydrazyl as an organic analytical reagent: Analysis of amines, *Anal. Chem. 37*: 899 (1965).
132. D. B. Hunsacker, Jr., and G. H. Schenk, The determination of thiols with diphenylpicrylhydrazyl as a spectrophotometric reagent, *Talanta 30*: 475 (1983).
133. S. Salman and N. Bayraktar, Spectrophotometric assay of some amino group-containing drugs, *Sci. Pharm. 57*: 139 (1989); *Chem. Abstr. 111*: 201734b (1989).
134. K. M. Emara, H. F. Askal, and G. A. Saleh, Spectrophotometric determination of tetracycline and oxytetracycline in pharmaceutical preparations, *Talanta 38*: 1219 (1991).
135. K. E. Emara, Application of diphenylpicrylhydrazyl as a spectrophotometric reagent in the determination of some phenothiazines, *Anal. Lett. 25*: 99 (1992).
136. K. M. Emara, A.-M. I. Mohamed, H. F. Askal, and I. A. Darwish, Spectrophotometric determination of some pharmaceutical compounds using 2,2-diphenyl-1-picrylhydrazyl (DPPH), *Anal. Lett. 26*: 2385 (1993).
137. T. S. Ma and W. L. Nazimowitz, Colorimetric methods for determining organic functional groups, *Mikrochim. Acta*: 345 (1969).
138. R. D. Sauerbrunn and E. B. Sandell, Determination of submicrogram quantities of osmium by catalysis, *Mikrochim. Acta*: 22 (1953).
139. F. Vydra and R. Přibil, New redox systems, *Talanta 9*: 1009 (1962).
140. P. B. Issopoulos, Analytical investigations of isonicotinic acid hydrazide

(isoniazid). VI. Sensitive colorimetric determination of micro amounts of isoniazid using an indirect redox method, *Int. J. Pharm. 70*: 201 (1991).

141. P. B. Issopoulos, High-sensitivity spectrophotometric determination of trace amounts of levodopa, carbidopa and α-methyldopa, *Fres. J. Anal. Chem. 336*: 124 (1990).

142. M. I. Farooqui, J. Anwar, A. Khan, R. M. Ali, and R. Mahmood, A new sensitive method for the microdetermination of ascorbic acid, *J. Chem. Soc. Pakistan 12*: 333 (1990); *Chem. Abstr. 115*: 99428y (1991).

143. F. M. Ashour, F. M. Salama, and M. A. E. Aziza, A colorimetric method for the determination of captopril, *J. Drug. Res. 19*: 323 (1990).

144. I. L. Mattos, E. A. G. Zagatto, and A. O. Jacintho, Spectrophotometric flow injection determination of sucrose and total reducing sugar in sugarcane juice and molasses, *Anal. Chim. Acta 214*: 247 (1988).

145. C. S. P. Sastry, A. Sailaja, T. Thirupathi Rao, and D. Murali Krishna, Three simple spectrophotometric methods for the determination of sulphinpyrazone, *Talanta 39*: 709 (1992).

6

Formation of New Organic Chromogens

This chapter deals with photometric procedures that are based on the formation of new organic species, mostly dyes of defined structures created by the reaction (condensation) of an organic analyte usually with an organic reagent. The products formed are new chemical species and have characteristic absorption bands in the UV or visible region that make specific or selective spectrophotometric determinations of the original analyte possible. The classification in this chapter is done according to the type of the resulting chromogens (the basic chromophore is in parentheses): azo dyes $(-N{=}N-)$, quinoneimine dyes $(>C{=}N-)$ and aryl- (alkyl)-aminoquinones, azomethines $(-CH{=}N-)$ and hydrazones $(-CH{=}N{-}NH-)$, nitroso- $(-N{=}O)$ and nitrocompounds $(-NO_2)$, di- and triarylmethane dyes, and polymethines $(-CH{=}CH{-}CH{=}CH-)$. It is obvious however, that it is not possible to classify all spectrophotometrically usable color reactions according to this scheme, since their mechanism is complicated and in some cases still obscure. The classification used here covers the most important and typical reactions.

A. AZO DYES

The application of this group of chromogens is based on the so-called azocoupling reaction of diazonium salts with phenols in weak alkaline medium (pH < 8) or with aromatic amines in weak acid medium (pH 2–5), or by reaction with compounds containing a methylene group activated by one or two neighboring strongly electron-attracting groups. The diazonium salt is classified as the active coupling component, the phenol or amine as the passive coupling component. The electrophilic reagent, the aryldiazonium cation, attacks the *p*-position of the phenolic OH group or of the −NH$_2$, −NHR, or −NR$_1$R$_2$ groups [1,2] (Eq. 1). If the *p*-position is occupied (e.g, in *p*-cresol, 2-naphthol), the reaction takes place in the *o*-position (Eq. 2). Certain substituents in the *o*- and *p*-positions can be split off and replaced by the azo group. Such reactions are dependent on the reactivity of the diazonium compound. If diazonium salts are used as the reagent in excess to the analyte, the formation of disazodyes can be observed.

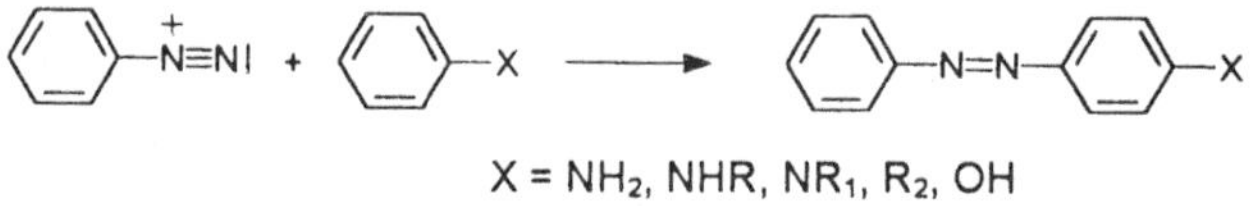

$$X = NH_2, NHR, NR_1, R_2, OH$$

If at least one of the reaction components contains a solubilizing group (−SO$_3$H, −COOH), the resulting azodye is water soluble; if not, the chromogen can be extracted into an organic solvent. Many of the hydroxyazo- and aminoazocompounds formed undergo acid–base equilibria accompanied by color changes. Nitrohydroxyazo dyes are converted into deeply colored quinonoid salts in alkali metal hydroxide solutions, whereas aminoazodyes are protonated in strong acid media (see also Chapter 2, Section C).

1. Diazonium Salts as Reagents

These reagents are prepared by diazotization of an appropriate primary aromatic amine freshly before use (one of the oldest used reagents is diazotized sulfanilic acid, known as Pauli reagent; or if stabilized diazonium salts (1,5-naphthalenedisulfonates, ZnCl$_2$ complexes, fluoroborates etc.) are used, they are simply dissolved in water, methylcellosolve, or dimethylsulfoxide before use. Many of these reagents are commercially available (Fast Salts) in forms appropriate for specific analytical purposes. The choice of the proper reagent is made on the basis of its coupling reactivity, sensitivity, color shade, and stability and according to the reproducibility of the determination [3]. The reactivity of the aryldiazonium cation can be altered by substitution of the aromatic nucleus. The substitution by nitrogroups brings about an increased reactivity. Thus, benzenediazonium chloride reacts with phenol but does not react with anisole (methoxybenzene). 2,4-Dinitrobenzenediazonium chloride reacts with anisole, and 2,4,6-trinitrobenzenediazonium chloride couples even with mesitylene.

The azocoupling reaction has been used for the determination of a great variety of both simple phenols and more complicated aromatic hydroxy compounds. Many particular determinations are described in handbooks [4–6] and in original articles [7,8]. Besides phenols and aromatic amines, diazonium compounds couple with compounds containing activated methylene groups, especially with ethyl acetoacetate, aldehydes and ketones, pyruvic acid, oxalacetic acid, ascorbic acid, 2-ketoglutaric acid, and some heterocyclic compounds [e.g., imidazoles, (histamine, histidine), pyridoxine, thiazoles, pyrazolone derivatives, theophylline, and aromatic diisocyanates]. The determination of thiamine is based on the theorized coupling of the thiazole part of the molecule. The reaction of diazonium compounds with the oxocompounds mentioned above results in the formation of the corresponding 1,5-diaryl-formazanes, hydrazones, and so on; the reaction with hexosamines in the formation of triazenes. Thus, the azocoupling reaction can be used for the spectrophotometric determination of all these compounds, [4–6,9–15]. The structure of the products can be affected by the properties of the heterocyclic analyte [16].

The azocoupling reaction of phenols and amines can be applied also in all cases where these compounds are obtained by decomposition reactions (e.g., hydrolysis) or chemical transformations of the original analyte. The hydrolysis of *N*-arylcarbaminates, by which phenols are set free and hydrolysis of arylisocyanates, by which arylamines result, serve as examples of decomposition reactions, and the fusion of aromatic sulfonic acids with alkali metal hydroxides as an example of analyte transformation [5,17–19].

2. Passive Coupling Components as Reagents

The formation of colored azocompounds can be utilized for the determination of diazonium compounds [20,21] and of all types of compounds from which the diazo compounds can be generated.

The choice of passive coupling component depends on the reactivity of both the diazo compounds and the coupling partner: 1- and 2-naphthols, resorcinol, orcinol, phloroglucinol, barbituric acid, 1-naphthylamine, *N*-ethyl-1-naphthylamine, *N*-(1-naphthyl)-ethylenediamine (Bratton-Marshall reagent) and many others have been used. It is advantageous to select a passive coupling component giving a uniform reaction product and not a mixture of isomers of different λ_{max}.

Furthermore, primary aromatic amines can be easily converted by the action of nitrite in acid media into the corresponding diazonium compounds [22]. The reactive agent is the cation NO^+, which appears in concentrated acids and in the course of conversions with complex NO^+ salts. Reactivity decreases in the series $NO^+ > H_2O^+{-}NO > Br{-}NO > Cl{-}NO$. The diazotization reaction proceeds in several steps, which can be in a simplified manner illustrated by Eq. (3):

$$\text{ArNH}_2 \longrightarrow \underset{\underset{H^+}{|}}{\overset{\overset{H}{|}}{\text{Ar}-\text{N}-\text{NO}}} \longrightarrow \underset{\underset{H}{|}}{\text{Ar}-\text{N}-\text{NO}} + \overset{+}{H} \longrightarrow \text{Ar}\overset{+}{N}{\equiv}\text{N} + \text{H}_2\text{O} \qquad (3)$$

Mostly, diazotation of aromatic amines with nitrous acid is carried out with solutions or suspensions of the amine in dilute hydrochloric acid (approximately 2.5–3 mol acid per 1 mol amine) by treating with the molar amount of sodium nitrite in concentrated aqueous solution at 0–5 °C. The conversion to the diazonium salt solution proceeds rapidly and almost quantitatively, and can be followed with iodine–starch indicator paper, which shows a violet coloration with the unreacted nitrous acid. Before subsequent reactions occur (especially prior to coupling with amines), the excess should be removed by the addition of urea or sulfamic acid. Weakly basic amines bearing strongly electron-attracting substituents on the aromatic ring are diazotized in concentrated acids to avoid hydrolysis of the diazonium salts in aqueous solution. Heterocyclic amines are diazotized similarly as the aniline derivatives. In some cases, however, the reaction may be complicated by the properties of the heteroaromatic ring.

Many examples of primary arylamine determination can be found [4–6,23–26]. The procedure can also be used in cases when the arylamine is obtained by a decomposition reaction (hydrolysis, e.g.) of more complex molecules [27–30]. 1,4-Benzodiazepines, which are easily hydrolyzed to the corresponding aminobenzophenones, are a typical example [28–30].

The easy reduction of aromatic nitro compounds yielding primary aromatic amines can be the basis of procedures for their determination [31,32]. Since aromatic compounds are easily nitrated (HNO_3, KNO_3 + H_2SO_4), the pathway of aromatic compound → nitro compound → primary aromatic amine → diazonium salt → azo dye, known as the Guerbet reaction, makes possible the determination of compounds representing the individual steps. For some possible complications see e.g. in ref. [33].

Arylhydrazines can be converted into diazonium salts by oxidation with selenium dioxide [Eq. (4), the Postowsky reaction] [34], which enables their determination in the form of azodyes [35]. The color reaction can also serve for the catalytic determination of selenium(IV) [36].

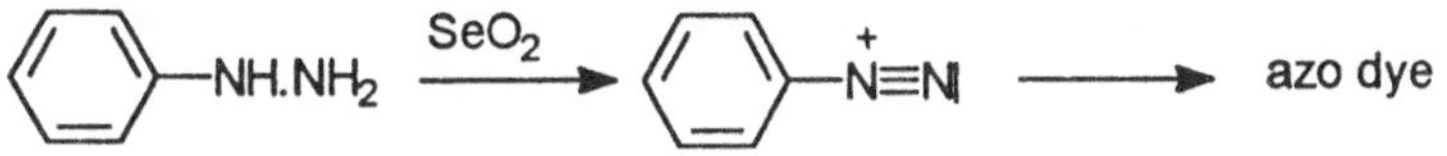

$$\langle\!\!\!\bigcirc\!\!\!\rangle\text{—NH.NH}_2 \xrightarrow{\text{SeO}_2} \langle\!\!\!\bigcirc\!\!\!\rangle\text{—}\overset{+}{\text{N}}\!\!\equiv\!\!\text{N} \longrightarrow \text{azo dye} \qquad (4)$$

3. Amines and Passive Coupling Components as Reagents (Determination of Nitrite)

The azocoupling reaction can also be used for the determination of nitrite. In this case, a suitable primary aromatic amine is used as a reagent to prepare the diazonium compound, which is coupled using an appropriate passive coupling component as the second reagent (Griess reaction) [37–40]. The procedure is applicable in those cases where an appropriate reaction can be used to release nitrite from more complicated molecules [41–43].

The Griess reaction can also be used for the determination of nitrates after their reduction with hydrazine, zinc, or modified cadmium metal or alloys [44–46]. The determination of both nitrite and nitrate in admixture is possible.

The diazotization-coupling reaction can be applied also to the determination of thiols, which when treated with nitrite are S-nitrosated. The excess nitrite is destroyed by the addition of sulfamic acid, and the S-nitroso derivatives upon acid hydrolysis in the presence of mercury(II) chloride release nitrous acid, which is used to diazotize sulfanilamide. The procedure is completed by coupling the diazonium salt formed with N-(1-naphthyl)-ethylene diamine [47].

4. Reagents Forming Azo Dyes by Other Reactions

Azo dyes can also be formed by the reaction of aromatic nitroso compounds with primary aromatic amines [Eq. (5)]. Bromnitrosol (2-nitroso-3-bromo-1-naphthol), 4-nitroso-benzoic acid and nitroso-*R*-salt (1-nitroso-2-naphthol-3,6-disulfonic acid) are examples of the reagents used [5].

$$\text{Ar—NO} + \text{H}_2\text{N—Ar}' \rightarrow \text{H}_2\text{O} + \text{Ar—N}\!=\!\text{N—Ar}' \qquad (5)$$

Another reaction resulting in the formation of azo compounds is the condensation of 1,2-napththoquinones with 2,4-dinitrophenylhydrazine [5], which is used for the determination of corresponding quinones [Eq. (6)].

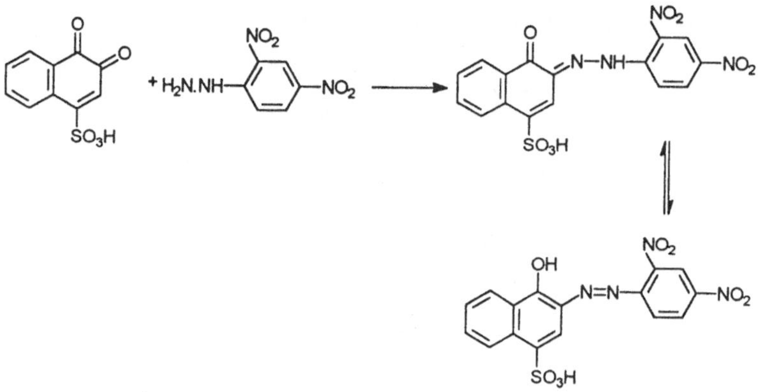

$$(6)$$

B.   QUINONEIMINE DYES AND AMINOQUINONES

1.   Quinonemine Dyes

The most common reaction resulting in the formation of quinonemines is oxidative coupling, which consists of the oxidation of a mixture of an aromatic *p*-diamine (*p*-phenylenediamine, *N,N*-dialkyl-*p*-phenylenedia-mine) or an aminophenol (*o*-, *m*-, *p*-aminophenol, metol, etc.) as one reaction component, a phenol (or an arylamine) as the second reaction compo-nent, and an appropriate oxidizing agent ($K_3Fe(CN)_6$, $K_2S_2O_8$, chloramine T, KIO_4, $FeCl_3$, 2-iodyl-benzoate, *N*-chloro- or bromosuccinimide, etc.) as the third one [4,5]. Indophenols (1), indoanilines (2), or indamines (3) are the products, depending on the character of the reaction components. The structure of the chromogens can be affected by further substituents of the reacting molecules [48]. The reaction proceeds via 1,4-benzoquinonedi-imine, which condenses under further oxidation with a molecule of phenol. When phenothiazines are used as one reacting component, Methylene Blue–type chromogens were identified as the reaction products [49].

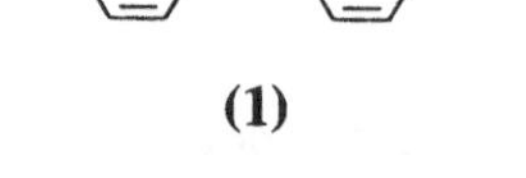

(1)    **(2)**

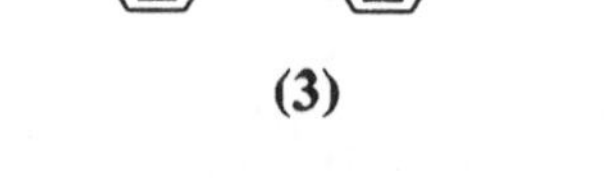

(3)

The reaction is used for the determination of either *p*-diamines (aminophenols) or phenols (arylamines) [4–6,50–56] and is also applicable in cases when one of the reacting components is obtainable by chemical degradation of more complicated molecules [57,58]. Arylhydrazines also give positive reactions [59,60].

This same reaction can also be used for the indirect determination of nitrite. The determination is based on the reaction of nitrite with *p*-sulfanilamide in excess, and the determination of the residual reagent is made by oxidative coupling with 4-*N*-methylaminophenol (metol) [61].

When used for the determination of ammonia, this reaction is known as the Berthelot reaction [62–66], which takes place in the presence of a further reaction component acting as catalyst (sodium nitroprusside, sodium dichloroisocyanurate). The same procedure is also applicable for the determination of amino acids [67,68].

The Berthelot reaction can also be used for the determination of nitrates after their reduction with Dewarda alloy in alkaline medium [69].

The reaction of guanidine and its derivatives with 1-naphthol or thymol and sodium hypochlorite, known as the Sakaguchi or Fearon reaction [4,5,70,71], belongs to this group of reactions. It has been shown, however, that chromogen formation involves the conversion of *N*-halogenoguanidine derivatives into the corresponding semicarbazones **(4,5)** [71].

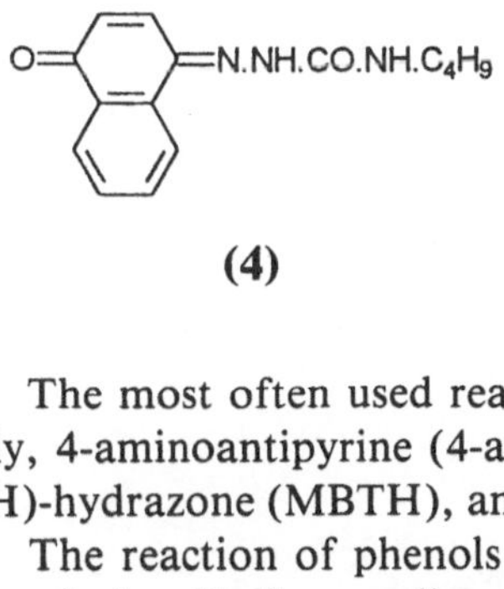

$$O = \text{(naphthalene ring)} = N.NH.CO.NH.C_4H_9$$

(4)

$$O = \text{(ring with } C_3H_7, Br, CH_3) = N.NH.CO.NH.C_4H_9$$

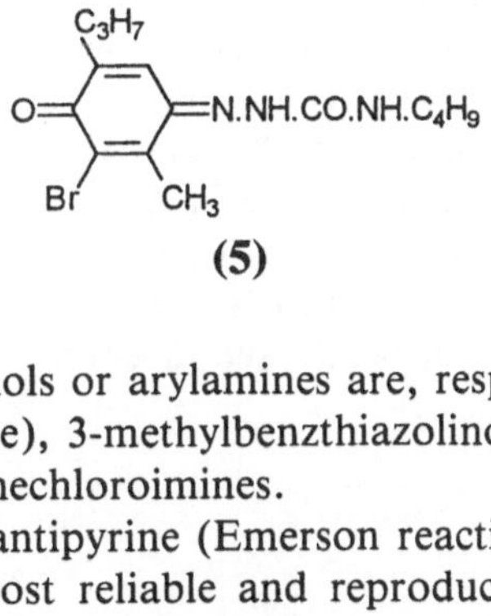

(5)

The most often used reagents for phenols or arylamines are, respectively, 4-aminoantipyrine (4-aminophenazone), 3-methylbenzthiazolinone-2(3H)-hydrazone (MBTH), and benzoquinonechloroimines.

The reaction of phenols with 4-aminoantipyrine (Emerson reaction) proceeds in alkaline media, yielding the most reliable and reproducible results [72] when a buffer solution of pH 7.5–9.5 and $K_3Fe(CN)_6$ or $K_2S_2O_8$ are used. A sufficient buffering capacity is required since there is a considerable decrease of pH during the reaction. The results are independent of the kind of buffer and oxidizing agent used, and of the ionic strength. It is necessary to work with at least a tenfold excess of the reagents, the ratio of 4-aminoantipyrine and the oxidizing agent being close to 1 : 1. The reaction takes place in the *p*-position [Eq. (7)] to the OH group. If this position is occupied by halogen or carboxy groups, the substituent is eliminated quantitatively and a product identical to that of the unsubstituted phenol results. Alkoxy groups in the *p*-position are eliminated slowly and the elimination is not quantitative. Phenols substituted in the *p*-position with alkyl, alkylcarboxy, and alkoxycarboxy groups react under the formation of only slightly colored yellow products. The concentration of these phenols and reagents, however, must be increased. Salicylic acid alone does not react at all; salicylamide, salicylaldehyde, and salicylic acid methyl and phenyl esters react very strongly. Their intensively colored products, however, are less stable [4–6,73–77]. The procedure can, of course, be used for the determination of phenols in all cases when these can be obtained from more complicated compounds by chemical degradation. Primary arylamines react in weak acid media [5]. *p*-Benzoquinone gives a positive reaction with 4-aminoantipyrine in acid media which can be made use of for the determination of certain *p*-quinones [5].

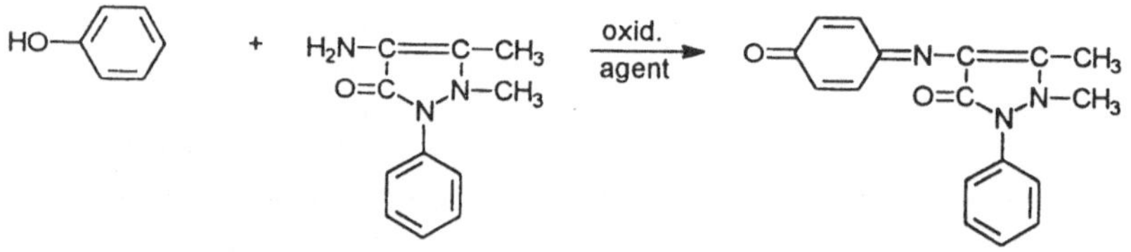

$$(7)$$

4-Dimethylaminoantipyrine (4-dimethylaminophenazone) has also been found to react with phenol because of its degradation to 4-aminoantipyrine under reaction conditions. The reaction is used for the determination of phenols [78] and has also been used for the determination of the reagent itself [79].

The phenol–4-aminoantipyrine reaction can also be used as the indicator reaction in the enzymatic determination of glucose in which hydrogen peroxide is released [80,81].

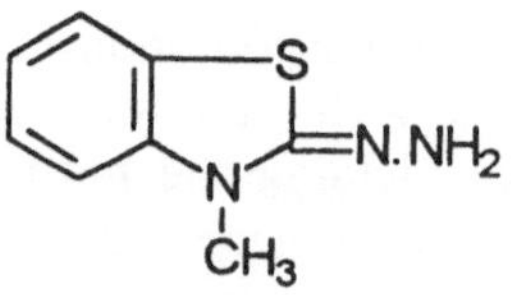

(6)

3-Methylbenzthiazolinone-2(3H)-hydrazone (MBTH) **(6)** was originally introduced as a reagent for aldehydes. Later its use was extended to a variety of organic compounds (e.g, phenols, arylamines, and different *N*- and *S*-heterocyclic compounds). The reagent is used in the form of an aqueous solution, and the reaction products can be extracted into chloroform if desired. The reaction with phenols proceeds in acid or alcoholic media and in the presence of an oxidizing agent, and takes place in the *p*-position (red dyes) or *o*-position (violet dyes) when these are free [82]. In the case of halogenated phenols, dehalogenation occurs in both *o*- and *p*-positions. The absorption maxima of different phenols are in the range 510–560 nm; arylamines in the range of 500–630 nm. Under reaction conditions, MBTH loses two electrons and one proton to form the electrophilic intermediate (7), which has been identified as the active coupling species that undergoes electrophilic substitution with phenol to form the colored product (8). The formation of the corresponding diazonium ion has not been observed.

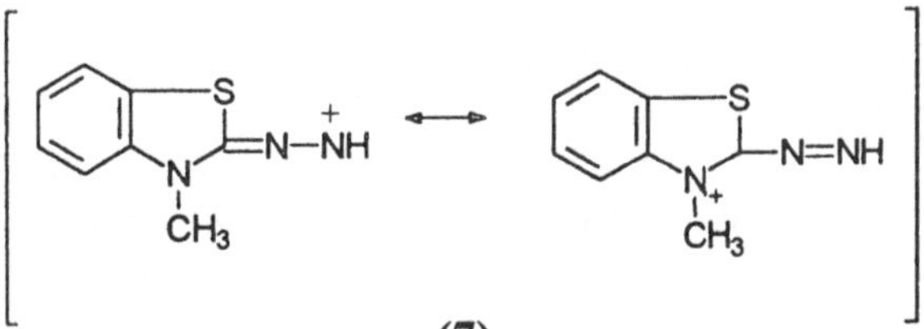

(7)

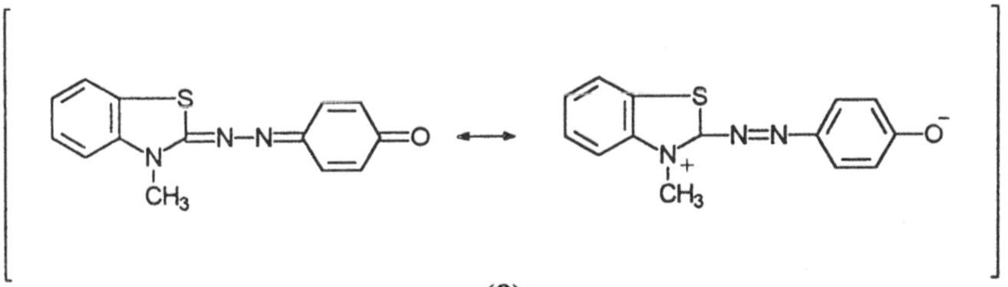

(8)

Many examples of the application of this reagent can be found [4–6,83–86]. In comparing the 4-aminoantipyrine and MBTH reagents for determination of phenols, the latter has been found to give better results [87]. The determination of benzodiazepine drugs can be based on the determination of their hydrolysis products (the corresponding aminobenzophenones) using the MBTH reagent [88].

Another reaction resulting in the formation of blue quinoneimines ($\lambda_{max} \sim 610$ nm) is the condensation of quinonechloroimines with phenols in aqueous alkaline medium. 2,6-Dichloro- and 2,6-dibromobenzoquinonechloroimines (Gibbs reagent) [4–6,92–94] and 1,4-benzoquinonedichlorodiimide are used most often [4,5] [Eq. (8)].

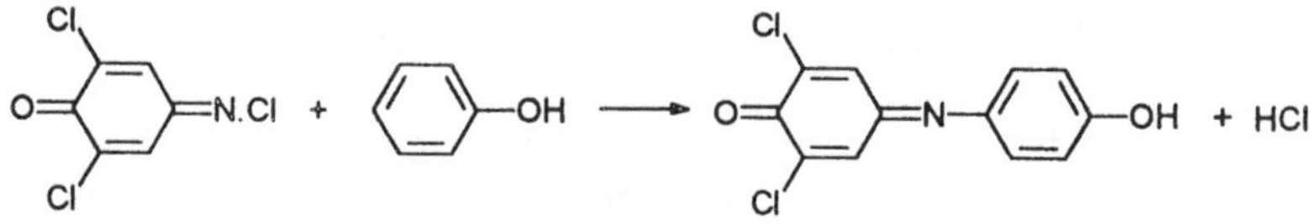

$$\text{(8)}$$

The reagents are used in the form of solutions in ethanol or acetone. Reproducible results are obtained [89,90] when 2×10^{-6} to 2×10^{-5} M phenol solutions react with a 30–50-fold reagent excess. The optimal pH is 8.5, and both phosphate and Britton-Robinson buffers are suitable; ammonia and glycocol buffers are inapplicable. 1-Butanol can be used when concentration of the colored product is desired.

The first step of the reaction is the formation of the corresponding quinoneimine, the two molecules of which react immediately with one molecule of phenol. For the mechanism of the reaction of quinonedichlorodiimide with secondary and tertiary aromatic amines, see Ref. [91].

Thiols react with the reagent in weakly alkaline medium according to Eq. (9), as do disulfides. With the latter, reductive cleavage is theorized to be the first reaction step [Eq. (10)] [5].

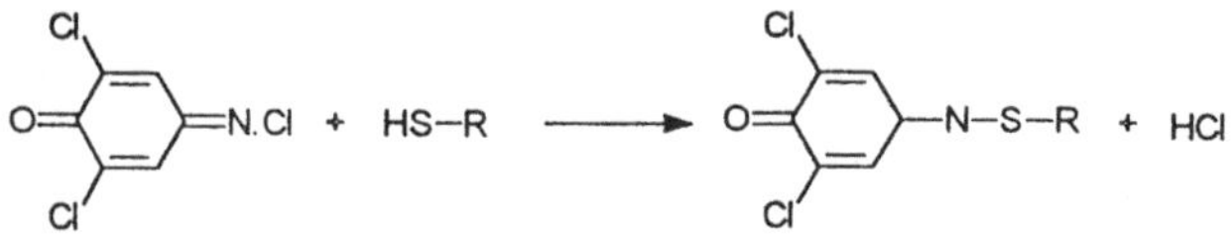

$$\text{(9)}$$

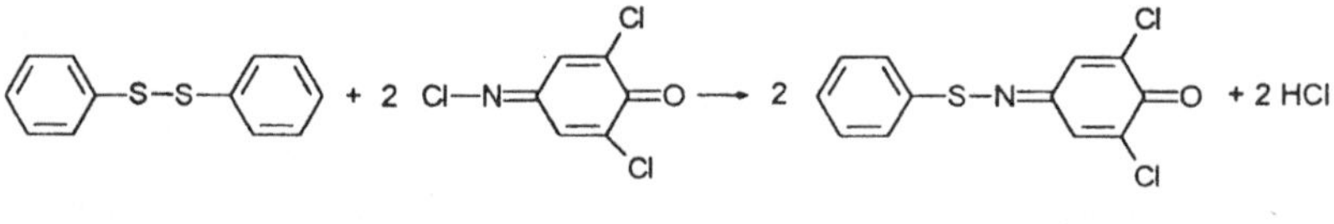

$$\text{(10)}$$

2. Aminoquinones

Primary and secondary amines and amino acids react with *p*-quinones in ethanolic solution at elevated temperatures (40–60°C) to form the corresponding alkyl(aryl)aminoquinones with new characteristic absorption bands [Eq. (11)]. *p*-Benzoquinone, chloranil, and 2,3-dichloro-1,4-naphthoquinone (diclone) are used as reagents [4–6,95–100]. Depending on the reaction conditions, the intermediate formation of corresponding π- and σ-complexes is theorized (see also Chapter 3, Section B).

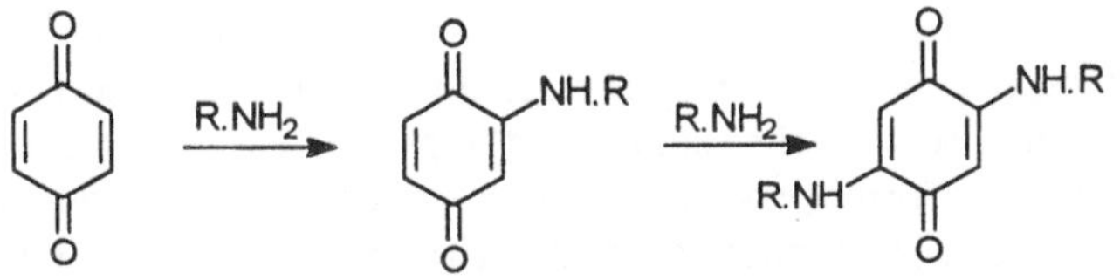

(11)

1,2-Naphthoquinone-4-sulfonic acid and its alkali salts (NQS, 3,4-dihydro-3,4-dioxonaphthalene-1-sulfonic acid) are very useful reagents (Folin reagent) and are frequently used for the determination of compounds containing two reactive hydrogen atoms on a nitrogen or carbon atom, that is, for primary and secondary amines, amino acids [Eq. (12)], hydrazines and compounds with active methylene groups (see Section F) [101–110]. The pale orange color of the reagent is changed to dark red of the chromogen (λ_{max} 450–510 nm). Two forms (9) and (10) of the reaction chromogen should be considered with primary amines:

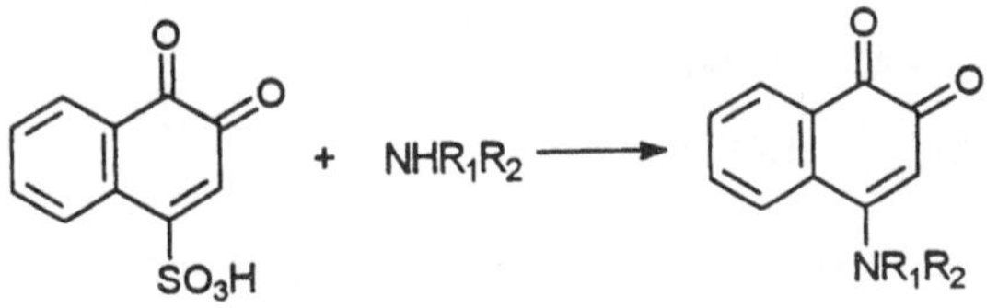

(12)

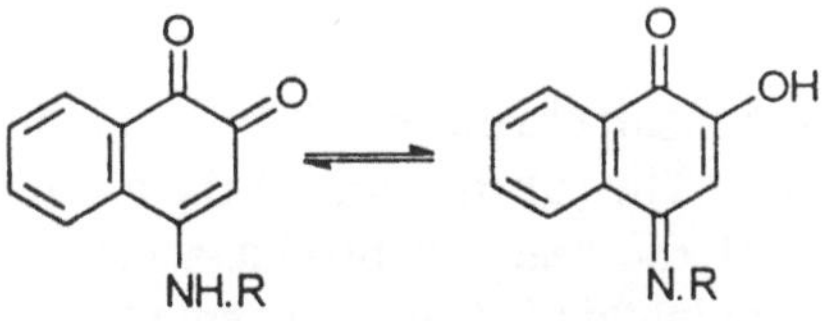

(9)   (10)

The reaction of NQS with the nucleophilic analyte proceeds in weak alkaline medium, usually under normal or slightly elevated temperatures. Chromogens derived for analytes containing no solubilizing groups are extractable with organic solvents. The difference in solubility can be used for the determination of sulfanilic acid in mixture with sulfonamide drugs [107,109]. There are potential problems with extractibility in the case of some ethanolamines, ephedrine, and bis-derivatives of diamines [106,107].

The reaction has also been used for the determination of aromatic nitro compounds in industrial wastewaters after their reduction with $TiCl_3$

[111]. Sulfides in sewage effluents have been determined indirectly based on reduction of NQS, the excess of which is reacted with sulfanilic acid [112].

Phenols condense with the reagent in the presence of sodium hydroxide and pyridine to form products with a structure analogous to 3'-hydroxy-5',6'-benzo-4,4'-diphenoquinone, the sodium salts of which are deeply colored (λ_{max} = 660–680 nm) [113].

C. AZOMETHINES AND HYDRAZONES

1. Azomethines

Spectrophotometrically important azomethines (Schiff bases) [114] are products of the reaction of aromatic aldehydes and primary arylamines under normal temperature in aqueous or ethanolic solution and acid media consisting mostly of hydrochloric or acetic acid [Eq. (13)]. Their protonated forms are the chromogens of the very sensitive color reaction that

$$Ar-CHO + H_2N-Ar' + HCl \rightarrow$$
$$Ar-CH=\overset{+}{N}H-Ar' + H_2O + Cl^- \tag{13}$$

enables the determination of both aromatic aldehydes and primary arylamines [4–6].

a. Aldehydes as reagents

Different substituted benzaldehydes and 2-furylaldehyde have been used as reagents [4–6,115–117]. The substitution of the aromatic ring brings about color intensification of the reaction products. The reagent used often is 4-dimethylaminobenzaldehyde (Ehrlich reagent) [4–6,118,119]. Its reaction products with most arylamines and urea derivatives are yellow to orange depending on the substituent in the molecule of the analyte.

4-Dimethylaminocinnamaldehyde is used very often in the same way. Its advantage is its ability to shift the chromogen absorption bands to longer wavelengths, for example, from yellow to red [120–122].

5-(p-Dimethylaminophenyl)-2,4-pentadienal, which contains one conjugated double bond more in its molecule, has also been proposed for the determination of primary arylamines. Their Schiff bases have absorption bands of $\lambda_{max} \sim 630$ nm [123].

The reaction of all these reagents is applicable in all cases in which arylamines are obtained by reduction [124–126], hydrolysis [127–129], or other decomposition reactions of more complicated organic molecules.

b. Arylamines as reagents

Primary arylamines are suitable reagents for the determination of aromatic and 2,3-unsaturated aldehydes based on the formation of the corresponding Schiff bases [4,5,130,131].

Besides the direct determination of those aldehydes for which *o*-dianisidine is often used as the reagent [4–6,130–131], the azomethine formation represents a very important final step in the determination of saccharides, pyridine derivatives, reactive monotopic alkyl di- or tri-halogenides, and cyanides.

The determination of analytes containing OH groups in appropriate positions (aminophenols) offers the opportunity of subsequent conversion of the resulting azomethines into metal chelates [132,133].

The determination of saccharides is based on the action of mineral acids of higher concentration at 100°C by which these compounds are converted into furaldehyde derivatives. 2-Furylaldehyde, for example, is formed from aldopentoses; 5-methyl-2-furylaldehyde from 6-deoxyaldoses; and 5-hydroxymethyl-2-furylaldehyde from aldohexoses. This last product, however, can undergo further degradation. These aldehydes react with an appropriate arylamine to form the desired Schiff base. It should be emphasized that pentoses and hexoses react with the amine also directly under mild conditions to form the corresponding Schiff bases without being degraded to furaldehyde derivatives [4,5,134].

The determination of pyridine derivatives is based on the Koenig reaction. Pyridine compounds when treated with cyanogen halides undergo opening of the pyridine ring under the formation of glutaconaldehyde derivatives, which react with a primary arylamine to form intensely colored Schiff bases [135–137]. Anthranilic acid or J-acid (6-amino-1-naphthol-3-sulfonic acid), respectively, or reagents containing reactive methylene groups (barbituric and thiobarbituric acids, e.g.) that condense with glutaconaldehyde to the corresponding polymethine dyes are used often instead of arylamines (see Section F).

The same reaction is also the basis of cyanide determination [138–139]. Cyanide ions are easily converted into cyanogen bromide by the addition of bromine. The addition of pyridine gives rise to glutaconaldehyde, which reacts with aromatic amines to form Schiff bases.

The reaction pathway can be broadened to include the determination of acrylonitrile, which on oxidation with $KMnO_4$ is converted into cyanide, which in turn is converted by bromination into cyanogen bromide. Pyridine added to cyanogen bromide is converted into glutaconaldehyde, and the determination is completed by the addition of 4-aminobenzoic acid, 4-

aminosalicylic acid, or 6-amino-1-naphthol-3-sulfonic acid as the corresponding Schiff base [140,141].

The determination of monotopic, di-, and trihalogeno derivatives is
based on the Fujiwara reaction, in which the pyridine ring is split by these
compounds to yield glutaconaldehyde derivatives, which form colored
Schiff bases with arylamines as in the preceding case [4,5,142–145]. For the
evaluation of the yellow coloration of glutaconaldehyde, see Section F.

A very important reagent of this group is ninhydrin (2,2-dihydroxy-
1H-indene-1,3(2H)-dione) (11), discovered by S. Ruhemann in 1911 [146].
Its reaction is used extensively in the determination of amines, amino acids,
peptides, and proteins [4–6,147–158], and is probably the most frequently
used color reaction of all. The reaction is performed in aqueous solution or
in alcohol, methylcellosolve, dimethylsulfoxide, and similar solvents under
elevated temperature. The reaction mechanism involves decarboxylation of
the amino acids to be determined, resulting in the formation of Ruhemann's
purple (12), and is accompanied by side reactions [159].

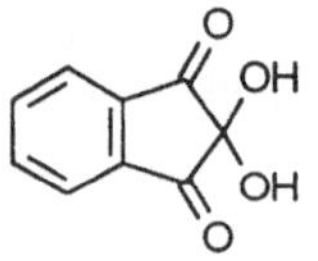

(11) 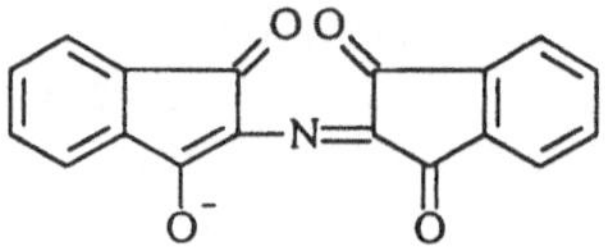   **(12)**

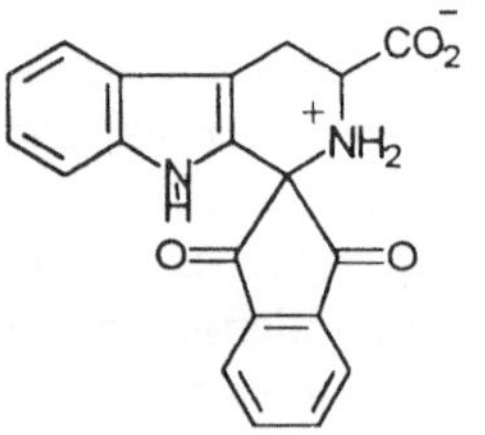

(13) 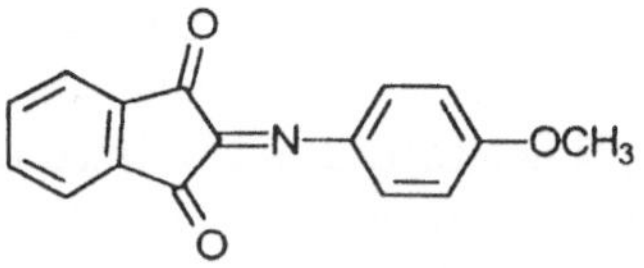   **(14)**

Chromogens have been isolated and their structure determined, for example, for tryptophan (13) [160,161] and p-anisidine (14) [162].

2. Arylhydrazones

Arylhydrazones result in the reaction of arylhydrazines having two hydrogen atoms on nitrogen atom with aldehydes and ketones [Eq. (14)].

$$Ar-CHO + H_2N \cdot NH-Ar' \rightarrow$$
$$Ar-CH=N \cdot NH-Ar' + H_2O \qquad (14)$$

a. Hydrazines as reagents

Besides being used for the direct determination of the oxocompounds, this reaction is a very important final step in the determination of all compounds that on oxidation, reduction, hydrolysis, or other chemical reaction can be converted into simple aldehydes or ketones, especially into formaldehyde, glyoxal, glyoxylic acid, 2-furaldehyde derivatives, and so on. Formaldehyde, for example, reacts with phenylhydrazine to form the corresponding phenylhydrazone. The excess of the reagent is oxidized by an appropriate oxidant to form with the phenylhydrazone an intensely colored 1,5-diphenylformazane. 1,2-Diols, 1,2-aminoalcohols, 1,2-ketols, and epoxides are readily oxidized by periodic acid to formaldehyde [Eq. (15)].

$$R-CH(OR)-CH(OH)-CH_2OH + HIO_4 \rightarrow$$
$$R-CH(OR)-CHO + HCHO + HIO_3 + H_2O \qquad (15)$$

and can therefore be determined using the reaction with phenylhydrazine. Oxalic acid can be reduced (Zn + HCl), yielding glyoxylic acid, and thus determined in the same way. 17,20-Dihydroxy-20-ketosteroids react with phenylhydrazine in ethanolic sulfuric acid and can be determined using this reaction (Porter-Silber reaction). Reducing sugars react with phenylhydrazine to form osazones ($\lambda_{max} \sim 400$ nm) [Eq. (16)].

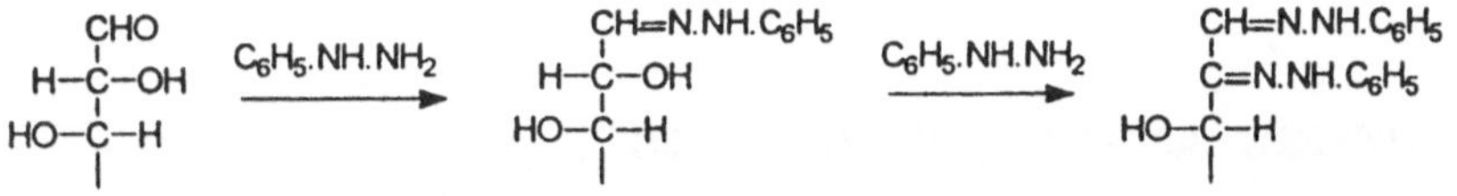

Arylhydrazines substituted with nitro groups react with oxocompounds to form the corresponding hydrazones, and react with acid derivatives (anhydrides, chlorides) to form hydrazides. The presence of nitro groups in the hydrazine molecule enables a sensitive color reaction of the hydrazone with alkali hydroxide, by which it is converted into a quinonoid structure, while the color formed with the reagent rapidly disappears [163–165]. This reaction can be used in combination with other reactions to converting the reaction products to derivatives with oxo groups. Thus, primary and secondary amines react with 1,2-naphthoquinone-4-sulfonate to form the corresponding aminoquinones (see Section B), and the excess reagent can be determined by the reaction with 2,4-dinitrophenylhydrazine to form the hydrazone tautomeric with the corresponding azodye (see Section A).

Isonicotinic acid hydrazide and especially 3-methyl-2-benzothiazolin-one hydrazone (**6**) are frequently used reagents [4–6,166–169]. The latter reagent was originally introduced for the determination of aliphatic alde-hydes. The reaction proceeds in aqueous medium in the presence of an appropriate oxidizing reagent (e.g., $FeCl_3$). Structure (**15**) has been attrib-uted to the blue chromogen (λ_{max} at 635 and 670 nm). The reaction is applicable in all cases where aliphatic aldehydes (especially formaldehyde) are obtained by oxidation of other compounds (1,2-diols, 1,2-aminoalco-hols, olefines, etc.). The compound has since become a useful reagent for phenols, arylamines, and similar compounds (see Section B).

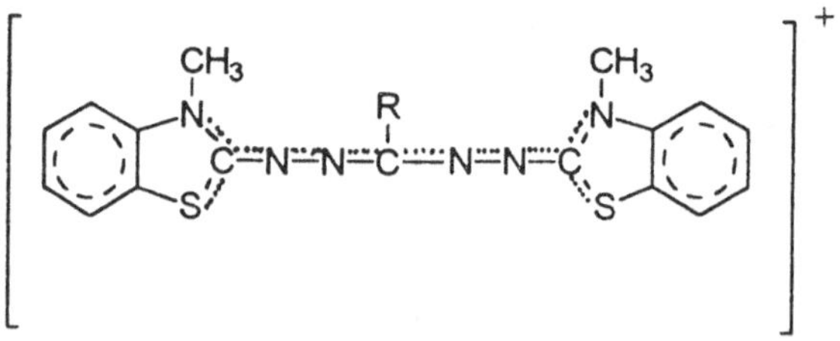

(**15**)

b. Aldehydes as reagents

Aromatic aldehydes, especially p-dimethylaminobenzaldehyde or -cinna-maldehyde, are the mostly used reagents for the determination of hydra-zines and acid hydrazides [4–6,170–178].

D. NITROSO- AND NITROCOMPOUNDS

These chromogens are formed by the action of nitrous or nitric acid on certain types of organic compounds. Depending on the structure of the analyte and the functional groups present, C-nitroso-, C-nitro- and N-nitrosocompounds result. The reaction products formed are spectrophoto-metrically evaluated directly or after conversion to alkali metal or ammo-nium salts. Both reagents can, of course, produce chromogens by oxidation or other reactions. Reactions involving both oxidation and nitrosation (ni-tration) are also possible.

1. Nitrous Acid as Reagent

Phenols react easily with nitrous acid ($NaNO_2$ + H^+) in acid media to form nitroso- or nitrophenols, depending on the reaction conditions (i.e., on the acid used and its concentration, temperature, reaction period, substituents on the phenol molecule, etc). The products are converted to their intensely colored alkali metal or ammonium salts by the addition of alkali metal hydroxide or ammonia [Eq. (17)].

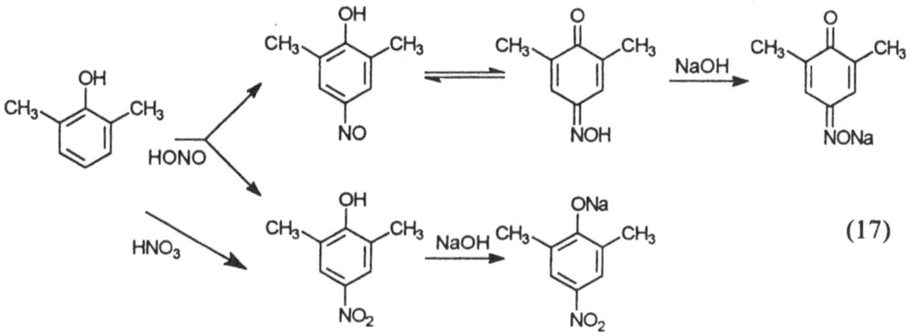

Most of the methods described are based on the procedure of Lykken, Tresseder, and Zahn [179], using acetate buffer medium (160 mL glacial acetic acid + 30 mL 15% KOH + 10 mL water) and sulfuric acid. The reaction products are converted into ammonium salts by adding ethanolic ammonia (225 mL ethanol + 125 mL water + 150 mL concentrated ammonia). The ammonium salts are yellow to orange and are suitable for photometric measurements. Although the formation of nitrosophenols was assumed originally, further investigations have shown that ammonium salts of nitrophenols are the chromogens [180]. The rate of color development and the proper reaction temperature differ for individual phenolic compounds. Nitro groups are introduced into ortho and para positions of the phenolic groups, and the number of nitro groups introduced is dependent on the reaction conditions and phenol structure. In the case of phenol, mono-, di-, or trinitrophenols are formed. When the *o*- and *p*-positions are substituted by certain functional groups, these groups are fully or partially replaced by the nitro groups: bromine, iodine, carboxylic and carbaldehydic groups; sulfo groups; the terc.butyl and tert.octyl groups, and so forth. Thus, 2,4,6-triiodophenol reacts with the reagent under the formation of successively nitrated products, the final chromogen being 2,4,6-trinitrophenol [Eq. (18)]. Fluorine and chlorine are not replaced. *p*-Hydroxybenzoic acid does not react. The reaction of 2,4,6-trichlorophenol results in the formation of 2,6-dichlorobenzoquinone. In the case of Dian (4,4′-isopropylidene-bis-phenol), the fission of the aliphatic part of the molecule and the formation of 2,4,6-trinitrophenol has been observed [181].

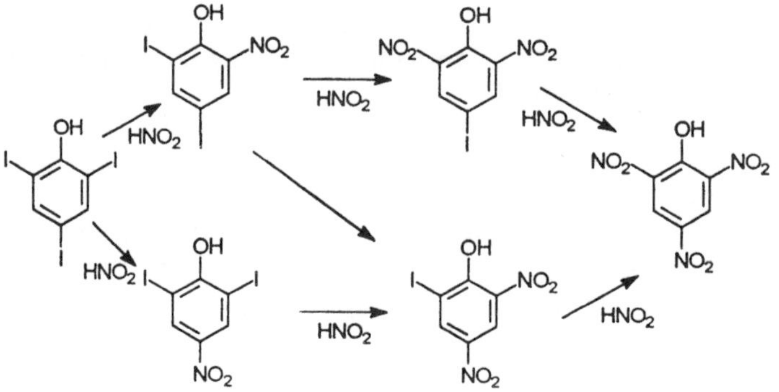

(18)

If the reaction is carried out in dilute hydrochloric or acetic acid, nitrosophenols are formed, sometimes accompanied by small amounts of the corresponding nitrocompounds. Thus, in the reaction of resorcinol with nitrite under such conditions (1 M HCl, e.g.), the 2,4-dinitrosocompound, or its sodium salts after alkalization, is the chromogen [182]. 2-Methyl-resorcinol is converted into 4-nitroso-2-methylresorcinol under the same conditions. When these reactions are carried out using the procedure according to Lykken et al., the corresponding nitrocompounds are the chromogens. The reaction of resorcinol with nitrite in HCl (1 : 1) at 100°C results in the formation of 2,4,6-trinitroresorcinol [183]. Phenolic acids react positively in 1 M HCl medium at 100°C. The resulting chromogens are the corresponding nitro acids [183].

5-Nitrosalicylic acid, or its sodium salt, respectively, is also the chromogen in the reaction of tryptophan with nitrous acid at elevated temperature. First, the heterocyclic nucleus in tryptophan is oxidized by nitrous acid to produce salicylic acid, and then nitration takes place [184].

The nitrosation reaction of paracetamol results in the formation of the 2-nitroderivative [185], and diethylstilboestrol is converted to nitroso- or nitrocompounds according to the conditions used [183,186]. For other applications see Refs. [187–189].

If the reaction of phenols with nitrite is carried out in the presence of heavy metal ions (Cu^{2+}, Ni^{2+}, Co^{2+}, Fe^{2+}, etc.), complexes of these metals with the corresponding nitrosophenols are formed. These can be used for spectrophotometric determination of phenols [190], the corresponding metals, or the nitrite ions (see Chapter 3, Section A). Sodium cobaltnitrite in dilute acetic acid had been proposed as a nitrosation agent forming cobalt chelates of nitrosophenols, although it has since been found that p-substituted phenols are nitrated when treated with this reagent [191,192].

C-Nitrosation also results when N-phenylpyrazolones with a free hydrogen in the 4-position are treated with sodium nitrite [4,5]. For the determination of antipyrine, see Ref. [193].

The reaction of nitrite with barbituric and thiobarbituric acids results in the formation of violuric acids, the alkali metal salts of which are deeply colored [4,5].

Conversely, the same reaction of phenolic compounds with nitrous acid is used to determine nitrite. 2,6-Dimethylphenol (2,6-xylenol) is used as the reagent and the 4-nitrosoderviative (λ_{max} = 307 nm) is the chromogen. The reaction medium of a 5 : 4 : 1 sulfuric acid–water–acetic acid mixture works best because the formation of byproducts is suppressed to a minimum. 3,3′,5,5′-Tetramethyldiphenoquinone is formed in weakly acid solutions (10% acetic acid); 3,3′,5,5′-tetramethylindophenol appears slowly in strongly acid solutions [194]. *N*-nitrosocompounds are the chromogens in the reaction of nitrite with secondary and tertiary aromatic amines and secondary aliphatic amines. Indol has been used as a reagent for the determination of nitrites as well [195].

The interaction of adrenaline with nitrous acid in acid medium serves as an example of contemporary *C*-nitrosation, *N*-nitrosation, and oxidation: 6-nitroadrenaline, 6-nitro-*N*-nitrosoadrenaline, and adrenochrom have been identified as chromogens [5]. Oxidation is the basis of the reaction of phenothiazines with nitrous acid [5].

It should be emphasized that the reaction of nitrous acid with primary aromatic amines results in the formation of diazonium salts (see Section A), some of which are colored or can be decomposed to colored products. The reaction of nitrite with rivanol (2-ethoxy-6,9-diaminoacridinium lactate) has been proposed for the determination of nitrite and/or nitrate (after reduction) [196].

2. Nitric Acid as Reagent

Nitrocompounds can also be obtained by the treatment of different aromatic compounds with nitric acid or nitrate in acidic medium [4,5]. Examples include phenacetine, which yields 3-nitro-4-acetaminophenetole, and vanillin which yields a picric acid derivative [4,5]. The so-called xanthoproteic reaction is based on nitration of aromatic amino acids (tryptophan, tyrosine, phenylalanine) in proteins and is used for the spectrophotometric determination [197].

As with the interaction of nitrous acid, nitric acid can oxidize phenothiazines to colored products [4,5]. Phenoxyacetic acids are nitrated and at the same time undergo oxidation to 2,4-dinitrophenol [4,5]. Pentachlorophenol is oxidized to tetrachlorobenzoquinone [4,5].

Conversely, the same reaction with phenols can be used to determine nitrate. Phenoldisulfonic acid, 2,4-dimethylphenol, 3,4-dimethylphenol, 2,6-dimethylphenol, and chromotropic acid have been proposed as reagents, with the last but one being recommended [198]. The solvent composition 6 : 3 : 1 sulfuric acid–water–acetic acid at room temperature is required for the fast formation of 4-nitro-2,6-xylenol (λ_{max} = 320–324 nm). To prevent interference (formation of colored products), nitrite present in the sample to be analyzed must be destroyed by the addition of sulfamic acid. Chlorides do not interfere at concentrations up to ten times higher than that of nitrate. At higher concentrations, mercuric perchlorate must be added.

E. DI- AND TRIARYLMETHANES

The initial step of this type of reactions is the attack of an electrophilic reagent (aldehyde) on an electron-rich (hetero)aromate under the formation of colorless di- or triaryl- or heteroarylmethanes [Eq. (19)]. The first step is followed by the oxidation step, in which the corresponding colored carbonium ions are formed. These are the chromogens of the spectrophotometric reaction [199–201]. The oxidation can also be caused by atmospheric oxygen and can be catalyzed by traces of metals. The chromogens formed can undergo further structural changes in the reaction medium, for example, cyclization to xanthilium compounds enabled by OH− groups in appropriate position on the benzene or naphthalene nucleus, in the case of resorcinol, phloroglucinol, chromotropic acid, and others.

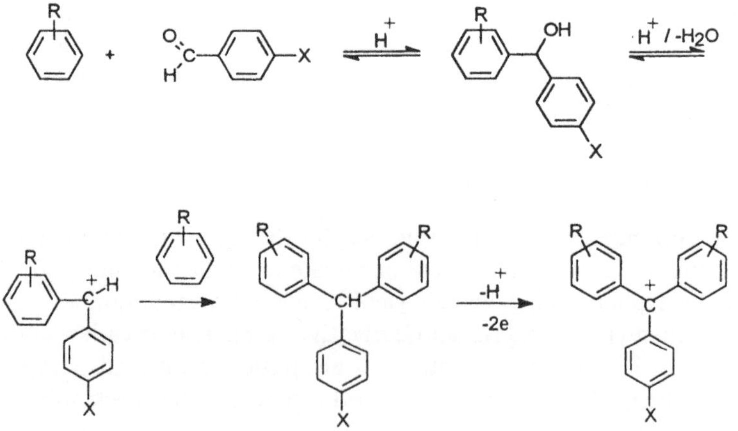

$$(19)$$

1. Aldehydes as Reagents

Formaldehyde, glyoxylic acid, furaldehyde, and substituted aromatic aldehydes (*p*-hydroxy-, *p*-methoxy-, *p*-nitro-, *p*-dimethylaminobenzaldehydes, vanillin, etc.) are the most often used reagents [4–6]. The condensation proceeds in acid media established by acetic acid, 50% to concentrated sulfuric acid, perchloric acid, and trichloracetic acid under addition of a respective oxidizing agent. Derivatives of phenols, alkylarylamines, pyrrole, and indole are the reaction partners that can be determined by this reaction [4–6,202–205].

Formaldehyde in concentrated sulfuric acid is known as the Marquis reagent, the reaction of glyoxylic acid formed in situ by oxidation of acetic acid by $FeCl_3$ in H_2SO_4 as the Hopkins–Cole reaction, and the reaction of *p*-dimethylaminobenzaldehyde with indole derivatives as the van Urk reaction. It has been shown that aldehydes as reagents can be substituted by triethylorthoformiate [206].

Fructose is decomposed in acid reaction medium at 100°C to 5-hydroxymethylfuraldehyde, which on dehydration yields 5,5′-diformyl-difurylether [Eq. (20)]. The resulting ether undergoes condensation with diarylamines with at least one free *p*-position and formation of the corresponding triarylmethanes. The reaction is used for the determination of diphenylamine derivatives [4,5]

$$2 \quad \underset{\text{OHC}}{\overset{\text{O}}{\bigsqcup}}\text{CH}_2\text{OH} \quad \xrightarrow{-\text{H}_2\text{O}} \quad \text{OHC}\underset{}{\overset{\text{O}}{\bigsqcup}}\text{CH}_2\text{.O.CH}_2\underset{}{\overset{\text{O}}{\bigsqcup}}\text{CHO} \quad (20)$$

2. (Hetero)arene Derivatives as Reagents

Phenolic compounds, secondary and tertiary arylamines, and compounds with pyrrole and indole structures in their molecules are useful reagents for the determination of aldehydes [4–6,207,208].

Chromotropic acid (1,8-dihydroxynaphthalene-3,6-disulfonic acid), J-acid (6-amino-1-naphthol-3-sulfonic acid), and phenyl-J-acid (6-phenylamino-1-naphthol-3-sulfonic acid) are reagents for formaldehyde [4–6,167,209–211]. The reagents can also be used for the determination of all compounds that release formaldehyde on heating with the reagent in a medium of highly concentrated sulfuric acid (methylenedioxy compounds, methylenediamines, methylazo-alkanes, glycolic acid, and *N*-methanesulfonates) or by other reactions, for example, on oxidation with periodic acid (primary α-aminoalcohols), $KMnO_4$ (methanol), or chloramine T (glycine) [4,5]. The highly colored reaction product has been found to have mono-cationic dibenzoxanthilium structure [212,213].

This type of reactions is of special importance for the determination of carbohydrates. 1-Naphthol in sulfuric acid, the so-called Mollisch reagent, is used for the determination of different sugars. Many other phenols (phenol, thymol, 3- or 4-hydroxybiphenyl, orcinol, naphthoresorcinol, etc.) as well as indole, carbazole, and their derivatives (3-indolyl-acetic acid, tryptophan) are used as reagents [4–6,214–220]. The reaction path involves the hydrolytic decomposition of sugar molecules to the corresponding furaldehyde derivatives (or to formaldehyde, respectively), which condense with the reagent to corresponding triarylmethanes and on oxidation form the intensely colored carbonium ions. The optimal acid concentration, temperature, and extraction conditions should be determined for each type of analytes and analyzed material.

In some cases the reaction products have been found to be not uniform. Many of these chromogens have been isolated and their structure elucidated [221–224], as in the case of the reaction of 4-ethylresorcinol with fructose (Seliwanoff reaction) [225], where the chromogen is of structure **(16)**.

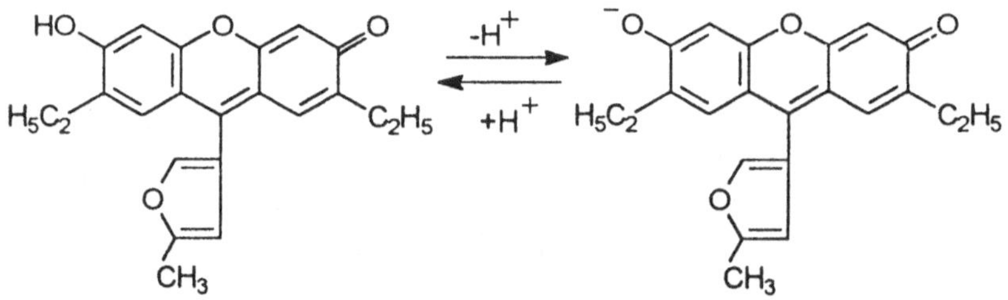

(16)

1,3-Naphthalenediol (naphthoresorcinol), carbazole, and 2- or 3-hydroxybiphenyl are the most often used reagents for the determination of glucuronic acids and glucuronides [226]. Naphthoresorcinol is used in hydrochloric acid medium and the coloration is extracted into 1-pentanol (λ_{max} = 615–620 nm), ethyl acetate, or toluene (λ_{max} = 580 nm). Ether is not a suitable extractant since the peroxides present decompose and decolorize the chromogen. The determination of bound glucuronic acid is achieved by selective oxidation of the free acid with sodium hypoiodite at pH 10 or by reduction with sodium borohydride at pH 9 to form products that do not give a positive reaction. The reaction with carbazole is carried out with concentrated sulfuric acid. The reproducibility for particular types of material analyzed is dependent on the heating period, acid concentration, and presence of salts. The reaction can be improved by the addition of sodium tetraborate. For further applications, see Refs. [227–231].

F. POLYMETHINE DYES AND SIMILAR CHROMOGENS

Aromatic aldehydes condense easily with different types of compounds possessing active methylene groups. The nucleophilic addition to the carbonyl group is represented by the reaction of aromatic aldehydes with barbituric or 2-thiobarbituric acid [Eq. (21)].

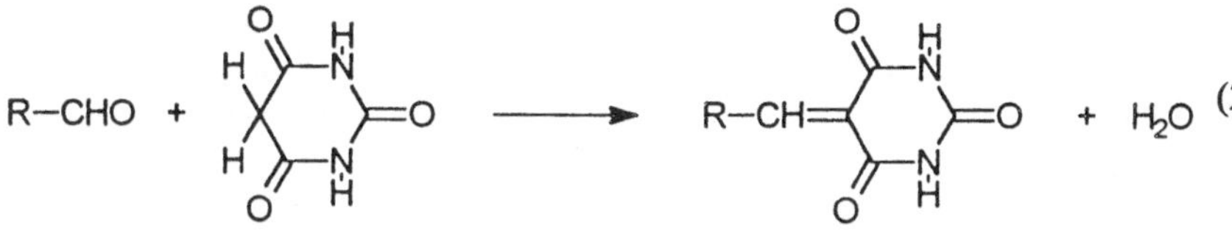

$$R-CHO + \text{(barbituric acid structure)} \longrightarrow R-CH= \text{(chromogen structure)} + H_2O \quad (21)$$

The reaction used for the determination of aromatic aldehydes proceeds in alcoholic media at slightly elevated temperatures [232]. Conversely, barbituric acid can be determined using p-dimethylaminobenzaldehyde in glacial acetic acid [233]. Glutaconaldehyde (2-pentenedial), malonaldehyde, and similar reactive aldehydes also react with barbituric and thiobarbituric acids to form colored derivatives. Procedures for the determination of pyridine, cyanide ions, and acrylonitrile mentioned in Section C can be carried out with 2-thiobarbituric acid as the reagent under the formation of a poly-methine chromogen [234–237]. Some researchers complaining of the insta-bility of this polymethine dye, prefer to evaluate the coloration of glutaco-naldehyde released by its decomposition [142–144]. Some researchers have succeeded in measuring the intermediate chromogen [236].

Compounds that can be oxidized, for instance, by periodic acid to reactive aldehydes (1,2-diols, 2-deoxysaccharides, quinic acid, etc.) can be determined by reacting the resulting aldehyde with thiobarbituric acid [4,5]. 3-Formylglutaric acid obtained by oxidation of sialic acids forms a chromo-gen (17) [5].

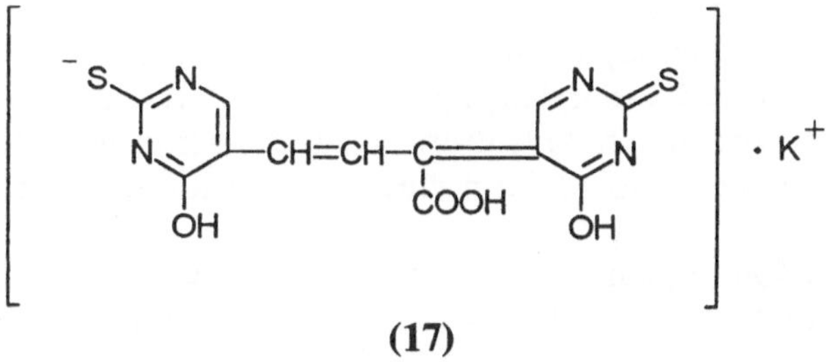

(17)

2-Thiobarbituric acid is also an important reagent for the determina-tion of different saccharides after their decomposition in acid media [4,5,238–244].

Reactive arylhalogenides (2,4-dinitrochlorobenzene, 2,4-dinitroflu-orobenzene, etc.) react with pyridine under the formation of a pyridine salt, which is converted in alkaline medium into a Schiff base of glutaconalde-hyde. The red glutaconaldehyde derivative, however, is unstable, and is hydrolyzed to yellow glutaconaldehyde and 2,4-dinitroaniline [Eq. (22)] [245]. The reaction is used for the determination of pyridine [245] and of arylhalogenides, which can be nitrated to more reactive nitro- or dinitrohalogenides [5].

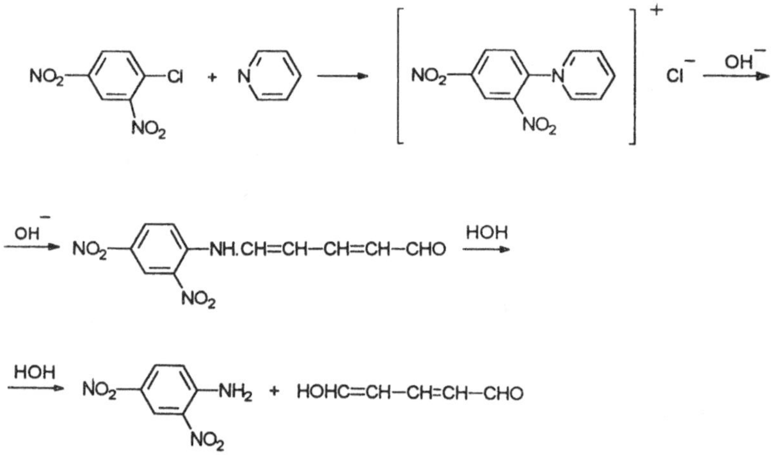

(22)

Anthrone in concentrated sulfuric acid is a favorite reagent for the determination of saccharides. Pentoses; methylpentoses; hexoses; hexuronic acids; di-, oligo-, and polysaccharides; dexoysaccharides; and glycosides; as well as aliphatic aldehydes and steroids give a positive reaction [4–6,246–251]. The reaction conditions (acid concentration, reaction temperature, and reaction period) should be modified for each particular analyte and material analyzed. 2-Furylaldehyde derivatives, which are formed from saccharides by the action of sulfuric acid, condense with anthrone at its C-10 methylene group to form chromogens with $\lambda_{max} = 610$–630 nm [252,253].

The reaction of guanidine derivatives with 1-naphthol and diketones of the common formula $R\text{-}CO.CO.CH_3$ is known as the Voges–Proskauer reaction. Its reaction products have been found to be not uniform [5].

Fluorene reacts rapidly with the sodium salt of 1,2-naphthoquinone-4-sulfonic acid in alkaline dimethyl sulfoxide medium containing sodium methoxide and methanol to give a colored product [254]. The reaction is used for the determination of fluorene.

4-(4-Nitrobenzyl)pyridine dissolved in acetone, methylethyl ketone, or acetophenone easily undergoes alkylation and acylation on the pyridine nitrogen. On addition of alkali hydroxide or an organic base, the products are converted to the intensely colored corresponding dihydropyridine derivatives [e.g., **(18)**]. The reaction enables the determination of alkylating (λ_{max} = 550–575 nm) and acylating agents (λ_{max} = 450–480 nm) and of organic phosphorus, arsenium, and silicium compounds [4–6,255]. For new regents of this type, see Ref. [256].

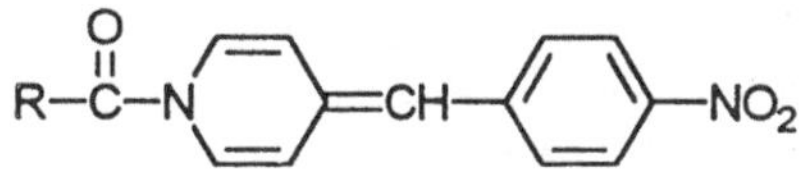

$$R-\overset{\overset{\displaystyle O}{\|}}{C}-N\diagdown\diagup=CH-\diagdown\diagup-NO_2$$

(18)

Aromatic amines react with formaldehyde and acetylacetone to form colored dihydrolutidine derivatives [Eq. (23)]. The procedure is suitable for the determination of formaldehyde, the diketones, and amines [257–259].

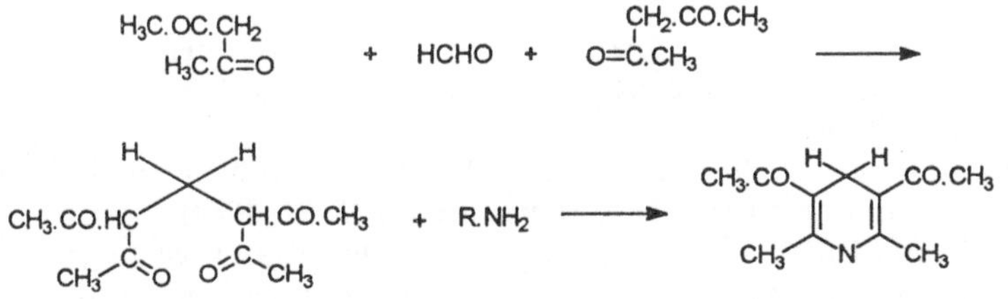

$$\begin{array}{c} H_3C.OC.CH_2 \\ | \\ H_3C.C=O \end{array} + HCHO + \begin{array}{c} CH_2.CO.CH_3 \\ | \\ O=C.CH_3 \end{array} \longrightarrow$$

$$CH_3.CO.HC \underset{\substack{| \\ CH_3 \diagup C \diagdown O}}{\overset{\substack{H \diagdown \diagup H}}{\diagup \diagdown}} CH.CO.CH_3 + R.NH_2 \longrightarrow \underset{CH_3 \diagup \overset{|}{N} \diagdown CH_3}{CH_3.CO \overset{H \diagdown \diagup H}{\diagup \diagdown} CO.CH_3} \quad (23)$$

The Komarowski reaction is based on the action of aromatic aldehydes (*p*-hydroxybenzaldehyde, vanillin, *p*-dimethylaminobenzaldehyde, etc.) on aliphatic alcohols in concentrated sulfuric acid at elevated temperatures [4,5,260,261]. The first reaction step involves dehydratation of the alcohols and formation of reactive olefins, which react with the aldehydes under formation of the corresponding colored oxonium salts [4,5].

REFERENCES

1. K. H. Saunders and R. L. M. Allen, *Aromatic Diazo Compounds*, 3rd edition, Arnold, London, 1985.
2. H. Zollinger, *Diazo and Azo Chemistry*, Interscience, New York, 1961.
3. T. Urbányi and J. A. Mollica, Potential diazo reagents for photometric determinations, *J. Pharm. Sci. 57*: 1257 (1968).
4. Z. J. Vejdělek and B. Kakáč, *Farbreaktionen in der spektrophotometrischen Analyse organischer Verbindungen. Band I. Organische Farbreagenzien*, G. Fischer Verlag, Jena, 1969.
5. Z. J. Vejdělek and B. Kakáč, *Farbreaktionen in der spektrophotometrischen Analyse organischer Verbindungen. Suppl. Vol. I. Organische Farbreagenzien*, G. Fischer Verlag, Jena, 1980.

6. M. Pesez and J. Bartos, *Colorimetric and Fluorimetric Analysis of Organic Compounds and Drugs*, Marcel Dekker, New York, 1974.

7. G. S. Sadana and G. G. Parikh, A simple colorimetric method for the determination of chlorquinaldol from pharmaceutical preparations, *Indian Drugs 24*: 531 (1987).

8. A. A. Al-Hatim and B. B. Ibraheem, Spectrophotometric determination of 1-naphthylamine in aqueous solution by coupling with diazotized 4-aminobenzophenone, *Anal. Lett. 22*: 2091 (1989).

9. S. A. Rahim, B. B. Ibraheem, and W. A. Bashir, Spectrophotometric determination of trace amounts of acetylacetone in aqueous solutions, *Microchem. J. 36*: 297 (1987).

10. S.-Y. Lin, K.-J. Duan, and T.-C. Lin, Variables affecting the spectrophotometric determination of ascorbic acid based on the reaction with Fast Red AL Salt, *Pharm. Acta Helv. 69*: 39 (1994).

11. S. Flamerz and W. A. Bashir, Spectrophotometric determination of acetaldehyde in aqueous solution with diazotized orthanilic acid, *Anal. Chem. 54*: 1734 (1982).

12. W. A. Bashir and S. Flamerz, Spectrophotometric determination of α-ketoglutaric acid with diazotized *p*-aminobenzoic acid, *Fresenius' Z. Anal. Chem. 309*: 401 (1981).

13. W. A. Bashir, G. W. Hagop, and S. Flamerz, Spectrophotometric determination of acetone in acetic acid, *Microchem. J. 28*: 77 (1983).

14. A. K. Ahmad, Y. I. Hassan, and W. A. Bashir, Diazotized 4-nitroaniline as a chromogenic reagent for the determination of trace amounts of pyrrole in aqueous solution, *Analyst 112*: 97 (1987).

15. A. P. Latawiec, Spectrophotometric determination of common aromatic diisocyanates using 4-nitrobenzenediazonium tetrafluoroborate in dimethyl sulphoxide, *Analyst 116*: 749 (1991).

16. G. Waldheim, H. Moehrle, and S. Ruediger, The coloring constituent of the spectrophotometric determination of phenylbutazone after reaction with *p*-nitrophenyl-diazonium chloride (in German), *Pharmazie 42*: 11 (1987).

17. K. M. Appaiah, U. C. Nag, and O. P. Kapur, Colorimetric determination of carbaryl and its residue in vegetables, *J. Food Sci. Technol. 20*: 252 (1983).

18. M. M. Ayad, S. F. Belal, and S. M. Al Adel, Determination of some pharmaceutical carbonyl compounds, *Acta Pol. Pharm. 42*: 561 (1985).

19. Das Joyce Vanisha, K. N. Ramachandran, and V. K. Gupta, Spectrophotometric determination of baygon and carbaryl, *Fresenius' J. Anal. Chem. 348*: 840 (1994).

20. H. Kuboki and S. Kikuchi, Spectrophotometric determination of diazonium salts and its applications, *Japan Analyst 13*: 66 (1964).

21. D. A. Ben-Ephreim, Detection and determination of diazo and diazonium groups, in *The Chemistry of Diazonium and Diazo Groups* Part I, (S. Patai, ed.), Wiley, New York, 1978, p. 149.

22. K. Schank, Preparation of diazonium groups, in *The Chemistry of Diazonium and Diazo Groups* Part I, (S. Patai, ed.), Wiley, New York, 1978, p. 647.

23. G. Norwitz and P. N. Keliher, Spectrophotometric determination of aromatic

amines by the diazotization-coupling technique with *N*-(1-naphthyl)ethylene-diamine as the coupling agents, *Anal. Chem. 54*: 807 (1982).

24. V. Kratochvíl and J. Kroupa, Spectrophotometric determination of benzidine, diphenyline, *o*-benzidine, *o*-tolidine and *o*-dianisidine by a coupling method (in Czech), *Chem. Prum. 37*: 311 (1987).

25. M. M. Abdel-Khalek, M. S. Mahrous, H. G. Daabees, and Y. A. Beltagy, Spectrophotometric determination of aminoglutethimide by diazotization and subsequent coupling, *Anal. Lett. 26*: 1109 (1993).

26. B. Gala, A. Goméz-Hens, and D. Pérez-Bendito, Direct kinetic determination of bromhexine hydrochloride in pharmaceutical formulations, *Anal. Lett. 26*: 2607 (1993).

27. J. S. Esteve-Romero, E. F. Simo-Alfonso, M. C. Garcia-Alvarez-Coque, and G. Ramis-Ramos, Conventional and thermal lens spectrophotometric determination of *p*-aminobenzoic acid and arylamine diuretics after previous azodye formation in a micellar medium, *Talanta 40*: 1711 (1993).

28. M. Carmona, M. Silva, and D. Perez-Bendito, Automatic kinetic determination of oxazepam by the continuous addition of reagent technique, *Talanta 39*: 1175 (1992).

29. H. Schütz, *Benzodiazepines: A Handbook*, Springer-Verlag, Berlin-Heidelberg-New York, 1982.

30. H. Schütz, *Benzodiazepines II: A Handbook*, Springer-Verlag, Berlin-Heidelberg-New York, 1989.

31. A. K. Sanyal, Rapid and selective method for quantitation of metronidazole in pharmaceuticals, *J. Assoc. Off. Anal. Chem. 71*: 849 (1988).

32. S. M. Hassan, F. Belal, M. Sharaf El-Din, and M. Sultan, Spectrophotometric determination of some pharmaceutically important nitro compounds in their dosage forms, *Analyst 113*: 1087 (1988).

33. H. A. Dingjan and H. A. Crezée, The Guerbet reaction applied to diodone, *Mikrochim. Acta II*: 395 (1978).

34. J. Gasparič, J. Skutil, and M. Teclová, The color reaction of arylhydrazines according to Postowsky (in German), *Mikrochim. Acta II*: 17 (1980).

35. M. Večeřa and M. Jureček, Analysis of pharmaceuticals. IV. Colorimetric determination of phenylhydrazine and phenylhydrazones in antipyrine (in Czech), *Chem. Listy 45*: 475 (1951).

36. S. Chen, S. Peng, and D. Yuan, Spectrophotometric determination of trace selenium(IV) with a new catalytic reaction (in Chinese), *Fenxi Huaxue 12*: 913 (1984); *Chem. Abstr. 102*: 39081b (1985).

37. E. Sawicki, T. W. Stanley, J. Pfaff, and A. D'Amico, Comparison of fifty-two spectrophotometric methods for the determination of nitrite, *Talanta 10*: 641 (1963).

38. A. A. Al-Hatim, Spectrophotometric determination of nitrite in aqueous solution by the diazotation-coupling method with *p*-aminobenzophenone-*N*-(1-naphthyl)ethylenediamine, *Int. J. Environ. Anal. Chem. 38*: 617 (1990).

39. R. Kaveeshwar, L. Cherian, and V. K. Gupta, Extraction-spectrophotometric determination of nitrite using 1-aminonapthalene-2-sulphonic acid, *Analyst 116*: 667 (1991).

40. G. Norwitz and P. N. Keliher, Interference of ascorbic and isoascorbic acids in the spectrophotometric determination of nitrite by the diazotisation-coupling technique, *Analyst 112*: 903 (1987).

41. N. Z. Muradov, Spectrophotometric method for nitroglycerine analysis in air, *Environ. Sci. Technol. 28*: 388 (1994).

42. T. Ohta, N. Goto, and S. Takitani, Spectrophotometric determination of *N*-nitroso compounds by flow-injection analysis, *Analyst 113*: 1333 (1988).

43. V. Mejstřík, I. Hronková, L. Držková, F. Krampera, J. Jandera, and Z. Ságner, Photometric determination of *N*-nitrosamines (in Czech), *Chem. Prum. 38*: 321 (1988).

44. K. N. Ramachandran, L. Cherian, and V. K. Gupta, Extraction spectrophotometric determination of nitrate in environmental samples, *Asian Environ. 15*: 53 (1993); *Chem. Abstr. 120*; 207559f (1994).

45. T. I. Ivkova, R. P. Pantaler, and L. I. Gorodilova, Photometric methods for determination of nitrate and nitrite ions in oxides of cadmium and tungsten, *Vysokochist. Veshchestava 6*(2): 127 (1993); *Chem. Abstr. 120*: 94290r (1994).

46. A.-M. Sjoeberg and T. A. Alanko, Spectrophotometric determination of nitrate in baby foods: Collaborative study, *J. AOAC Int. 77*: 425 (1994).

47. B. Saville, A scheme for the colorimetric determination of microgram amounts of thiols, *Analyst 83*: 670 (1958).

48. H.-J. Kallmayer and J. Weiten, The colored product of the reaction of sulfanilamide with thymol and NaClO (in German), *Pharmazie 43*: 130 (1988).

49. F. A. Mohamed, A. I. Mohamed, and S. R. El-Shabouri, Visible spectrophotometric determination of sulfonamides, *J. Pharm. Biomed. Anal. 6*: 175 (1988).

50. T. T. Ngo and C. F. Yam, Sensitive and selective spectrophotometric determination of *p*- and *o*-aminophenol with hypochlorite-alkaline phenol (Berthelot) reagents, *Anal. Lett. 17 (A15)*: 1771 (1984).

51. F. Belal, Spectrophotometric determination of halogenated 8-hydroxyquinoline derivatives, *Talanta 31*: 648 (1984).

52. E. E Rao and C. S. P. Sastry, New spectrophotometric determination of terbutaline sulfate, *Microchem. J. 32*: 293 (1985).

53. C. S. P. Sastry, P. L. Kumari, and B. G. Rao, New colorimetric method for the assay of procainamide, metoclopramide, and nitrazepam, *Chem. Anal. (Warsaw) 30*: 461 (1985).

54. D. G. Sankar, C. S. P. Sastry, M. N. Reddy, and N. R. P. Singh, Spectrophotometric determination of some adrenergic drugs with *p-N,N*-dimethylphenylenediamine, *Indian Drugs 24*: 410 (1987); *Chem. Abstr. 107*: 64987n (1987).

55. K. K. Verma, S. K. Snaghi, and A. Jain, Spectrophotometric determination of primary aromatic amines with 4-*N*-methylaminophenol and 2-iodyl benzoate, *Talanta 35*: 409 (1988).

56. S. Amlathe and V. K. Gupta, Spectrophotometric determination of aniline in air and water via oxidative coupling with guaiacol, *Microchem. J. 43*: 208 (1991).

57. C. S. P. Sastry, T. T. Rao, and A. Sailaja, Spectrophotometric determination

of nicoumalone, acebutolol hydrochloride and procainamide hydrochloride, *Talanta 38*: 1057 (1991).

58. K. K. Verma and A. Jain, Spectrophotometric determination of oxyphenbutazone in drug formulations using the indophenol reaction, *Analyst 110*: 997 (1985).

59. H. Sakurai and M. Umeda, Color reaction of phenols with oxidizing agents and phenylhydrazine and its derivatives. I. Color reaction of phenols with chloramine T and phenylhydrazine or its derivatives, *Yakugaku Zasshi 82*: 1282 (1962); *Chem. Abstr. 58*: 13828f (1963).

60. P. Siraj, S. S. N. Murthy, B. S. Reddy, B. G. Rao, and C. S. Prakasasastry, Spectrophotometric determination of isoniazid using potassium iodate–metol reagent, *Natl. Acad. Sci. Lett. (India) 4*: 203 (1981); *Chem. Abstr. 96*: 129852r (1982).

61. K. K. Verma and K. K. Stewart, Spectrophotometric determination of aromatic primary amines and nitrite by flow-injection analysis, *Anal. Chim. Acta 214*: 207 (1988).

62. W. Leithe and G. Petschl, Determination of ammonia in air via the indophenol reaction (in German), *Z. Anal. Chem. 230*: 344 (1967).

63. C. J. Patton and S. R. Crouch, Spectrophotometric and kinetic investigation of the Berthelot reaction for the determination of ammonia, *Anal. Chem. 49*; 464 (1977).

64. M. D. Krom, Spectrophotometric determination of ammonia: A study of a modified Berthelot reaction using salicylate and dichloroisocyanurate, *Analyst 105*: 305 (1980).

65. Xing-Chu Qiu, Guo-Ping Liu, and Ying-Quan Zhu, Determination of water-soluble ammonium ion in soil by spectrophotometry, *Analyst 112*: 909 (1987).

66. C. Low and B. G. Adams, Spectrophotometric determination of urea-ammonia in the urea degradation pathway of *Saccharomyces cerevisiae*, *J. Microbiol. Methods 11*: 229 (1990).

67. M. K. Tummuru, K. E. Rao, and C. S. P. Sastry, Spectrophotometric determination of amino acids using Metol and sodium hypochlorite or chloramine-T, *Mikrochim. Acta*: 199 (1984).

68. Y. Z. Ahmed, M. Abd-Elmottalb, and I. M. Abd-Ellah, Spectrophotometric determination of microgram amounts of amino acids by indophenol reagents, *Acta Chim. Hung. 122*: 111 (1986).

69. R. Oliver, Determination of nitrates by automatic colorimetry after reduction by Dewarda alloy, *Analusis 6*: 126 (1978).

70. M. A. Arustamyan, L. E. Zel'tser, D. Kh. Yunusov, and N. Suleimanova, Spectrophotometric determination of guanidine (in Russian), *Zh. Anal. Khim. 38*: 129 (1983).

71. A. Hessing and K. Hoppe, Sakaguchi and Fearon reaction: Structure of the dyes, their mechanism of formation, and the specificity of the reaction, *Chem. Ber. 100*: 3649 (1967).

72. D. Svobodová and J. Gasparič, Investigation of the colour reaction of phenols with 4-aminoantipyrine, *Mikrochim. Acta*: 384 (1971).

73. A. Abou Ouf, M. I. Walash, S. M. Hassan, and S. M. El-Sayed, Spectrophotometric determination of oxyphenbutazone in pharmaceutical preparations, *Analyst 105*: 169 (1980).

74. Ya. I. Korenman, A. T.Alymova, N. S. Kobeleva, I. A. Belen'kaya, E. F. Goritskaya, and G. A. Markus, Adsorption-photometric determination of phenols in natural and purified wastewaters (in Russian), *Zavod. Lab. 50*: 5 (1984).

75. Ya. I. Kornenman, T. N. Ermolaeva, T. A. Kuchmenko, and A. V. Michina, Photometric determination of phenol traces in hydrophilic extracts (in Russian), *Zh. Prikl. Khim. 66*: 2043 (1993).

76. M. Rodriguez-Alcala, P. Yanez-Sedeno, and L. M. Polo-Diez, Determination of pentachlorophenol by FIA and spectrophotometry. *Talanta 35*: 601 (1988).

77. R. Karlíček and P. Solich, FIA determination of tetracycline antibiotics, *Anal. Chim. Acta 285*: 9 (1994).

78. D. Svobodová, M. Fraenkl, and J. Gasparič, On the mechanism of the colour reaction of phenols with 4-dimethylaminoantipyrine, *Mikrochim. Acta I*: 285 (1977).

79. S. Ono, R. Onishi, and K. Kawamura, Analysis of mixed anticold pharmaceutical preparations. II. Colorimetric determination of aminopyrine, *Yakugaku Zasshi 85*: 245 (1965); *Chem. Abstr. 63*: 436g (1965).

80. C. L. M. Stults, A. P. Wade, and S. R. Crouch, Computer-assisted optimization of an immobilized-enzyme flow-injection system for the determination of glucose, *Anal. Chim. Acta 192*: 155 (1987).

81. I. Goedicke and W. Goedicke, Vanillinic acid, a new chromogen for the enzymic determination of glucose, *Z. Med. Laboratoriumsdiagn. 28*: 247 (1987); *Chem. Abstr. 107*: 194360p (1987).

82. J. Gasparič, D. Svobodová, and M. Pospíšilová, Investigation of the colour reaction of phenols with the MBTH reagent, *Mikrochim. Acta I*: 241 (1977).

83. M. E. El-Kommos and K. M. Emara, Spectrophotometric determination of some phenothiazine drugs using 3-methylbenzothiazolin-2-one hydrazone, *Analyst 113*: 1267 (1988).

84. K. R. Raju, T. N. Parthasarathy, S. R. K. M. Akella, and U. T. Bhalerao, Spectrophotometric assay of Isoproduron and Metoxuron in technical grade and formulation samples using 3-methyl-benzothiazolin-2-one hydrazone hydrochloride, *Talanta 39*: 1387 (1992).

85. J. N. Rodriguez-Lopez, J. Escribano, and F. Garcia-Canovas, A continuous spectrophotometric method for the determination of monophenolase activity of tyrosinase using 3-methyl-2-benzothiazolinone hydrazone, *Anal. Biochem. 216*: 205 (1994).

86. J. J. Berzas Nevado, J. M. Lemus Gallego, and P. Buitrago Laguna, Determination of sulfamethazine and sulfathiazole in mixtures using 3-methyl-2-benzothiazolone hydrazone in the presence of Fe(III) by ratio spectra derivatization, *Analusis 22*: 226 (1994).

87. H. O. Friestad, D. E. Ott, and F. A. Gunther, Automated colorimetric

microdetermination of phenols by oxidative coupling with 3-methyl-2-benzothiazolinone hydrazone, *Anal. Chem. 41*: 1750 (1969).

88. K. M. Emara and N. M. Mahfouz, Colorimetric determination of certain benzodiazepine drugs, *Egypt. J. Pharm. Sci. 34*: 267 (1993); *Chem. Abstr. 121*: 308480z (1994).

89. D. Svobodová, P. Křenek, M. Fraenkl, and J. Gasparič, Colour reaction of phenols with the Gibbs reagent: The reaction mechanism and decomposition and stabilisation of the reagent, *Mikrochim. Acta I*: 251 (1977).

90. D. Svobodová, P. Křenek, M. Fraenkl, and J. Gasparič, Colour reaction of phenols with the Gibbs reagent: The properties of the coloured product and the optimum reaction conditions, *Mikrochim. Acta II*: 197 (1978).

91. M. Yamamoto and T. Uno, Mechanism of color reaction of aromatic secondary and tertiary amines with quinonedichlorodiimide, *Chem. Pharm. Bull. 27*: 688 (1979).

92. D. M. Shingbal and S. D. Naik, Colorimetric determination of salbutamol sulfate, *Can. J. Pharm. Sci. 16*: 65 (1981).

93. D. G. Sankar, C. S. P. Sastry, M. N. Reddy, and S. N. R. Prasad, Spectrophotometric determination of some adrenergic drugs using 2,6-dichloroquinone-chloroimide, *Indian J. Pharm. Sci. 49*: 69 (1987); *Chem. Abstr. 107*: 161782z (1987).

94. S. S. Artemchenko, V. V. Petrenko, N. P. Zhovna, and V. A. Tsilinko, Application of 1,4-benzoquinone chloroimine for determination of drugs (in Russian), *Zh. Anal. Khim. 40*: 744 (1985).

95. K. T. Finley, The addition and substitution chemistry of quinones, in *The Chemistry of Quinonoid Compounds*, Part 2, (S. Patai, ed.), Wiley, New York, 1974, p. 877.

96. M. A. Korany, A. M. Wahbi, and M. H. Abdel-Hay, *p*-Benzoquinone as a reagent for determining some catecholamines, *J. Pharm. Biomed. Anal. 2*: 537 (1984).

97. M. L. Iskander, H. A. A. Medien, and S. Nashed, Spectrophotometric determination of aromatic amines by reaction with *p*-benzoquinone, *Microchem. J. 36*: 368 (1987).

98. M. L. Iskander, H. A. A. Medien, and S. Nashed, Kinetics and determination of sulfa drugs via interaction with *p*-benzoquinone, *Microchem. J. 39*: 43 (1989).

99. M. L. Iskander and H. A. A. Medien, Some observations on the spectrophotometric determination of amino acids via interaction with *p*-benzoquinone, *Microchem. J. 41*: 172 (1990).

100. D. A. M. Zaia, W. J. Barreto, N. J. Santos, and A. S. Endo, Spectrophotometric method for the simultaneous determination of proteins and amino acids with *p*-benzoquinone, *Anal. Chim. Acta 277*: 89 (1993).

101. M. C. Gutierrez, A. Gomez-Hens, and D. Perez-Bendito, Kinetic-photometric determination of spermine, spermidine, and their mixtures by the stopped-flow technique, *Fresenius' Z. Anal. Chem. 331*: 642 (1988).

102. S. A. Filipeva, L. N. Strelets, V. V. Petrenko, and V. P. Buryak, Use of

sodium 1,2-naphthoquinone-1-sulfonate for spectrophotometric determination of some pharmaceuticals (in Russian), *Zh. Anal. Khim. 44*: 131 (1989).

103. J. Saurina and S. Hernandez-Cassou, Continuous-flow spectrophotometric determination of amino acids with 1,2-naphthoquinone-4-sulfonate reagent, *Anal. Chim. Acta 283*: 414 (1993).

104. C. Molins Legua, P. Campins Falcó, and A. Sevillano Cabeza, Extractive-spectrophotometric determination of amphetamine in urine samples with sodium 1,2-naphthoquinone-4-sulfonate, *Anal. Chim. Acta 283*: 635 (1993).

105. A. Sevillano Cabeza, P. Campins Falcó, and C. Molins Legua, Kinetic-spectrophotometric determination of primary and secondary amines by reaction with 1,2-naphthoquinone-4-sulfonate, *Anal. Lett. 27*: 1095 (1994).

106. P. Campins Falcó, A. Sevillano Cabeza, and C. Molins Legua, Extractive-spectrophotometric determination of ephedrine with sodium 1,2-naphthoquinone-4-sulfonate in pharmaceutical formulations, *Anal. Lett. 27*: 531 (1994).

107. J. R. Lindsay Smith, A. U. Smart, F. E. Hancock, and M. V. Twigg, High-performance liquid chromatographic determination of low levels of primary and secondary amines in aqueous media via derivatisation with 1,2-naphthoquinone-4-sulfonate, *J. Chromatog. 483*: 341 (1989).

108. A. Salman, Spectrophotometric determination of dihydralazine sulfate with 1,2-naphthoquinone-4-sulfonate, *Sci. Pharm. 55*: 255 (1987).

109. A. Punta Cordero, F. J. Barragán de la Rosa, and A. Guiraum, Spectrophotometric determination of sulfanilic acid and sulfonamides in pharmaceutical samples with potassium 1,2-naphthoquinone-4-sulfonate, *Can. J. Chem. 67*: 1599 (1989).

110. C. S. P. Sastry, A. Sailaja, and T. T. Rao, Sensitive spectrophotometric methods for the determination of pindolol, *Mikrochim. Acta 107*: 1 (1992).

111. J. Jeník, F. Renger, and M. Smetanová, Spectrophotometric determination of organic nitro compounds in industrial wastewaters, *Vodni Hospod. B 31*: 325 (1981); *Chem. Abstr. 97*: 28320f (1982).

112. A. Punta, F. J. Barragan, M. Ternero, and A. Guiraum, Determination of sulfide in sewage effluents using a new spectrophotometric method, *J. Environ. Anal. Chem. 43*: 91 (1991).

113. M. Umeda, Color reaction of phenols. I. Color reaction mechanism of o-cresol and resorcinol with 1,2-naphthoquinone-4-sulfonate (in Japanese), *Yakugaku Zasshi 85*: 28 (1965).

114. S. Dayagi and Y. Degani, Methods of formation of the carbon-nitrogen double bond, in *The Chemistry of the Carbon–Nitrogen Double Bond* (S. Patai, ed.), Interscience, New York, 1970, p. 61.

115. O. S. Kamalapurkar and G. J. Kamat, Spectrophotometric evaluation of sulfamethoxazole, *Indian Drugs 22*: 95 (1984).

116. F. B. Salem and M. I. Walash, Spectrophotometric determination of certain sympathomimetic amines, *Analyst 110*: 1125 (1985).

117. C. S. P. Sastry, B. G. Rao, and B. S. Sastry, Spectrophotometric determination of primary aromatic amines using syringaldehyde, *J. Indian Chem. Soc. 63*: 1006 (1986).

118. J. Vachek, Spectrophotometric determination of diaveridine in the presence of sulfadimidine (in Czech), *Cesk. Farm. 34*: 367 (1985).
119. A. A. Belyakov, L. V. Mel'nikova, and L. T. Kurenko, Unified standard substance for photometric determination of aromatic amines and isocyanates (in Russian), *Zavod. Lab. 53*(4): 6 (1987).
120. M. Qureshi and I. A. Khan, Detection and spectrophotometric determination of some aromatic nitrogen compounds with *p*-dimethylaminocinnamaldehyde, *Anal. Chim. Acta 86*: 309 (1976).
121. M. I. Walash, A. Abou Ouf, and F. B. Salem, Colorimetric determination of sympathomimetic amines methyldopa and noradrenaline, *J. Assoc. Off. Anal. Chem. 68*: 91 (1985).
122. P. Parimoo and S. Kumar, Determination of bromhexine hydrochloride by colorimetry, *Indian Drugs 31*: 41 (1994).
123. S. Nakatsuji, R. Nakano, M. Kawano, K. Nakashima, and S. Akiyama, 5-(*p*-Dimethylaminophenyl)-2,4-pentadienal as an analytical reagent: A simple preparation of the reagent and its application to the colorimetric determination of primary aromatic amines, *Chem. Pharm. Bull. 30*: 2467 (1982).
124. D. M. Shingbal and A. S. Khandeparkar, Colorimetric estimation of nifedipine and its dosage forms, *Indian Drugs 24*: 415 (1987).
125. K. R. Mahadik, G. B. Byale, H. N. More, and S. S. Kadam, A spectrophotometric method for estimation of nifedipine and its formulations, *J. Inst. Chem. (India) 63*: 218 (1991).
126. P. Y. Khashaba, Spectrophotometric determination of azidoamphenicol in pure form and in combination with other drugs, *Alexandria J. Pharm. Sci. 8*: 145 (1994); *Chem. Abstr. 121*: 308471x (1994).
127. A. Saeed, S. Haque, and S. Z. Qureshi, Resin bead detection and spectrophotometric determination of oxyphenbutazone with *p*-dimethylaminocinnamaldehyde: Application to bulk drug and dosage forms, *Talanta 40*: 1867 (1993).
128. K. K. Verma, A. Jain, N. Patel, and S. K. Sanghi, Spectrophotometric determination of dipyrone, phenylbutazone and oxyphenbutazone by their hydrolysis and Schiff bases formation with 4-dimethylaminobenzaldehyde, *Farmaco, Ed. Prat. 42*: 185 (1987).
129. T. Kitagawa, K. Iwakura, M. Ohsugi, and E. Hirai, Analytical studies on isoxazoles. V. Colorimetric determination of isouron and its isomer with *p*-dimethylaminocinnamaldehyde, *Chem. Pharm. Bull. 32*: 2736 (1984).
130. M. Lopez-Nieves, P. D. Wentzell, and S. R. Crouch, Continuous flow method for the determination of aromatic aldehydes, *Anal. Chim. Acta 258*: 253 (1992).
131. F. Yang and S. Li, Determination of furfural in wastewater by colorimetry, *Huanjing Kexue, 13*: 80 (1992); *Chem. Abstr. 118*: 131606e (1993).
132. M. S. Mayadeo and R. K. Banavali, Spectrophotometric determination of *p*-aminophenol and acetaminophen through Schiff base formation and subsequent chelation, *Indian J. Chem., Sect. A, 25A*: 789 (1986); *Chem. Abstr. 106*: 12057u (1987).
133. I. E. Kalinichenko and E. Ya. Mateeva, Simultaneous spectrophotometric

determination of polyethylene-polyamines (in Russian), *Zh. Anal. Khim. 49*: 230 (1994).

134. V. Nirmalchandar, K. E. John, and N. Balasubramanian, Spectrophotometric determination of ascorbic acid, *Indian Drugs 25*: 309 (1988).

135. S. Amlathe, S. Upadhyay, and V. K. Gupta, A sensitive spectrophotometric determination of traces of pyridine with anthranilic acid in environmental samples, *Microchem. J. 37*: 225 (1988).

136. K. N. Ramachandran and V. K. Gupta, Spectrophotometric determination of pyridine with 6-amino-1-naphthol-3-sulfonic acid, *Microchem. J. 44*: 272 (1991).

137. M. Bhattacharjee, S. Amlathe, and V. K. Gupta, An extractive spectrophotometric method for the determination of pyridine in air and environmental samples, *Int. J. Environ. Anal. Chem. 45*: 127 (1991).

138. S. Upadhyay and V. K. Gupta, Spectrophotometric method for the determination of cyanide and its application to biological fluids, *Analyst 109*: 1619 (1984).

139. K. N. Ramachandran and V. K. Gupta, A new spectrophotometric method for determination of cyanide with J-acid, *Chem. Anal. (Warsaw) 37*: 485 (1992).

140. K. N. Ramachandran and V. K. Gupta, Studies on extraction spectrophotometric determination of acrylonitrile by Koenig reaction, *Chem. Anal. (Warsaw) 38*: 491 (1993).

141. V. Agrawal and V. K. Gupta, An extractive sensitive spectrophotometric method for the determination of acrylonitrile via Koenig reaction, *Asian Environ. 15*: 61 (1993); *Chem. Abstr. 120*: 235146n (1994).

142. A. M. Taha, N. A. El-Rabbat, and M. E. El-Kommos, Novel modification of the Fujiwara reaction, *J. Pharm. Belg. 35*: 107 (1980).

143. O. M. Polishchuk and T. V. Gorbonos, Determination of 2-trichloromethyl-6-chloropyridine in wastewaters (in Russian), *Khim. Tekhnol. Vody 5*: 229 (1983); *Chem. Abstr. 99*: 110420p (1983).

144. J. Y. C. Huang and G. C. Smith, Spectrophotometric determination of total trihalomethanes in finished waters, *J. Am. Water Works Assoc. 76*: 168 (1984); *Chem. Abstr. 101*: 27980k (1984).

145. S. P. Pande, Spectrophotometric determination of chloroform in drinking water, *J. Inst. Chem. (India) 59*: 151 (1987); *Chem. Abstr. 107*: 222925x (1987).

146. R. West, Siegfried Ruhemann and the discovery of ninhydrin, *J. Chem. Educ. 42*: 386 (1965).

147. S. Ruan, Improvement of colorimetric estimation of amino acids with ninhydrin (in Chinese), *Huaxue Shijie 27*: 357 (1986); *Chem. Abstr. 105*: 183157k (1986).

148. G. Yang and S. Na, Determination of amino acids in mixed alcohol medium by ninhydrin spectrophotometry (in Chinese), *Zhiwu Shenglixue Tongxun* 1989 (4), 46; *Chem. Abstr. 111*: 190775n (1989).

149. C. Magne and F. Larher, High sugar content of extracts interferes with

colorimetric determination of amino acids and free proline, *Anal. Biochem.* *200*: 115 (1992).

150. T. Yao, G. Zhang, Z. Liu, and T. Chen, Preparation of ninhydrin reagent for automatic amino acid analyzer, *Zhejiang Yike Daxue Xuebao 21*: 25 (1992); *Chem. Abstr. 117*: 123752p (1992).

151. G. Yang and S. Na, Determination of amino acids in mixed alcohol medium by ninhydrin spectrophotometry, *Zhiwu Shenglixue Tongxun 1989(4)*: 46; *Chem. Abstr. 111*: 190775n (1989).

152. S. Itoh, S. Saitoh, E. Arai, and Y. Nakanishi, The study of amino acid analysis of proteins (in Japanese), *Toso Kenkyu Hokoku 35*: 105 (1991); *Chem. Abstr. 115*: 115051c (1991).

153. E. De Tinguy-Moreaud, S. Cicirello, Nguyen Ba Chanh, and E. Neuzil, Reactivity of indole amino acids with ninhydrin. I. Tryptophan, *Bull. Soc. Pharm. Bordeaux 128*: 19 (1989).

154. M. Szakacs Pinter, and I. Perl Molnar, Spectrophotometric determination of tryptophan in acid solutions with ninhydrin (in Hungarian), *Magy. Kem. Foly. 95*: 462 (1989).

155. N. A. Zakhari, S. M. Hassan, and Y. El-Shabrawy, Spectrophotometric determination of ergot alkaloids with ninhydrin, *Acta Pharm. Nord. 3*: 151 (1991); *Chem. Abstr. 116*: 46400x (1992).

156. I. A. Biryuk, V. V. Petrenko, and B. P. Zorya, Spectrophotometric determination of dopamine by its reaction with ninhydrin (in Russian), *Farm. Zh. (Kiev)*:(2), 57 (1992).

157. G. R. Rao, S. S. N. Murthy, and I. R. K. Raju, Spectrophotometric determination of piperazine and its pharmaceutical dosage forms, *Indian Drugs 24*: 558 (1987).

158. I. A. Biryuk and V. V. Petrenko, Spectrophotometric determination of hexamethylene tetramine by its reaction with ninhydrin (in Russian), *Zh. Anal. Khim. 48*: 187 (1993).

159. Z. Khan and A. A. Khan, Kinetics and mechanism of decarboxylation of α-amino acids by ninhydrin, *J. Indian Chem. Soc. 67*: 963 (1990).

160. A. Hessing, R. Müller-Matthesius, and H. Rose, The reaction of ninhydrin with tryptophan (in German), *Ann. Chem. 735*: 72 (1970).

161. E. Neuzil, M. Malgat, and A.-M. Lacoste, Reaction of some β-amino acids with ninhydrin and with two related triketones, *Biochem. Soc. Trans. 20*: 180s (1992).

162. H. J. Roth and W. Kok, Investigation of the ninhydrin reaction. I. Reactions with 2- and 4-methyoxyaniline (in German), *Arch. Pharm. 308*: 401 (1975).

163. M. H. Abdel-Hay, M. A. Korany, M. M. Bedair, and A. A. Gazy, Colorimetric determination of seven nonsteroidal antiinflammatory drugs using 2-nitrophenylhydrazine hydrochloride, *Anal. Lett. 23*: 281 (1990).

164. Y. Murata, S. Kawashima, E. Miyamoto, N. Masauji, and A. Honda, Colorimetric determination of alginates and fragments thereof as the 2-nitrophenylhydrazides, *Carbohydr. Res. 208*: 289 (1990).

165. P. K. Dasgupta, G. Zhang, S. Schulze, and J. N. Marx, Measurement of

carbonyl compounds as the 2,4-dinitrophenylhydrazonate anion: Reaction mechanism and an automated measurement system, *Anal. Chem. 66*: 1965 (1994).

166. A. G. Davidson and T. O. Dawodu, Difference spectrophotometric assay of 5-hydroxymethylfurfuraldehyde in hydrolyzed pharmaceutical syrups. II. Isoniazid reagent, *J. Pharm. Biomed. Anal. 6*: 61 (1988).

167. A. D. Pickard and E. R. Clark, The determination of traces of formaldehyde, *Talanta 31 (10A)*: 763 (1984).

168. R. Goebel, A. Krug, and R. Kellner, Spectrophotometric flow-injection analysis for formaldehyde in aqueous solutions using 3-methyl-2-benzothiazolinone hydrazone, *Fresenius' J. Anal. Chem. 347*: 491 (1993).

169. J. D. Pakulski and R. Benner, An improved method for the hydrolysis and MBTH analysis of dissolved and particulate carbohydrates in seawater, *Mar. Chem. 40*: 143 (1992); *Chem. Abstr. 118*: 87250h (1993).

170. L. K. Maslii, M. N. Umetskaya, A. A. Tikhomolov, and T. N. Timofeeva, Comparative study of the efficiency of *p*-dimethylaminobenzaldehyde as an analytical reagent for hydrazine and phenylhydrazine (in Russian), *Zh. Anal. Khim. 38*: 120 (1983).

171. V. Ya. Veselov, L. F. Urovskii, and A. P. Grekov, 4-Dimethylaminobenzaldehyde as a photometric reagent for determination of carboxylic acid hydrazides (in Russian), *Zh. Anal. Khim. 38*: 115 (1983).

172. S. Amlathe and V. K. Gupta, Spectrophotometric determination of trace amounts of hydrazine in polluted water, *Analyst 113*: 1481 (1988).

173. S. Fan, Determination of trace hydrazine in wastewater (in Chinese), *Fenxi Ceshi Tongbao 5*: (6) 35 (1986); *Chem. Abstr. 107*: 12524q (1987).

174. N. E. Nechaeva, M. A. Lukashevich, and V. V. Lukachina, Determination of hydrazine using *p*-dimethylaminobenzaldehyde in the presence of vanadate (in Russian), *Ukrain. Khim. Zhur. 40*: 1207 (1974).

175. A. Steup, J. Metzner, and A. Voll, Simple photometric method for the determination of L-dopa and carbidopa in drug preparations (in German), *Pharmazie 41*: 739 (1986).

176. K. Nakashima, K. Shimada, and S. Akiyama, Properties of hydrazones of hydralazine and colorimetric determination of hydralazine hydrochloride with *p*-dimethylaminocinnamaldehyde based on solvent extraction, *Chem. Pharm. Bull. 33*: 1515 (1985).

177. A. M. DiPietra, P. Roveri, R. Gotti, and V. Cavrini, Spectrophotometric and chromatographic (HPLC) analysis of hydralazine, dihydralazine and hydrazine after derivatization with 2-nitrocinnamaldehyde, *Farmaco 48*: 1555 (1993).

178. L. C. Bailey and T. Medwick, Spectrophotometric determination of hydrazine and 1,1-dimethylhydrazine, separately or in admixture, *Anal. Chim. Acta 35*: 330 (1966).

179. L. Lykken, R. S. Treseder, and V. Zahn, Colorimetric determination of phenols, *Anal. Chem. 18*: 103 (1946).

180. J. Gasparič, The reaction of phenols with nitrous acid (in German), *Z. Anal. Chem. 199*: 276 (1964).

181. J. Gasparič and V. Koula, Photometric determination of 4,4'-isopropyli-dene-bis-phenol (Dian) with nitrous acid: Identification of the chromogen, *Mikrochim. Acta II*: 61 (1989).

182. V. Gervay and I. Kelemen, Colorimetric and gravimetric determination of resorcinol, *Acta Pharm. Hung. 25*: 58 (1955).

183. J. Gasparič, to be published.

184. K. K. Verma, A. Jain, and J. Gasparič, Spectrophotometric determination of tryptophan by reaction with nitrous acid, *Talanta 35*: 35 (1988).

185. L. Chafetz, R. E. Daly, H. Schriftman, and J. L. Lomner, Selective colori-metric determination of acetaminophen, *J. Pharm. Sci. 60*: 463 (1971).

186. R. B. Patel, A. A. Patel, M. R. Patel, and U. Pattani, Spectrophotometric estimation of diethylstilboestrol in pharmaceutical formulations, *Ind. J. Pharm. Sci. 49*: 61 (1987).

187. M. E. El-Sadek, H. E. A. Latef, and A. A. A. Khier, Colorimetric determina-tion of terbutaline sulfate and orciprenaline sulfate via nitrosation and differ-ence spectrophotometry, *J. Assoc. Off. Anal. Chem. 70*: 568 (1987).

188. A. A. El Kheir, S. F. Belal, and A. El Shanwani, Spectrophotometric analysis of mixtures of acetaminophen, salicylamide, and codeine phosphate in tab-lets, *J. Assoc. Off. Anal. Chem. 68*: 1048 (1985).

189. R. B. Patel, A. A. Patel, and U. Pattani, Spectrophotometric determination of salbutamol sulphate and its combinations in pharmaceutical dosage forms, *Indian Drugs 24*: 298 (1987).

190. I. Johannes, L. Molder, J. Sidoruk, and L. Tiikma, Selective colorimetric method for the determination of resorcinol and its alkyl derivatives (in Esto-nian), *Eesti Tead. Akad. Toim. Keem. 43*: 91 (1994); *Chem. Abstr. 121*: 314923n (1994).

191. R. V. Smith and M. J. Garst, The chemistry and quantitative utility of sodium cobaltnitrite in the determination of phenols, *Anal. Chim. Acta 65*: 69 (1973).

192. A. M. Wahbi, H. Abdine, M. Korany, and M. A. Abdel-Hay, Sodium co-baltnitrite as a reagent for determining some phenolic drugs, *J. Assoc. Off. Anal. Chem. 61*: 1113 (1978).

193. M. Jimenez, R. M. Gonzales, L. Crovetto, and Y. V. Jara, Determination of anti-pyrine in commercial forms (in Spanish), *An. Real. Acad. Farm. 53*: 269 (1987).

194. A. M. Hartley and R. I. Asai, Spectrophotometric determination of nitrite as 4-nitroso-2,6-xylenol, *Anal. Chem. 35*: 1214 (1963).

195. S. A. Rahim, N. A. Fakhri, and W. A. Bashir, Indole as a reagent for nitrite in aqueous solution, *Microchem. J. 28*: 479 (1983).

196. Š. Hošková and I. Němcová, A spectrophotometric study of the reaction of nitrite with rivanol: Determination of nitrite and nitrate, *Microchem. J. 41*: 296 (1990).

197. L. M. Buruiana, The determination of proteins by means of the xanthopro-teic reaction (in French), *Naturwisseschaften 45*: 339 (1958).

198. A. M. Hartley and R. I. Assai, Spectrophotometric determination of nitrate with 2,6-xylenol, *Anal. Chem. 35*: 1207 (1963).

199. U. Pindur, Structural investigation of the van-Urk-color-reaction of ergot alkaloids (in german), *Pharm. Acta Helv. 57*: 112 (1982).

200. U. Pindur, 2,2′-Diindolylmethanes. VII: Influence of diindolylmethane leu-cobases on the course of the van Urk reaction with physiologically active indoles (in German), *Arch. Pharm. 317*: 502 (1984).

201. H. Auterhoff and H.-J. Riethmüller, The color reaction of diethylstilboes-trole with vanillin (in German), *Arch. Pharm. 305*: 386 (1972).

202. S. P. Agarwal, N. Chandrasekhara, and A. U. Madu, A colorimetric method for the determination of amphetamines, *Acta Pharm. Technol. 27*: 181 (1981); *Chem. Abstr. 96*: 24858u (1982).

203. G. Waldheim, H. Möhrle, and S. Rudzky, Condensation of drugs of the series of *N,N*-disubstituted aniline derivatives with 4-*N*-dimethylaminoben-zaldehyde (in German), *Pharm. Acta Helv. 60*: 71 (1985).

204. A. Ahmad, M. Qureshi, I. A. Khan, and K. Z. Alam, Spectrophotometric determination of diphenylamine, pyrrole and indole: A kinetic study, *Anal. Lett. 23*: 1139 (1990).

205. M. S. Mahrous and M. M. Abdel-Khalek, Spectrophotometric determination of some thiophene-containing cephalosporins using different aldehydes, *Egypt. J. Pharm. Sci. 34*: 47 (1993); *Chem. Abstr. 121*: 308445s (1994).

206. U. Pindur and H. Witzel, The reactions of pindolol, meprindolol, carazolol, and related model compounds with triethylorthoformate: Pathway and prod-ucts of a new color reaction, *Arch. Pharm. 323*: 427 (1990).

207. H. A. A. Medien, Spectrophotometric determination of some aromatic alde-hydes, *Anal. Lett. 27*: 983 (1994).

208. S. Yang and T. Zhou, Colorimetric determination of aromatic aldehydes by phloroglucinol–sulfuric acid reagents, *Huaxue Shiji*: 237 (1981); *Chem. Ab-str. 96*: 96799y (1982).

209. P. Bachhausen, N. Buchholz, and H. Hartkamp, Determination of formalde-hyde in air by means of chromotropic acid. 1. Stability of the reagent, *Z. Anal. Chem. 320*: 347 (1985).

210. J. Chrastil and R. M. Reinhardt, Direct colorimetric determination of form-aldehyde in textile fabrics and other materials, *Anal. Chem. 58*: 2848 (1986).

211. W. Xu and G. Yang, Colorimetric determination of chloroethanol in air with chromotropic acid (in Chinese), *Zhonghua Yufangyixue Zazhi 16*: 309 (1982); *Chem. Abstr. 98*: 39847e (1983).

212. K. Rehse and H.-G. Kawerau, Mechanism of the reaction of aromatic com-pounds with formaldehyde in concentrated sulfuric acid (Marquis reagent) (in German), *Arch. Pharm. 307*: 934 (1974).

213. P. E. Georghiou and C. K. Ho, The chemistry of the chromotropic acid method for the analysis of formaldehyde, *Can. J. Chem. 67*: 871 (1989).

214. S. Honda, H. Chiba, and K. Kakehi, Selective determination of pentoses, hexoses and deoxyhexoses by dual-wave-length spectrophotometry with the cysteine-phenol-sulfuric acid system, *Anal. Chim. Acta 131*: 205 (1981).

215. P. Rao and T. N. Pattabiraman, Reevaluation of the phenol–sulfuric acid reaction for the estimation of hexoses and pentoses, *Anal. Biochem. 181*: 18 (1989).

216. M. Monsigny, C. Petit, and A. C. Roche, Colorimetric determination of

neutral sugars by a resorcinol-sulfuric acid micromethod, *Anal. Biochem. 175*: 525 (1988).

217. M. Koleva, S. Ninov, and S. Kolev, Spectrophotometric determination of monosaccharides with thymol in concentrated sulfuric acid, *Dokl. Bolg. Akad. Nauk 41*: 39 (1988); *Chem. Abstr. 110*: 107364p (1989).

218. Y. Jiang, Z. Huang, K. Ma, and X. Chen, Resorcinol-spectrophotometric method for determination of fructose in fructose-glucose syrup, *Huaxue Shijie 30*: (1) 18 (1989); *Chem. Abstr. 111*: 22286k (1989).

219. V. Yu. Andreev, Quantitative spectrophotometric method of determination of monosaccharide composition of plant polysaccharides with *o*-toluidine reagent, *S-kh. Biol. 1992* (1), 154; *Chem. Abstr. 117*: 3618c (1992).

220. J. Buysse and R. Merckx, An improved colorimetric method to quantify sugar content of plant tissue, *J. Exp. Bot. 44*: 1627 (1993), *Chem. Abstr. 120*: 49319w (1994).

221. T. Momose and M. Nakamura, Organic analysis. XXXIII. Mechanism of the color reaction between fructose and *N*-methyldiphenylamine, *Chem. Pharm. Bull. 10*: 544 (1962).

222. T. Momose, Y. Ueda, and M. Iwasaki, Organic analysis. XXXIV. Reaction mechanism of D-glucuronic acid with 1,2-naphthalenediol, *Chem. Pharm. Bull. 10*: 546 (1962).

223. T. Momose and Y. Ohokura, Organic analysis. XXXV. Mechanism of the color reaction between hexose and thymol, *Chem. Pharm. Bull. 10*: 550 (1962).

224. T. Momose, Y. Ueda, and M. Iwasaki, Organic analysis. XXXVII. Reaction mechanism of D-mannuronic acid and D-galacturonic acid with 1,3-naphthalenediol, *Chem. Pharm. Bull. 10*: 663 (1962).

225. M. Ohta, M. Iwasaki, K. Kuono, and Y. Ueda, Study on the color reaction mechanism of the Seliwanoff reaction with 4-ethylresorcinol as the color-developing reagent, *Chem. Pharm. Bull. 33*: 2421 (1985).

226. M. Vosmanská, Spectrophotometric determination of glucuronic acid and glucuronides (in Czech), *Chem. Listy 88*: 636 (1994).

227. J. T. Galambos, J. R. McCain, The reaction of carbazole with carbohydrates I. Effect of borate and sulfamate on the carbazole color of sugars, *Anal. Biochem. 19*: 119 (1967).

228. M. Kosakai and Z. Yoshizawa, A partial modification of the carbazole method of Bitter and Muir for quantitation of hexuronic acids, *Anal. Biochem. 93*: 295 (1979).

229. S. Matsuhashi and C. Hatanaka, Difference between the free and conjugated galacturonate residues in their color reaction with carbazole or *m*-hydroxybiphenyl reagents, *Biosci. Biotechnol. Biochem. 56*: 1142 (1992); *Chem. Abstr. 117*: 149601p (1992).

230. K. A. Taylor and J. G. Buchanan-Smith, A colorimetric method for the quantitation of uronic acids and a specific assay for galacturonic acid, *Anal. Biochem. 201*: 190 (1992).

231. K. A. C. C. Taylor, A colorimetric method for the quantitation of galacturonic acid, *Appl. Biochem. Biotechnol. 43*: 51 (1993).

232. H. A. A. Medien, Simple and rapid spectrophotometric method for determination of some aromatic aldehydes: A kinetic study, *Anal. Lett. 27*: 2727 (1994).

233. E. M. Volynskaya, P. V. Prisyazhnyuk, and N. G. Prodanchuk, Photometric determination of barbituric acid and its Na salt in the air (in Russian), *Gig. Sanit. 1990 (4)*, 86; *Chem. Abstr. 114*: 11345q (1991).

234. J. L. Lambert, J. Ramasamy, and J. V. Paukstelis, Stable reagents for the colorimetric determination of cyanide by modified Koenig reaction, *Anal. Chem. 47*: 916 (1975).

235. A. Tanaka, K. Deguchi, and T. Deguchi, Spectrophotometric determination of cyanide and thiocyanate based on a modified Koenig reaction in a flow-injection system, *Anal. Chim. Acta 261*: 281 (1992).

236. H. Ma and J. Liu, Flow injection determination of cyanide by detecting an intermediate of the pyridine-barbituric acid chromogenic reaction, *Anal. Chim. Acta 261*: 247 (1992).

237. S. Amlathe and V. K. Gupta, A new spectrophotometric method for the determination of acrylonitrile in traces and its application to biological samples, *J. Indian Chem. Soc. 66*: 359 (1989).

238. L. Lilov, Quantitative determination of lactulose, *Pharmazie 42*: 138 (1987).

239. L. Lilov, Thiobarbituric method for analysis of sugars: L-Rhamnose, 2-deoxy-D-glucose, and D-glucose, *Dokl. Bolg. Akad. Nauk 42*: 103 (1989); *Chem. Abstr. 111*: 190700j (1989).

240. L. Lilov, Thiobarbituric method for analysis of sugar fructose, *Dokl. Bolg. Akad. Nauk 42*: 139 (1989); *Chem. Abstr. 111*: 36071y (1989).

241. M. I. R. M. Santoro, E. R. M. Hackmann, J. F. Magalhaes, and M. J. Vernengo, Determination of 5-hydroxymethylfurfural on oral rehydration salts containing glucose, *Rev. Farm. Bioquim. Univ. Sao Paolo 22*: 77 (1986); *Chem. Abstr. 107*: 84000p (1987).

242. D. Tu, S. Xue, C. Meng, A. Espinosa-Mansilla, A. Munoz de la Pena, and F. Salinas Lopez, Simultaneous determination of 2-furfuraldehyde and 5-hydroxymethyl-2-furfuraldehyde by derivative spectrophotometry, *J. Agr. Food Chem. 40*: 1022 (1992).

243. A. Espinosa-Mansilla, A. Munoz de la Pena, F. Salinas, and M. Martinez Galera, Elimination of spectral interferences in the reaction of 2-thiobarbituric acid with malonaldehyde, 2-furfuraldehyde and 5-hydroxymethyl-2-furfuraldehyde by partial least squares multivariate calibration (PLS), *Fresenius' J. Anal. Chem. 347*: 371 (1993).

244. A. Espinosa-Mansilla, A. Munoz de la Pena, F. Salinas, and M. Martinez Galera, Simultaneous determination of 2-furfuraldehyde, 5-hydroxymethyl-furfuraldehyde and malonaldehyde in mixtures by derivative spectrophotometry and partial least-squares analysis, *Anal. Chim. Acta 276*: 141 (1993).

245. R. Zalewski, E. Zakrzewska, and P. Tomasik, Studies on the quantitative determination of pyridine in coal tars. I. The alkaline hydrolysis of the products of the reactions of some chloronitrocompounds with pyridine, *Chem. Anal. (Warsaw) 21*: 73 (1976).

246. Y. Lin, Comparison and improvement of several common anthrone-

colorimetric methods for sugar content in plants, *Zhiwu Shenglixue Tongxun* (4), 53 (1989); *Chem. Abstr. 111*: 190776p (1989).

247. J. Jakovljevic and Z. Boskov, Modified anthrone method for selective D-fructose determination (in Serbo-Croatian), *Hem. Ind. 42*: (Suppl. 3–4) 86 (1988); *Chem. Abstr. 110*: 93645s (1989).

248. J. R. Brooks, V. K. Griffin, and M. Kattan, A modified method for total carbohydrate analysis of glucose syrup, maltodextrins, and other starch hydrolysis products, *Cereal Chem. 63*: 465 (1986).

249. B. L. Somani, J. Khanade, and R. Sinha, A modified anthrone-sulfuric acid method for the determination of fructose in the presence of certain proteins, *Anal. Biochem. 167*: 327 (1987).

250. A. Silins, Comparison of methods for determination of sugars in plant material (in Russian), *Zh. Anal. Khim. 43*: 308 (1988).

251. M. Ranjbar-Hamghawandi, Photometric determination of biopolysaccharides in brine, *Fresenius' J. Anal. Chem. 336*: 41 (1990).

252. T. Momose, Y. Ohkura, and K. Hirauchi, Organic analysis. XLVII. Color reaction mechanism of anthrone with sugars. 2. Reaction products of furfural with anthrone, *Chem. Pharm. Bull. 11*: 1364 (1963).

253. H. Hörmann and I. A. Siddiqui, Color reactions of carbohydrates. IV. Products of the color reaction of fructose and glucose with anthrone/sulfuric acid (in German), *Ann. Chem. 714*: 174 (1968).

254. M. Tachibana and M. Furusawa, Rapid spectrophotometric method for the determination of fluorene, *Analyst 115*: 1495 (1990).

255. D. Yu. Rogov'skii, Determination of bromophos in a chemico-toxicological examination of biological material (in Russian), *Farm. Zh. (Kiev)*: (2) 45 (1993).

256. W. Stuecker, E. Golovinski, G. Mueller, M. Oberst, K. Maher, and K. Norpoth, New color reagents for the determination of alkylating agents, *Z. Anal. Chem. 330*: 42 (1988).

257. R. S. Shah, S. A. Shah, M. B. Devani, and K. P. Soni, Spectrophotometric determination of mesalamine and its dosage forms, *Indian Drugs 31*: 34 (1994).

258. O. V. Kaisina, N. A. Krylova, and O. A. Chumicheva, Determination of formaldehyde with an acetylacetone reagent (in Russian), *Gig. Sanit*: (6) 43 (1987); *Chem Abstr. 107*: 72192q (1987).

259. M. B. Devani, I. T. Patel, and T. M. Patel, Spectrophotometric determination of ampicillin and its dosage forms, *Talanta 39*: 1391 (1992).

260. S. Igarashi, Spectrophotometric determination of tertiary alcohols with vanillin and sulfuric acid, *Bunseki Kagaku 31*: 610 (1982); *Chem Abstr. 98*: 10889j (1983).

261. D. Bose, K. N. Ramachandran, and V. K. Gupta, New method for spectrophotometric determination of allyl alcohol using aryl aldehydes as reagents, *Chem. Anal. (Warsaw) 38*: 645 (1993).

Index